SOCIETY FOR EXPERIMENTAL BIOLOGY

SEMINAR SERIES · 6

NEURONES WITHOUT IMPULSES

NEURONES WITHOUT IMPULSES:

their significance for vertebrate and invertebrate nervous systems

Edited by

ALAN ROBERTS

Department of Zoology, University of Bristol

AND

BRIAN M. H. BUSH

Department of Physiology, University of Bristol

CAMBRIDGE UNIVERSITY PRESS

Cambridge

London New York New Rochelle

Melbourne Sydney

Published by the Press Syndicate of the University of Cambridge
The Pitt Building, Trumpington Street, Cambridge CB2 1RP
32 East 57th Street, New York, NY 10022, USA
296 Beaconsfield Parade, Middle Park, Melbourne 3206, Australia

First published 1981

Bitish Library Cataloguing in Publication Data

Neurones without impulses. – (Society for Experimental Biology.
Seminar series; 6).

1. Neural transmission – Congresses
2. Neurons – Congresses
I. Roberts, Alan II. Bush, Brian M H
III. Society for Experimental Biology.
Neurobiology Group.
591.1′88 QP364.5 79-42672

ISBN 0 521 23364 X hard covers
ISBN 0 521 29935 7 paperback

Transferred to digital printing 2002

CONTENTS

CONTRIBUTORS

Bullock, T. H.
Neurobiology Unit, Scripps Institution of Oceanography and Department of Neurosciences, School of Medicine, University of California San Diego, La Jolla, California 92093, USA.

Burrows, M.
Department of Zoology, University of Cambridge, Downing Street, Cambridge CB2 3EJ, UK.

Bush, B. M. H.
Department of Physiology, University of Bristol, School of Veterinary Science, Park Row, Bristol BS1 5LS, UK.

Fain, G. L.
Jules Stein Eye Institute, UCLA School of Medicine, Los Angeles, California 90024, USA.

Rall, W.
National Institutes of Health, Bethesda, Maryland 20205, USA.

Russell, I. J.
Ethology and Neurophysiology Group, School of Biological Sciences, University of Sussex, Brighton, Sussex, BN1 9QG, UK.

Shaw, S. R.
Department of Neurobiology, Research School of Biological Sciences, Australian National University, P.O. Box 475, Canberra City, ACT 2601, Australia.

Shepherd, G. M.
Section of Neuroanatomy, Yale University Medical School, 333 Cedar Street, New Haven, Connecticut 06510, USA.

Simmers, A. J.
Department of Physiology, University of Bristol, School of Veterinary Science, Park Row, Bristol BS1 5LS, UK.

PREFACE

During the last fifteen years studies in a large variety of nervous tissues have revealed a number of sensory and central neurones which appear normally to function without impulses. We feel that this type of neurone may prove to be central to the way nervous systems operate. This book considers all cases of neurones without impulses that are currently known in both vertebrate and invertebrate animals. We hope that a full review of the occurrence and properties of neurones without impulses might allow an overall view of their probable significance in normal nervous function. As Dr Shepherd points out in his introduction, studies on neurones which lack this property raise many questions about the normal role of impulses themselves.

Like so many changes of viewpoint in science, the re-evaluation of the relative importance of impulses and non-impulsive, graded communication between neurones has depended on technical advances. Impulses are easier to record electrically than graded potentials which require stable intracellular recordings from individual neurones. These can now be made, with difficulty, from small neurones and neuronal processes often less than 10 μm in diameter. A second requirement is that the cells recorded must be identified. This is now possible as dyes can be passed from recording micropipettes to allow subsequent histological identification of the recorded cells, as Dr Simmers' and Dr Burrows' chapters show. It is these technical advances that have allowed elegant studies like those on vertebrate retina (which Dr Fain describes), insect eyes (Dr Shaw's contribution), vertebrate acoustic hair cells (Dr Russell's chapter) and invertebrate central nervous ganglia (Dr Simmers' and Dr Burrows' chapters). Despite these successes it is clear that the techniques are harder to apply to vertebrate central nervous systems. It was in the vertebrate olfactory bulb that extensive evidence of graded interactions between neurone dendrites was first found in the mammalian brain. However, these studies depended on extracellular recording and detailed modelling more than on direct evidence from intracellular recording, as Dr Rall's and Dr

Shepherd's discussions demonstrate. It seems very likely that many other local graded interactions will be found in the vertebrate central nervous system, and very probable that many cells will lack impulses. We must wait for direct evidence which might well appear in the next decade as microelectrode recording techniques are applied to isolated slices of vertebrate central nervous tissue. In our final chapter Dr Bullock tries to look into this future prospect.

Our title 'Neurones without impulses', raises some slight problems of definitions. Are cochlear hair cells and retinal rods, cones and horizontal cells really neurones? We leave the question open, but we felt their properties were so closely related to the undisputed neurones that it would be foolish to omit them. For some receptors there is no doubt about their neuronal identity. The large non-impulsive stretch receptors in crabs, which Dr Bush describes, have provided a very convenient preparation where, for a change, intracellular recording is easy so the membrane properties of a neurone without impulses can be studied (see also the Introduction).

This book is based on a meeting of the Neurobiology Group of the Society for Experimental Biology held at the University of York in April 1979. We would like to thank the officers of that Society for their help. We would also like to give special thanks to Dr Jack W. Hannay, Professor J. W. S. Pringle and Dr Gordon M. Shepherd for their early encouragement of this venture.

November 1979

A. R., B.M.H.B.

Editors for the Society for Experimental Biology

GORDON M. SHEPHERD

Introduction: the nerve impulse and the nature of nervous function

The nerve impulse has occupied a central position in our traditional concepts of neurone function. As an expression of 'irritability' it traces its lineage back to the first suggestions by Haller and his contemporaries in the eighteenth century that what we now call 'nervous' functions – sensations, emotions, thinking and motor behaviour – are mediated exclusively by nervous tissue. As the basic 'excitable' property of peripheral nerves and muscles, its role has been apparent since the time of Galvani, and its study has lain at the core of almost two centuries of work since then on the physiology of nervous tissue and nerve membrane.

Although it has thus appeared that nervous function and nerve excitability are ineradicably bound together, work over the past several decades has revealed an increasing number of exceptions to this rule. It is now apparent that what would generally be regarded as the basic functions of nerve cells – the receiving, transmitting, generation, integration and emission of nerve signals – may be carried out in the absence of impulses. Any of the functions can be mediated by the graded potentials arising from actions at the synaptic junctions between nerve cells. In some cases these potentials spread the relatively short distances through the nerve cell branches (dendrites) to activate synaptic outputs from the branches themselves. In other cases the branches may be differentiated into impulse generating compartments or loci. A wealth of diversity throughout the nervous system is becoming apparent in this respect.

If nervous functions are thus capable of being mediated by non-impulse types of activity, it is also becoming apparent that non-nervous functions in other parts of the body may be mediated by 'nervous' ones, i.e., impulse activity. The ability of non-nervous cells (other than muscle) to generate impulses is unfamiliar to many neuroscientists, but in fact this is a robust field of research with a history almost as venerable as that concerned with the nerve impulse. In recent years the instances of cells showing this type of activity have increased dramatically, as single unit recording techniques have improved. Thus, any reassessment of the role of the impulse in the nervous system must take account of the fact that impulse activity is a much more general and widespread cellular property than heretofore realised.

We will review briefly these two lines of research, and see that they give a new perspective on excitability as an expression of voltage-dependent ionic conductances in cell membranes. Nerve cells employ a range of excitable and non-excitable properties in carrying out their many different functions within multicellular ensembles and circuits. From these considerations we can tentatively suggest a new conceptual framework for defining the neurone and characterising the fundamental cellular nature of nervous activity.

The nerve impulse

The early work on nerve excitability, by du Bois Reymond and Helmholtz in the 1840s, was carried out in the frog nerve–muscle preparation, where it was established that excitability is associated with an impulse propagating at a finite velocity and giving rise to a quick muscle response (see Boring, 1950, for historical review). For a century this preparation provided almost the only direct evidence for the functional properties of nervous tissue. When the works of Sherrington (1906) and others began to engender speculation about the nature of integrative processes in the central nervous system, Lucas (1917) countered with the proposition that central phenomena could be explained solely on the basis of properties that could be experimentally identified in peripheral nerve. Forbes (1922) carried this further, and foresaw the day when 'the elements of neural activity underlying consciousness and behaviour could be reduced to the single basis of the nerve impulse'. He conceived of central processes such as inhibition, summation, and long-lasting activity as being mediated solely by impulses in axonal and dendritic branches. Lorente de No, drawing on the richness of Golgi-impregnated neuronal structures, elaborated numerous examples of how impulse traffic might circulate and reverberate in neural circuits and account thereby for nervous function (Lorente de No, 1938).

The mechanism of the impulse itself was elucidated in the sequence of brilliant studies of Cole, Marmont, Goldman, Hodgkin, Huxley and Katz, culminating in the model of Hodgkin & Huxley in 1952. This provided a quantitative and explicit mechanism for the generation of the action potential, in terms of variables controlling sodium conductance, sodium inactivation, and potassium conductance (Fig. 1*a*). The importance of this model, for the development of neurobiology in general and the mechanisms controlling membrane potential changes in nerve cells in particular, cannot be overemphasised; it has been the solid touchstone for all subsequent research. Additional fast and slow conductances have been added over the years, appropriate for different cells and conditions of repetitive discharge, so that a fairly complete description is now available for several temporal sequences of impulse firing (Fig. 1*b*).

An essential feature of the mechanism of the action potential is the voltage dependence of the ionic conductances. The relation between sodium (or calcium) conductance and membrane depolarisation is regenerative, and this is the property that imparts 'excitability' to the membrane. The conductances are in turn determined by permeability channels for their respective ions. The

Fig. 1. Functional models of the nerve impulse. *a*. Model of Hodgkin & Huxley (1952) for single impulse in squid giant axon. 1. Membrane action potential. 2. Time courses of ionic current (I_i), capacity current $\left(-C_M \frac{dV}{dt}\right)$ and total membrane current $\left(I = \frac{C_M}{K}\frac{d^2V}{dt^2}\right)$ 3. Time courses of ionic current (I_i) and its two main components, sodium current (I_{Na}) and potassium current (I_K). *b*. Model for repetitive impulse firing in ink-gland motor neurone of *Aplysia* (Byrne, 1980). Memb pot, membrane potential; IFK, fast potassium current; IDK, delayed potassium current; INA, sodium current; ICA, calcium current; IS, stimulating current, in nanoamperes.

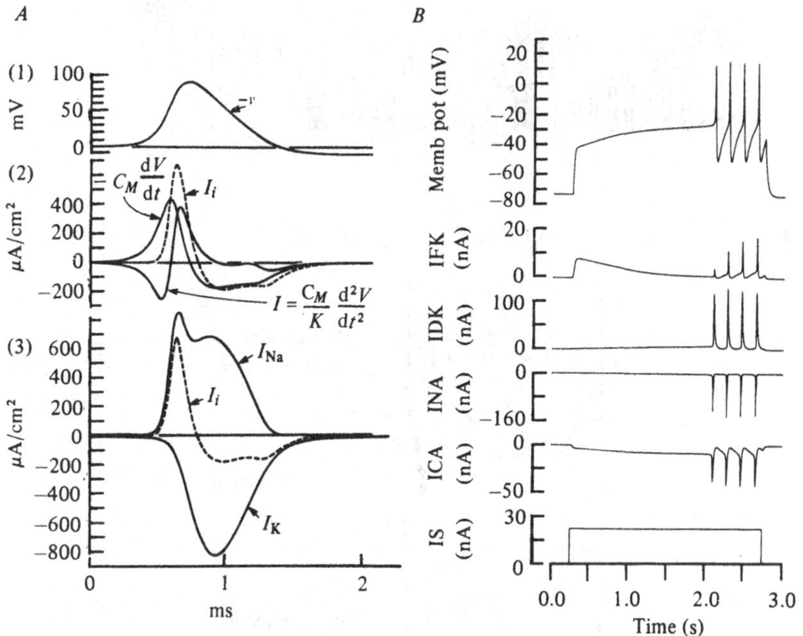

mechanism of formation of the channels and the kinetics of their activation and deactivation have been the subject of intensive study. These properties have been incorporated into molecular models for the electrical excitability of lipid bilayer membranes, as illustrated in Fig 2*a*. In natural membranes, a variety of molecular probes are being used to analyse the nature of the ionic channels (Fig. 2*b*). There is thus the prospect of increasingly refined models for defining excitability in molecular terms.

Fig. 2. Theoretical models of impulse-generating membrane. *a*. Excitable channels in artificial lipid bilayer. The molecules forming the gating parts of the ionic channel are in the upper half of the bilayer, and link up with the ion-selector molecules of the lower half to permit ions to permeate the channel. The channels of the lower half are formed by treatment with such substances as amphotericin; the gating molecules aggregate and insert through the membrane under the influence of the membrane potential (Mueller, 1979). *b*. The sodium channel in the node of Ranvier of myelinated nerve fibres, showing sites of action of various agents used to analyse the molecular basis of excitability of the channel. R_{Tox}, receptor site with which indicated toxins interact; R_{LA}, receptor site with which local anaesthetics interact; m and h gates refer to parameters of the Hodgkin–Huxley model controlling sodium current activation and inactivation, respectively. (Ritchie, 1979).

A

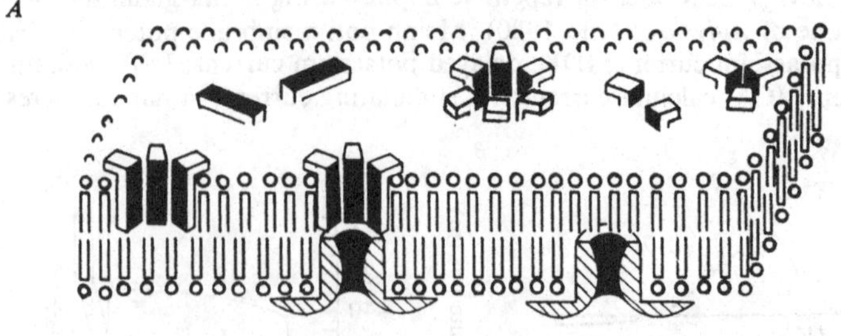

B

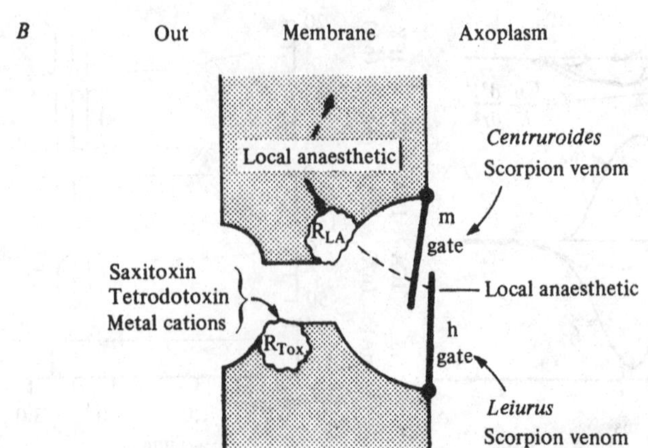

Graded potentials

The first intracellular recordings of graded non-impulse activity related to synaptic junctions were obtained from the neuromuscular junction (Fatt & Katz, 1951). Eccles and his co-workers obtained the first intracellular recordings of post-synaptic excitatory and inhibitory potentials in central neurons (Eccles, 1953; Fig. 3*a*). Eyzaguirre & Kuffler (1955) in their work on the crayfish stretch receptor cell provided a particularly elegant model of the relation between graded potentials, both sensory and synaptic, and the generation of impulses. During this period also, Clare & Bishop (1955) inferred the existence of graded potentials in the cerebral cortex, in the dendrites of cortical pyramidal cells.

These studies left little doubt about the importance of graded electrical activity in the integrative life of the nerve cell. In a classical review of the time, Bishop (1956) not only emphasised this point, but went further and proposed that 'the chief and most characteristic functions of nervous and other excitable tissue are performed by means of graded responses', and that the nerve impulse was 'a special case of the general property of excitability', developed in evolution chiefly for transmitting nervous signals over long distances. The

Fig. 3. Graded actions at synaptic junctions. *a*. Synaptic potentials in motoneurones. 1–4. Excitatory post-synaptic potentials. 2, recorded at normal resting potential (about −70 mV); 1, with membrane depolarized; 3, hyperpolarised; 4, synaptic potential is at threshold for eliciting action potentials (rapid upstrokes). (Coombs, Eccles & Fatt, 1955*a*). 5–9. Inhibitory post-synaptic potentials, with increasing strength of afferent volley, as shown in upper traces. (Coombs, Eccles & Fatt, 1955*b*). *b*. Junction potentials at muscle end-plate, showing grading of frequency and summation of miniature end-plate potentials (upper traces) due to increasing depolarisation of pre-synaptic nerve terminals. (del Castillo & Katz, 1954). Voltage scale: 1 mV.

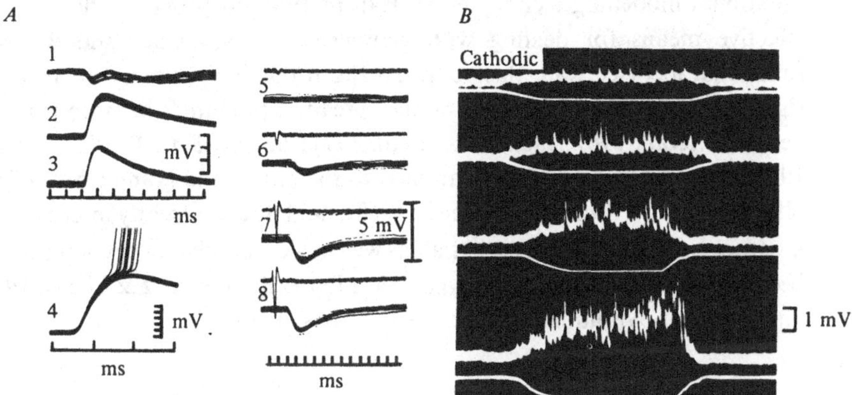

complexity of graded processes underlying impulse output was further discussed by Bullock (1959), who noted the 'quiet revolution' that this was bringing about in concepts of neuronal functions, and the revisions that were becoming necessary in the traditional views that had gathered about the 'neurone doctrine'. Among various possibilities he envisaged that there might be 'intercellular reactions mediated by graded activity without the intervention of all-or-none impulses'.

By this time the distinction between the all-or-nothing impulse, on the one hand, and graded receptor or synaptic potentials, on the other, seemed clear. In between was a middle ground of what appeared to be graded, prolonged potentials spreading by 'decremental conduction'. It was unclear to what extent these potentials involved 'active' membrane mechanisms, or could be accounted for solely by the passive spread of synaptic potentials through the complicated geometry of dentritic branching trees. In the case of cortical pyramidal cell dendrites, the interpretation was all the more uncertain because of the indirect nature of the evidence, being obtained as it was from summed extracellular recordings. Intracellular recordings in motoneurones indicated that synaptic potentials spread passively. However, the focus in these early studies was on synapses near the cell body, and the significance of synapses distributed over the dendritic tree and out on the most distal branches was either ignored or left enigmatic.

To deal with these questions it was necessary to develop rigorous methods for describing the spread of electrotonic potentials in dendritic trees. This can easily be seen in retrospect, though at the time the hope of adequately incorporating the complicated geometries of many types of dendritic tree seemed remote. There were two important steps. One was the extension of analytical methods for cable analysis from the case of the axon to the case of dendrites. This began with Rall in 1957, in his analysis of electrotonic transients in motoneurones. The second step was the introduction of computational modelling techniques by Rall in 1964, to provide a flexible and effective means for dealing with geometrical complexities and different functional loci (see Rall, this volume). From this have evolved explicit and rigorous models for characterising the spread of potentials in motoneuronal dendrites (Rall, 1967; Jack & Redman, 1971; Barrett & Crill, 1974; Traub, 1977). The methods were early applied to the mitral and granule cells of the olfactory bulb (Rall & Shepherd, 1968), and more recently models that incorporate sites of action potentials as well as slow conductance changes have become available for other cell types (Purkinje cell: Pellionicz & Llinás, 1977; hippocampal pyramidal cell: Traub & Llinás, 1979).

Non-impulse activity controlling synaptic output of nerve cells

As long as the motoneurone and the neuromuscular junction were the main models for central synapses it was natural to assume that synaptic output took place exclusively at an axon terminal, and that a synapse was activated by an impulse propagating into the terminal. In fact, closer study of the neuromuscular junction revealed that the impulse itself was not necessary, and that the critical factor was the amount of depolarisation of the terminal membrane. The frequency of miniature end-plate potentials, and consequent amount of depolarisation of the post-synaptic membrane, was found to be continuously graded with the amount of presynaptic membrane depolarisation, at the neuromuscular junction (del Castillo & Katz, 1954) and also at the squid giant synapse (Takeuchi & Takeuchi, 1962; Bloedel, Cage, Llinás & Quastel, 1966; Katz & Miledi, 1967). Recordings from the neuromuscular junction are shown in Fig. 3*b*. This work had its own importance for the analysis of synaptic mechanisms at the membrane and molecular level, but it had the additional importance, for our present interest, in implying that any small, slow, graded depolarisation, due to a synaptic input on to a cell body or dendrite, would be similarly effective in controlling output from a synapse situated at or near that site.

The possibility that cell bodies or dendrites might have synaptic outputs as well as inputs, however, had never been entertained in all the theories and speculations about nerve cells and their functions. The only serious evidence that might have borne on this question had to do with the contention around 1900 that there were points of contact or continuity between nerve cells as seen in Golgi or other types of histological preparations (cf. Barker, 1896). When Cajal (1911, 1954) carried the day in his arguments for the neurone doctrine and the 'dynamic polarisation' of the nerve cell these contentions were quietly forgotten. An echo was heard in the report by van der Loos (1960) of contacts between dendrites of Golgi-impregnated cortical cells, but apart from the first use of the term 'dendro-dendritic' this echo too faded away. The discovery of presynaptic inhibition of axon terminals on motoneurones (Frank & Fuortes, 1957; Eccles, 1964), and the evidence of Gray (1962) for axoaxonic synapses to mediate this inhibition, provided the first real departure from the classical model of a neurone in which all synaptic inputs occur on dendrites and all synaptic outputs occur from axons. But it was not too difficult to incorporate this as an exception that proved the rule.

Physiological evidence for synaptic output from dendrites of vertebrate neurones was forthcoming in studies of the mitral and granule cells of the olfactory bulb (Phillips, Powell & Shepherd, 1963; Rall, Shepherd, Reese & Brightman, 1966; Rall & Shepherd, 1968; Rall, this volume). The dendro-dendritic synapses to mediate the interactions between these cells were demonstrated in several studies at about the same time (Hirata, 1964; Andres, 1965; Rall *et al.*, 1966). Concurrently, similar types of synaptic arrangement were seen between the processes of cells in the vertebrate retina (Dowling & Boycott, 1966). Intracellular recordings soon established that most of the types of retinal neurones respond to light with slow graded potentials (see Fig. 4*a*), and that the ganglion cell is the only type to show the kinds of synaptic potentials and impulse activity expected from the classical motoneurone model (Werblin & Dowling, 1969; Kaneko, 1970; Baylor, Fuortes & O'Bryan, 1971; see Fain, this volume).

Since these early studies several lines of work have emerged. One has been the identification of synapses made by cell bodies and dendrites in different parts of the nervous system. A variety of types of synaptic connection, including dendro-dendritic, somato-dendritic, and axo-axonic, have been identified, in addition to the classical types of axo-somatic and axo-dendritic. Table 1 lists some of the regions in which synapses arising from dendrites

Fig. 4. Non-spiking activity in different types of nerve cell. *a*. Mudpuppy retina, showing intracellular recordings of slow graded responses to a flash of light in a receptor (R), horizontal cell (H) and bipolar cell (B), as well as synaptic and impulse responses in an amacrine cell (A) and ganglion cell (G_T). (Werblin & Dowling, 1969). Voltage calibration: 1 vertical division = 1 mV(R); 2 mV (H, B, G); 5 mV(A). Time: 1 horizontal division = 200 ms. *b*. Stretch receptor of coxal joint of crab, showing intracellular recording of graded response to stretch of increasing amplitude, and comparison with impulse activity induced by the receptor in the motor nerve. (Bush & Roberts, 1968). *c*. Interneurone of suboesophageal ganglion of lobster. Top trace, intracellular recording of graded depolarisations in an oscillator neurone; lower trace, extracellular recordings of associated impulse activity in two motor nerve branches. (Mendelson, 1971). 1 s cycle in upper trace. Voltage, 5 mV.

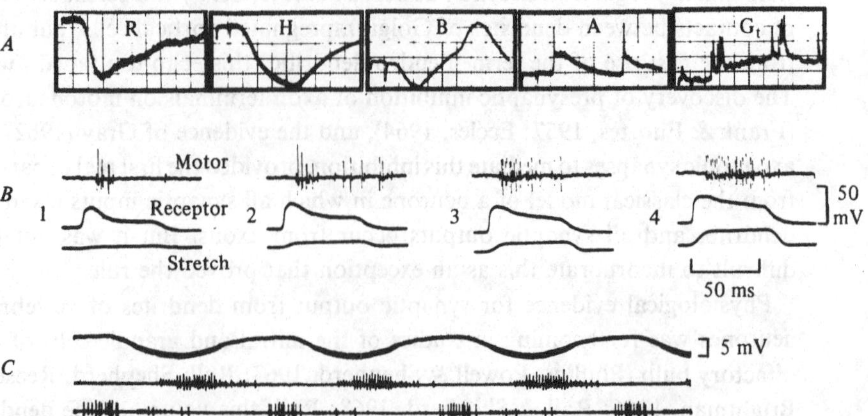

Table 1. *Regions with presynaptic dendrites or cell bodies*

Region	Location	Reference
Vertebrates[a]		
Olfactory bulb	Glomerular layer	Pinching & Powell, 1971
	External plexiform layer	Rall *et al.*, 1966; Jackowski *et al.*, 1978
Retina	Inner plexiform layer	Dowling & Boycott, 1966
	Outer plexiform layer	Stell, 1972
Thalamus	Lateral geniculate nucleus	Famiglietti & Peters, 1972
	Medial geniculate nucleus	Morest, 1971
	Ventrobasal complex	Ralston, 1971
	Ventrolateral nucleus	Harding, 1973
	Centre median	Harding, 1973
Cerebral cortex	Motor area	Sloper, 1972
Basal ganglia	Substantia nigra	Wilson *et al.*, 1978
Olfactory tubercle		Hosoya, 1973
Hypothalamus	Suprachiasmatic nucleus	Guldner, 1976
Superior colliculus		Lund, 1969
Optic tectum		Setalo & Szekely, 1967
Pons	Basilar nucleus	Mihailoff, 1978
Cerebellum	Deep nuclei	Hamori & Mezey, 1977
Trigeminal complex	Caudal nucleus	Gobel, 1974
Spinal cord	Dorsal horn	Ralston, 1968
Superior cervical ganglion		Williams & Palay, 1969
Carotid body		McDonald & Mitchell, 1975
Invertebrates[b]		
Coelenterates		
Jellyfish	Marginal ganglia	Horridge & MacKay, 1962
Annelids		
Leech	Segmental ganglia	Muller & McMahan, 1976
Arthropods		
Arachnids		
Horsehoe crab	Optic neuropil	Whitehead & Purple, 1970
Spider	Optic neuropil	Trujillo-Cenoz, 1965
Crustaceans		
Crab	Brain	Sandeman & Mendum, 1971
Crayfish	Abdominal ganglia	Stirling, 1972
Lobster	Optic neuropile	Hamori & Horridge, 1966
	Stomatogastric ganglion	King, 1976
Insects		
Bee	Corpora pedunculata	Schurmann, 1971
Cockroach	Thoracic ganglia	Castel *et al.*, 1976
Cricket	Corpora pedunculata	Schurmann, 1971
Dragonfly	Median ocellus	Dowling & Chapell, 1972
	Optic neuropil	Arnett-Kibel *et al.*, 1977
Fly	Optic neuropil	Braitenberg & Debbage, 1974; Strausfeld, 1970
	Thoracic ganglion	King, 1977
Molluscs		
Octopus	Optic lobe	Gray, 1974
	Vertical lobe	Gray, 1970
	Stellate ganglion	Gray, 1974

[a] Modified from Shepherd, 1977.

[b] From Pearson, 1979.

('presynaptic dendrites') are present. We have noted above the evidence from the neuromuscular junction that local graded depolarisation, such as might be due to nearby synaptic inputs, could control the output from a dendritic synapse. This was explicit in the model of the dendro-dendritic interactions in the olfactory bulb (Rall & Shepherd, 1968), and was suggested by Ralston (1971) to explain the function of presynaptic dendrites in thalamic neurones. Thus, the morphological demonstration of dendritic output synapses, such as those noted in Table 1, has by itself implied that these synapses could be activated by local graded synaptic potentials without the intervention of impulse activity, though of course it does not rule out the possible additional contribution of such activity.

With the development of intracellular recording and staining techniques it

Table 2. *Nerve cells without action potentials*[a]

System	Location	Cell type	Reference
Motor			
Cockroach walking	Thoracic ganglia	Interneurones	Pearson & Fourtner, 1975
Locust walking	Thoracic ganglia	Interneurones (16 types)	Siegler & Burrows, 1979
Crustacean ventilation	Suboesophageal ganglion	Interneurones	Mendelson, 1971
Sensory			
Vertebrate vision	Retina	Photoreceptors (10 species)	Tomita, 1970
		Horizontal cells (10 species)	Werblin & Dowling 1969
		Bipolar cells (3 species)	Werblin & Dowling, 1969
Mammalian olfaction	Olfactory bulb	Granule	Rall & Shepherd, 1968
Invertebrate vision	Insect retina	Retinula cells (6 species)	Naka & Eguchi, 1962
		Monopolar cells (3 species)	Shaw, 1968
	Crustacean and mollusc eyes	Photoreceptors (4 species)	Jacklet, 1969
Crustacean proprio-ception	Coxal joints (crabs)	Stretch receptors	Ripley, Bush & Roberts, 1968

[a] Modified from Pearson (1976) which may be consulted for references to different species.

has been found that the retina is not alone in having neurones which generate only slow potentials. Fig. 4*b*, and *c* show examples of recordings from a proprioceptor and a motor interneurone respectively, and Table 2 summarises some of the cells with this property that have been identified thus far and which are examined in detail in later chapters. Interestingly, much of the evidence has come from invertebrate species, as shown in Fig. 4*b*, *c*. In these studies it has become common to refer to a cell which shows only graded potentials as a 'non-spiking' ('non-impulse') neurone. Although the term has not been necessary in characterising retinal neurones, it has gained currency in referring to this whole class of cells, and can serve this purpose until a better, preferably non-negative, definition can agreed upon.

The prevalence of these properties will not be known until methods have improved so that the smallest cells and cell processes can be adequately studied. However, the significance of the findings to date has not gone unnoticed. Pearson (1976) has suggested that 'taken together, the discoveries of dendrodendritic synapses and graded interactions between neurons are... the beginnings of a revolution in our views about the integrative mechanisms within the nervous system'. It can now be stated that any part of a neurone (other than a myelinated segment) is a possible site of input or output, or both (Bodian, 1972). The classical view of the polarised neurone, receiving inputs in its dendrites and outputs through its axon, must be merged with a more generalised conceptual framework for the organisation of functional units on the basis of synaptic connections (Shepherd, 1972, 1979; see also below). In addition to local circuits formed by synaptic connections (Rakic, 1976) it is becoming clear that there is a wealth of molecular interactions between neurones that is non-synaptic and not directly dependent on electrical activity (Schmitt, Dev & Smith, 1976; Schmitt & Worden, 1979). In reviewing this evidence, Schmitt *et al.* have suggested that '...immensely complex aggregates of local circuits may indeed prove to be neural substrates of higher brain function' with respect to learning, memory or other cognitive processes.

The general concerns raised by the new findings are discussed in the reviews cited above. Here I wish to address the specific issue of the nerve impulse and the task of reassessing its special role in the function of the nerve cell. In order to do this, we shall first have to make a small detour, and consider a complementary question, that of impulse activity in non-nervous cells.

Impulses in non-nervous cells

Work on the impulse-generating properties of non-nervous cells (other than muscle) began more than a century ago; one of the earliest and best known studies was that of Burdon-Sanderson (1873) on the Venus fly-trap.

Fig. 5. Action potentials in different types of non-nervous cells. *a. Neurospora* (Slayman, Long & Gradmann, 1976). *b.* Pumpkin stem (Sinyuhkin & Gorchakov, 1966). Duration of trace: 80 s. *c. Paramecium* (Eckert, Naitoh & Friedman, 1972). *d.* Immature egg of annelid (Hagiwara & Miyazaki, 1977). Resting potential: −62 mV; duration of trace: 500 ms. *e.* Skin of *Xenopus* larva (Roberts & Stirling, 1971). Ordinate, 20 mV divisions; abscissa, 100 msec. *f.* Cells that secrete melanocyte-stimulating hormone (MSH) in the intermediate lobe of the rate pituitary (Douglas & Taraskevich, 1978). Resting potential: −58 mV; time step: 60 ms. *g.* Insulin-secreting (*β*) cells of the rat pancreas (Dean, Matthews & Sakamoto, 1975). Resting potential: −65 mV.

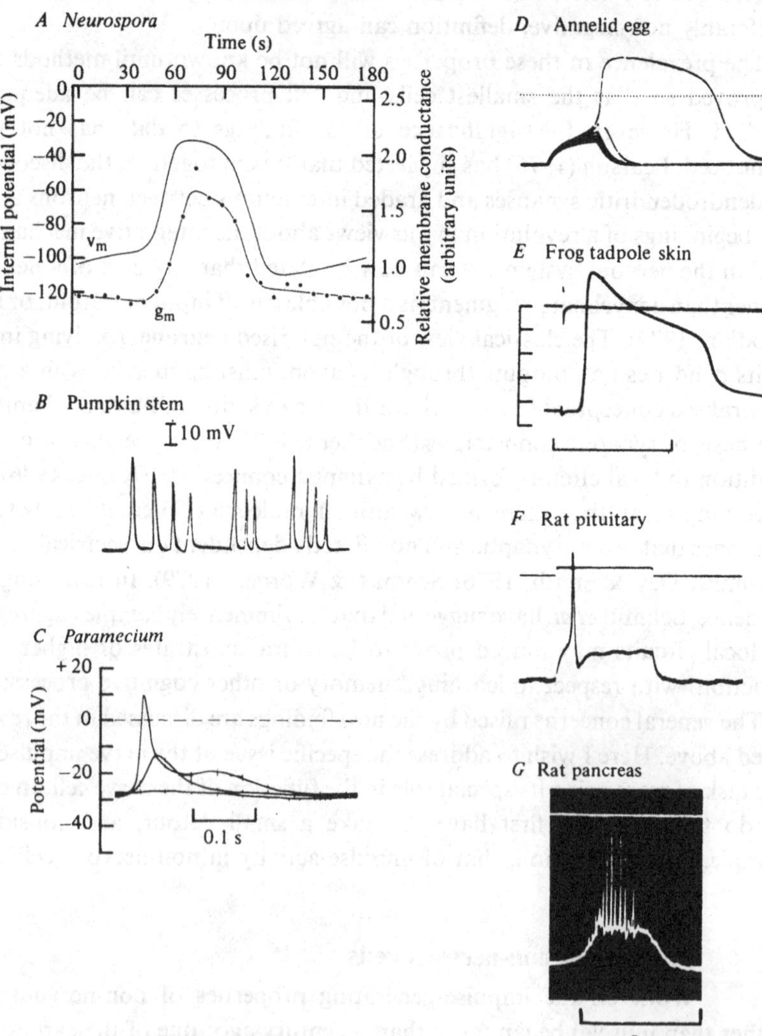

In the 1920s and 1930s, analysis of the action potential in the plant *Nitella* preceded that in the giant axon of the squid (Blinks, Harris & Osterhout, 1929). In recent years, single unit and intracellular recording techniques have revealed impulse activity in an extraordinary variety of cells.

It would be impossible to review this work within the scope of the present character, but some essential points may be noted. Examples of recordings of impulse activity in several types of non-nervous cells are illustrated in Fig. 5. Note the very slow impulse in the fungus *Neurospora* (a), lasting for approximately 90s, and the bursts of impulses in the stem of a higher plant, the pumpkin (*b*). In the unicellular *Paramecium* (*c*) impulses are a sensory response as well as mediators of the control of ciliary movement. Both the unfertilised and fertilised egg of an invertebrate, the polychaete annelid (*d*), are capable of generating action potentials. Frog tadpole skin (*e*) shows a propagating action potential present in several types of epithelial tissue. Finally, impulse activity has been demonstrated in several mammalian gland cells, such as the cells of the intermediate lobe of the pituitary (*f*) and the insulin-secreting β-cells of the pancreas (*g*).

Table 3 summarises examples of various species and their non-nervous cells in which evidence for impulse activity has been obtained. The list is not exhaustive; the attempt is simply to illustrate the diversity of cell types showing this activity. Also indicated is the evidence for the ionic mechanisms that may underlie the impulses, and the functions in which the impulses may be involved.

It seems safe to conclude that impulse generation is a property that is widely distributed in a variety of cells and tissues throughout the plant and animal kingdoms. It also appears that the ionic mechanisms for generating impulses vary in different cells. Finally, it seems clear that the functions of impulses, whether in nervous or non-nervous tissue, go considerably beyond the classical concept confined to signal propagation over long distances.

At this point we must return to the question of defining the impulse, for it is possible that the properties of impulses in nerve cells might be in some way unique, and differentiable in this respect from non-nervous impulses. Table 4 lists some of the general and widely-accepted properties of impulses, such as may be found in any textbook of biology or physiology. Most of these properties apply to non-nervous as well as nerve impulses; none as yet has been demonstrated to be unique to either type. One or two may be missing in any given cell type; several of those that have been found to be modified or absent in different cells are shown in parentheses in Table 4. Note that propagation, one of the classically-defining characteristics of the impulse, must now be regarded as a variable property in both nerve and non-nervous cells.

Table 3. *Non-nervous cells with impulse activity*[a]

Plants

Name	Species	Ions	Function	Reference
Algae	*Acetabularia*	Cl	Morphogenesis	Mummert & Gradmann, 1976
	Nitella	Cl		Mullins, 1962
Fungi	*Neurospora*	Cl	Ion movements	Slayman *et al.*, 1976
Angiospermae	*Cucurbita* (pumpkin)		Vascular	Sinyuhkin & Gorchakov, 1966
	Helianthus (sunflower)		Vascular	Sinyuhkin & Gorchakov, 1966
	Pisum (pea)			
	Mimosa			Sibaoka, 1962
	Biophytum			Sibaoka, 1973
	Dionaea (Venus fly-trap)		Movement: fast	Siboaka, 1966
	Drosera		Movement: slow	Williams & Pickard, 1972

Animals

Name	Cells	Ions	Function	Reference
Protozoa	*Paramecium*	Ca	Cilia movement	Eckert *et al.*, 1972
	Dinoflagellate		Bioluminescence	Eckert, 1965
Hydrozoa	Epithelium	Na	Secretion	Mackie, 1976
jellyfish	Secretory epithelium (rete mirabile)			
	Exumbrella epithelium		Sensory/propagation	Mackie, 1970
Annelida	Oocyte	Ca, Na		Hagiwara & Miyazaki, 1977
Echinodermata				
sea urchin	Oocyte	Ca	Activation	Scheer *et al.*, 1954
starfish	Oocyte	Ca		Miyazaki *et al.*, 1975

Tunicata	Oocyte	Ca (Na)		Miyazaki *et al.*, 1974
	Epithelium		Sensory/propagation	Mackie & Bone, 1977
Amphibia				
frog	Skin		Secretion	Finkelstein, 1964
toad	Bladder			Finkelstein, 1964
	Oocyte	Cl		
tadpoles	Skin	Na	Sensory/propagation	Roberts, 1971
Mammalia				
mouse	Oocyte	Ca		Okamoto *et al.*, 1977
rat	Pancreas (β cells)	Ca (Na)	Secretion	Dean & Matthews, 1970
	Adrenal medulla			
	(chromaffin)	Na (Ca)	Secretion	Brandt *et al.*, 1976
	Pituitary			
	(posterior)	Ca	Secretion	Douglas, 1974
	(intermediate)	Na	Secretion	Douglas & Taraskevich, 1978
	(anterior)		Secretion	Taraskevich & Douglas, 1977
many species	Muscle			
	skeletal	Na, K	Movement	
	cardiac	Na, K	Movement	
	(smooth)		Movement	
tumour cells	Pheochromocytoma			
	Thyroid			
	Bronchial			
	Pituitary adenoma			
Artificial				
Lipid bilayers				

[a] See also reviews by Mackie, 1970, and Spencer, 1975.

What can be said of the functions of impulses? Table 4 summarises, in a more coherent order, the functions noted previously in Table 3. It is apparent that impulse activity may be involved in a number of basic biological mechanisms in the life of the cell. Moreover, the functions are not necessarily mutually exclusive; impulses might serve different functions at successive developmental stages, or even contemporaneously at a given stage.

In view of these considerations, it appears that our understanding of the cellular functions mediated by impulses is undergoing a revolution every bit as profound as that surrounding the discovery of neuronal functions that are not mediated by impulses. The pervasiveness of the new view may be judged by the following observations of Slayman, Long & Gradmann (1976) in their study of action potentials in the fungus *Neurospora*.

> The number of biological systems (in addition to the classical excitable tissues of nerve, muscle, and the giant algae) that are now known to display reversible all-or-none changes in membrane permeability, giving rise to action potentials, is sufficiently large to raise the question of whether essentially all biological membranes may be inherently capable of producing action potentials, given appropriate environmental conditions.

The two lines of research we have been following have thus led us to what

Table 4. *Properties and functions of impulses*

Physiological properties of impulses	*Functions of impulses*
	Development
Transient time course	in fertilisation
(Threshold for activation)	in cell division
Regenerative relation to membrane potential	in morphogenesis
All-or-nothing	Ion transfers across
Depolarising	membrane
(Overshoot of membrane potential)	Bioluminescence
Change in membrane permeability and ionic	Secretion
conductance	hormonal
Ultimately dependent on energy	glandular
Refractory periods	synaptic
(Propagating)	Motility
	ciliary
	vascular
	muscle
	Nerve signalling
	propagation
	other functions

appear to be incompatible conclusions, that many nerve cells function without impulses *versus* the inherent capacity to generate impulses of many, and possibly all, cell membranes. In order to reconcile these views, we return to the nerve cell and consider more closely the nature of spiking and non-spiking activity.

Relations between spiking and non-spiking activity in nerve

Analysis of neuronal properties must take into account many factors. Among these is the technical problem that any experimental result is contingent on the recording conditions. One must always rule out, for example, that a normally excitable cell or cell process has not been rendered inexcitable, or, conversely, a normally inexcitable cell or cell process excitable, by the act of microelectrode recording or penetration or some other factor in the physiological state of the animal or tissue.

It must next be emphasised that when one characterises a neurone as 'non-spiking' it is only relative to the site of recording. One must take into account that there may be spiking activity in distal dendrites and small branches and terminals that is not registered at the site of recording in a cell body. This becomes all the more relevant as instances are reported of 'dendritic spikes' or 'hot spots' (cf. Spencer & Kandel, 1961; Llinás & Nicholson, 1971; Traub & Llinás, 1979). Thus, the fact that a cell is non-spiking in one part does not rule out that it may be spiking in another; conversely, the fact that a cell is spiking in one part does not rule out that it may have synaptic outputs controlled by non-spiking activity in other parts.

Many nerve cells thus appear to be morphological entities which nonetheless contain multiple functional loci. We must recognise that these loci may have different functional properties. If a locus has the property of excitability, its function is not restricted to the classical one of impulse propagation. This would apply, for example, to dendritic spikes as non-propagating boosters of signal transmission in dendrites. However, one can also conceive of other functions, such as boosting the intensity of local dendritic output, digitising non-propagating signals, prolonging the duration of a signal, or contributing to synaptic plasticity (cf. Baylor & Nicholls, 1969; Erulkar & Weight, 1977, Kandel, 1979). It is also appropriate to begin to consider the functions for impulses in non-nervous cells, such as are listed in Table 3, and to ask whether some of these may apply also to nerve cells. Of particular interest in this regard are the possible roles of impulses in differentiation and development; the maintenance of ionic concentrations across membranes; the non-synaptic release of substances and other non-neural secretory functions; the involvement in motility. Many of these functions may seem improbable for cell bodies and large axons. However, it is well known that in fine unmyelinated axons, down to diameters of 0.2 μm, the surface-to-volume relations force a much tighter coupling between impulse activity and metabolism (cf. Greengard & Ritchie, 1971), and similar considerations should apply to activity at the level of the finest processes in the neuropil.

Finally, the distinction between spiking and non-spiking activity may not be as hard and fast as envisaged in classical terms. This of course was recognised in the older literature on decrementing conduction. From recent research the problem can be formulated in terms of molecular mechanisms in the membrane, as represented in the models of Fig. 2*a*, *b*. It can be appreciated that the voltage-dependence of the ionic conductances is determined by several factors, such as the number and density of channels, and the kinetics of their formation. One can thus conceive that all cell membranes might contain the molecular mechanisms for forming voltage-dependent ionic channels, but some cells, such as non-spiking neurones or many types of non-nervous cells, would not show this property under normal conditions simply because they lack, as an example, sufficient density of channels (cf. Baumann & Mueller, 1974). Such a cell might nonetheless be induced to show some degree of excitability under extreme conditions (cf. the weak responses of frog skin to strong depolarising pulses; Finkelstein, 1964). Thus, in these respects, excitability itself might be regarded as a graded property. Similar considerations apply also to synaptic potentials; although these have classically been regarded as due to changes in ionic conductances that are independent of membrane potential, recent research has provided evidence for some degree of voltage sensitivity in a number of cases, such as the neuromuscular junction (Magleby & Stevens, 1972), cardiac ganglion cell (Hartzell, Kuffler, Stickgold & Yoshikami, 1977), olfactory cortex (Mori, Satou & Takagi, 1978) and olfactory bulb (Mori & Shepherd, 1979; see my chapter in this book). These and other mechanisms such as conductance-decrease synapses (Weight, 1974) and extremely prolonged synaptic potentials indicate that a wide variety of properties is associated with synaptic actions (cf. Shepherd, 1979 and this volume).

The cellular nature of the neurone

Our reassessment of the nerve impulse has thus provided us with a new perspective on the nerve cell as a most complex morphological substrate providing for multiple state-dependent functions. In addition, it is obvious that this new perspective requires a reconsideration of the question of how we define the neurone.

The classical view was simple: a neurone is a cell that conducts impulses.

Bishop was right, as we have seen, to insist that many nervous functions in fact depend on graded electrical signals. However, the idea that impulse transmission ultimately occurs in all nerve cells, and that it is mediated by the axon, has remained firmly rooted in our thinking. Bishop himself, for example, regarded the impulse as an evolutionary specialisation for long-distance transmission; he stated that 'The only known or even probable function of the axon is to conduct energy (i.e. impulses) between two regions of graded response tissue'. We have, however, reviewed considerable evidence for other functions for impulses. As for the morphological structure referred to as the axon, there is evidence that not only do the axons of horizontal cells not generate impulses, they can in fact serve to reduce or eliminate communication between the cell body and the axonal arborisation (Nelson, Lutzow, Kolb & Gouras, 1975). Finally, the idea that graded activity is the more general and primitive property seems irrelevant for most of the cells with this property, and also inconsistent with the prevalence of excitability in cells of different phyla and at all stages of morphological development, beginning with the ovum.

We thus find ourselves drawn to the conclusion that the classical concept of excitability – the capacity to generate action potentials – as the defining property of the neurone is no longer tenable. Rather, it appears that excitability should be regarded as a general property, found in a variety of cell types (possibly, to some degree, in most or all cell types) throughout the plant and animal kingdoms. One view would be that cells that clearly display this property be considered as having nervous, or 'neuroid', properties (cf. Spencer, 1975). But in view of the fact that many nerve cells function without impulses, a more reasonable view is that *the impulse in nervous tissue expresses general biological properties of excitability*. Some nerve cells may use this property for specific functions, others may not. Rather than engendering confusion or vagueness, this conclusion instead may impart a new unity to the study of nervous function. Just as the anatomical studies of the 1880s showed that the nerve cell shares in the morphological unity of the cell theory, so the physiological properties of nerve cells can now be seen to be shared, to a greater or lesser degree, with basic biological functions common to all cells. The analysis of non-spiking neurones thus provides a clue to the essentially cellular nature of the neurone. This general conclusion is not restricted to the example of excitability; consideration of the growing evidence for non-electrical and non-synaptic properties and interactions indicates that many other neuronal functions may be mediated by mechanisms shared with other cells of the body (cf. Schmitt & Worden, 1979). In this respect the study of the neurone is merging with the study of cell biology.

Is there left, then, any defining characteristic of the nerve cell? Here we must recognise that, although there can be agreement on regarding a motoneurone or a cortical pyramidal cell as a neurone, there is an enormous diversity in the morphological structure and physiological properties of cells in the nervous system; the more familiar we become with this diversity, the more it seems that there is virtually a continuum extending from cells like the motoneurone to the cells of muscles, glands or epithelia. In the retina, for example, the horizontal cells of teleost fish have some of the fine structural characteristics of glial cells, together with the physiological property of responding to light with only graded responses. Stell (1967) discussed the question of whether these cells are in fact neurones, and concluded that they are, on the basis that they 'are morphological parts of organized multicellular networks concerned with the processing of specific visual information'.

In slightly more general terms, one can suggest that *a neurone is a cell that is part of an interconnected multicellular system concerned with the processing of specific information of behavioural significance*. This serves to remind us that it is information that a nerve cell is concerned with; that information can be generated, transmitted or stored by a variety of physiological processes, including impulses where necessary; that it can be mediated by a variety of morphological interrelations with other cells, including, among the most characteristic, the synapse; and that a single cell has its significance as part of multicellular functional ensembles. These considerations do not by any means provide an unequivocal definition for all possible or probable nerve cells, but they at least provide a perspective that reflects the remarkable diversity of cellular mechanisms found in nerve cells, and the unique contributions that they make to the economy of the body.

References

Andres, K. H. (1965). Der Feinbau des Bulbus Olfactorius der Ratte unter besonderer Berucksichtigung der synaptischen Verbindungen. *Z. Zellforsch.* **65**, 530–61.

Arnett-Kibel, C., Meinertzhagen, I. A. & Dowling, J. E. (1977). Cellular and synaptic organization in the lamina of the dragon-fly, *Sympetrum rubicundulum. Proc. R. Soc. Lond. Series B*, **196**, 385–412.

Barker, L. F. (1896). *The Nervous System and its Constituent Neurons.* New York: Appleton.

Barrett, J. N. & Crill, W. W. (1974). Specific membrane properties of cat motoneurones. *J. Physiol., Lond.* **239**, 301–24

Baumann, G. & Mueller, P. (1974). A molecular model of membrane excitability. *J. Supramolec. Struct.* **2**, 538–57.

Baylor, D. A., Fuortes, M. G. F. & O'Bryan, P. M. (1971). Receptive fields of cones in the retina of the turtle. *J. Physiol., Lond.* **214**, 265–94.

Baylor, D. A. & Nicholls, J. G. (1969). After-effects of nerve impulses on signalling in the central nervous system of the leech. *J. Physiol., Lond.* **203**, 571–89.

Bishop, G. H. (1956). Natural history of the nerve impulse. *Physiol. Rev.* **36**, 379–99.

Blinks, L. R., Harris, E. S. & Osterhout, W. J. V. (1929). Studies on stimulation in *Nitella. Proc. Soc. Exptl. Biol. Med.* **26**, 836–8.

Bloedel, J., Gage, P. W., Llinás, R. & Quastel, D. M. J. (1966). Transmitter release at the squid giant synapse in the presence of tetrodotoxin. *Nature, Lond.* **212**, 49–50.

Bodian, D. (1972). Synaptic diversity and characterization by electron-microscopy. In *Structure and Function of Synapses* ed. G. D. Pappas & D. P. Purpura. New York: Raven.

Boring, E. G. (1950). *A History of Experimental Psychology*. New York: Appleton-Century-Crofts.

Braitenberg, V. & Debbage, P. (1974). A regular set of reciprocal synapses in the visual system of the fly, *Musca domestica. J. Comp. Physiol.* **90**, 25–31.

Brandt, B. L., Hagiwara, S., Kidokoro, Y. & Miyazaki, S. (1976). Action potentials in the rat chromatoffin cell and effects of acetylcholine. *J. Physiol., Lond.* **263**, 417–39.

Bullock, T. H. (1959). Neuron doctrine and electrophysiology. *Science*, **129**, 997–1002.

Burden-Sanderson, J. (1873). Note on the electrical phonemena which accompany stimulation of the leaf of *Dionaea muscipula. Proc. Roy. Soc.* **21**, 495–6.

Bush, B. M. H. & Roberts, A. (1968). Resistance reflexes from a crab muscle receptor without impulses. *Nature, Lond.* **218**, 1171–3.

Byrne, J. H. (1980). Analysis of ionic conductance mechanisms in motor cells mediating inking behavior in *Aplysia californica. J. Neurophysiol.* **43**, 630–50.

Castel, M., Spira, M. E., Parnas, I. & Yarom, Y. (1976). Ultrastructure of region of a low safety factor in inhomogeneous giant axon of the cockroach. *J. Neurophysiol.* **39**, 900–8.

Clare, M. H. & Bishop, G. H. (1955). Properties of dendrites: apical dendrites of cat cortex. *Electroencephalog. Clin. Electrophysiol.* **7**, 85–98.

Coombs, J. S., Eccles, J. C. & Fatt, P. (1955*a*). Excitatory synaptic action in motoneurones. *J. Physiol., London*, **130**, 374–95.

Coombs, J. S., Eccles, J. C. & Fatt, P. (1955*b*). The inhibitory suppression of reflex discharges from motoneurones. *J. Physiol., Lond.* **130**, 396–413.

Dean, P. M. & Matthews, E. K. (1970). Glucose-induced electrical activity in pancreatic islet cells. *J. Physiol., London*, **210**, 255–64.

Dean, P. M., Matthews, E. K. & Sakamoto, Y. (1975). Pancreatic islet cells: effects of monosaccharides, glycolytic intermediates and metabolic inhibitors on membrane potential and electrical activity. *J. Physiol., Lond.* **246**, 459–78.

del Castillo, J. & Katz, B. (1954). Changes in end-plate activity produced by presynaptic polarization. *J. Physiol., Lond.* **124**, 586–604.

Douglas, W. W. (1974). Mechanisms of release of neurohypophysiol hormones: stimulus-secretion coupling. In *Handbook of Physiology, Sect. F., Endocrinology, Vol. IV, The Pituitary Gland and its Endocrine Control, Part 1*, ed. E. Knobil & W. H. Sawyer, pp. 191–224. Washington: American Physiological Society.

Douglas, W. W. & Taraskevich, P. S. (1978). Action potentials in gland cells of rat pituitary pars intermedia: inhibition by dopamine, an inhibitor of MSH secretion. *J. Physiol., Lond.* **285**, 171–84.

Dowling, J. E. & Boycott, B. B. (1966). Organization of the primate retina: electron microscopy. *Proc. R. Soc. Ser. B*, **166**, 80–111.

Dowling, J. E. & Chappell, R. L. (1972). Neural organization of the median occellus of the dragonfly. II. Synaptic structure. *J. Gen. Physiol.* **60**, 121–65.

Eccles, J. C. (1953). *The Neurophysiological Basis of Mind.* Oxford: Clarendon.

Eccles, J. C. (1964). *The Physiology of Synapses.* Berlin: Springer.

Eckert, R. (1965). Bioelectric control of bioluminescence in the dinoflagellate noctiluca. *Science*, **147**, 1140–5.

Eckert, R., Naitoh, Y. & Friedman, K. (1972). Sensory mechanisms in *Paramecium*. I. Two components of the electrical responses to mechanical stimulation of the anterior surface. *J. Exp. Biol.*, **56**, 683–94.

Erulkar, S. D. & Weight, F. F. (1977). Extracellular potassium and transmitter release at the giant synapse of squid. *J. Physiol., Lond.* **266**, 209–18.

Eyzaguirre, C. & Kuffler, S. W. (1955). Processes of excitation in the dendrites and in the soma of single isolated sensory nerve cells of the lobster and crayfish. *J. Gen. Physiol.* **39**, 87–119.

Famiglietti, E. V., Jr. & Peters, A. (1972). The synaptic glomerulus and the intrinsic neuron in the dorsal lateral geniculate nucleus of the cat. *J. Comp. Neurol.* **144**, 285–334.

Fatt, P. & Katz, B. (1951). An analysis of the end-plate potential recorded with an intracellular electrode.. *J. Physiol., Lond.* **115**, 320–70.

Finkelstein, A. (1964). Electrical excitability of isolated frog skin and toad bladder. *J. Gen. Physiol.* **47**, 545–65.

Forbes, A. (1922). The interpretation of spinal reflexes in terms of present knowledge of nerve conduction. *Physiol. Rev.* **2**, 301–414.

Frank, K. & Fuortes, M. G. F. (1957). Presynaptic and post-synaptic inhibition of monosynaptic reflex. *Fed. Proc.* **16**, 39–40.

Gobel, S. (1974). Synaptic organization of the substantia gelatinosa glomeruli in the spinal trigeminal nucleus of the adult cat. *J. Neurocytol.* **3**, 219–43.

Gray, E. G. (1962). A morphological basis for presynaptic inhibition? *Nature, Lond.* **193**, 82–3.

Gray, E. G. (1970). The fine structure of the vertical lobe of the octopus brain. *Philos. Trans. R. Soc. Lond., Series B*, **258**, 379–95.

Gray, E. G. (1974). Synaptic morphology with special reference to microneurones. In *Essays on the Nervous System*, ed. R. Bellairs & E. G. Gray, pp. 155–78. Oxford: Clarendon Press.

Greengard, P. & Ritchie, J. M. (1971). Metabolism and function in nerve fibers. In *Handbook of Neurochemistry*, Vol. 5, Part A, ed. A. Lajtha, pp. 317–35. New York: Plenum.

Guldner, F.-H. (1976). Synaptology of the rat suprachiasmatic nucleus. *Cell and Tissue Res.* **165**, 509–44.

Hagiwara, S. & Miyasaki, S. (1977). Changes in excitability of the cell membrane during differentiation without cleavage in the egg of the annelid, *Chaetopterus pergamentaceus*.. *J. Physiol., Lond.* **272**, 197–216.

Hamori, J. & Horridge, G. A. (1966). The lobster optic lamina. II. Types of synapse. *J. Cell Sci.* **1**, 257–70.

Hamori, J. & Mezey, E. (1977). Serial and triadic synapses in the cerebellar nuclei of the cat. *Exp. Brain Res.* **30**, 259–73.

Harding, B. N. (1973). An ultrastructural study of the centre median and ventrolateral thalamic nuclei of the monkey. *Brain Res.* **54**, 335–40.

Hartzell, H. C., Kuffler, S. W., Stickgold, R. & Yoshikami, D. (1977). Synaptic excitation and inhibition resulting from direct action of acetylcholine on two types of chemoreceptors on individual amphibian parasympathetic neurones.. *J. Physiol., Lond.* **271**, 817–46.

Hirata, Y. (1964). Some observations on the fine structure of the synapses in the olfactory bulb of the mouse, with particular reference to the atypical synaptic configuration. *Arch. Histol. Japan*, **24**, 293–302.

Hodgkin, A. L. & Huxley, A. F. (1952). A quantitative description of membrane current and its application to conduction and excitation in nerve. *J. Physiol., Lond.* **117**, 500–44.

Horridge, G. A. & MacKay, B. (1962). Naked axons and symmetrical synapses in coelenterates. *Q. J. Microsc. Sci.* **103**, 531–41.

Hosoya, Y. (1973). Electron microscopic observations of the granule cells (Calleja's island) in the olfactory tubercle of rats. *Brain Res.* **54**, 330–4.

Jack, J. J. B. & Redman, S. J. (1971). An electrical description of the motoneurone, and its application to the analysis of synaptic potentials. *J. Physiol., Lond.* **215**, 321–52.

Jacklet, J. W. (1969). Electrophysiological organization of the eye of *Aplysia. J. Gen. Physiol.* **53**, 21–42.

Jackowski, A., Parnevalas, J. G. & Lieberman, A. R. (1978). The reciprocal synapse in the external plexiform layer of the mammalian olfactory bulb. *Brain Res.* **159**, 17–28.

Kandel, E. R. (1979). Cellular insights into behavior and learning. *Harvey Lectures*. pp. 19–92. New York: Academic Press.

Kaneko, A. (1970). Physiological and morphological identification of horizontal, bipolar and amacrine cells in goldfish retina. *J. Physiol., Lond.* **207**, 623–33.

Katz, B. & Miledi, R. (1967). A study of synaptic transmission in the absence of nerve impulses. *J. Physiol., Lond.* **192**, 407–36.

King, D. G. (1976). Organization of crustacean neuropil. I. Patterns of synaptic connections in lobster stomatogastric ganglion. *J. Neurocytol.* **5**, 207–37.

King, D. G. (1977). An interneuron in *Drosophila* synapses within a peripheral nerve onto the dorsal longitudinal muscle motor neurons. *Neurosci. Abst.* **3**, 180.

Llinás, R. & Nicholson, C. (1971). Electrophysiological properties of dendrites and somata in alligator Purkinje cells. *J. Neurophysiol.* **34**, 532–51.

Lorente de Nó, R. (1938). The cerebral cortex: architecture, intracortical connections and motor projections. In *Physiology of the Nervous System* ed. J. F. Fulton, pp. 291–325. London: Oxford.

Lucas, K. (1917). *The Conduction of the Nervous Impulse*. Cambridge University Press.

Lund, R. D. (1969). Synaptic patterns of the superficial layers of the superior colliculus of the rat. *J. Comp. Neurol.* **135**, 179–208.

McDonald, D. M. & Mitchell, R. A. (1975). The innervation of glomus cells, ganglion cells and blood vessels in the rat carotid body: a quantitative ultrastructural analysis. *J. Neurocytol.* **4**, 177–230.

Mackie, G. O. (1970). Neuroid conduction and the evolution of conducting tissues. *Quart. Rev. Biol.* **45**, 319–32.

Mackie, G. O. (1976). Propagated spikes and secretion in a coelenterate glandular epithelium. *J. Gen Physiol.* **68**, 313–25.
Mackie, G. O. & Bone, Q. (1977). Locomotion and propagated skin impulses in salps (*Tunicata: Thaliacae*). *Biol. Bull.* **153**, 180–97.
Magleby, K. L. & Stevens, C. G. (1972). The effects of voltage on the time course of end-plate currents. *J. Physiol., Lond.* **223**, 151–71.
Mendelson, M. (1971). Oscillator neurons in crustacean ganglia. *Science*, **171**, 1170–3.
Mihailoff, G. A. (1978). Anatomic evidence suggestive of dendrodendritic synapses in the opossum basilar pons. *Brain Res. Bull.* **3**, 333–40.
Miyazaki, S., Ohmori, H. & Sasaki, S. (1975). Action potential and non-linear current–voltage relation in starfish oocytes. *J. Physiol., Lond.* **246**, 37–54.
Miyazaki, S., Takahashi, K., Tsuda, K. & Yoshi, M. (1974). Electrical excitability in the egg cell membrane of tunicate. *J. Physiol., Lond.* **238**, 37–54.
Morest, D. K. (1971). Dendrodendritic synapses of cells that have axons: the fine structure of the Golgi type II cell in the medial geniculate body of the cat. *Z. Anat. Entwicklungsgeschichte*, **133**, 216–46.
Mori, K., Satou, M. & Takagi, S. F. (1978). Fast and slow inhibitory postsynaptic potentials in the piriform cortex neurons. *Proc. Jap. Akad.* **54B**, 484–9.
Mori, K. & Shepherd, G. M. (1979). Synaptic excitation and longlasting inhibition of mitral cells in the *in vitro* turtle olfactory bulb. *Brain Res.* **172**, 155–9.
Mueller, P. (1979). The mechanism of electrical excitation in lipid bilayers and cell membranes. In *The Neurosciences: Fourth Study Program*, ed. F. O. Schmitt & F. G. Worden, pp. 641–58. Cambridge, Massachusetts: MIT Press.
Muller, K. J. & McMahan, V. J. (1976). The shapes of sensory and motor neurones and distribution of their synapses in ganglia of the leech: A study using intracellular injections of horseradish peroxidase. *Proc. R. Soc. Lond. Ser. B*, **194**, 481–99.
Mullins, L. J. (1962). Efflux of chloride ions during the action potential of *Nitella. Nature, Lond.* **196**, 986.
Mummert, H. & Gradmann, D. (1976). Voltage dependent potassium fluxes and the significance of action potentials in *Acetabularia. Biochim. Biophys. Acta*, **443**, 443–50.
Naka, K. & Eguchi, E. (1962). Spike potentials recorded from insect photoreceptors. *J. Gen. Physiol.* **45**, 663–80.
Nelson, R., von Lutzow, A., Kolb, H. & Gouras, P. (1975). Horizontal cells in cat retina with independent dendritic systems. *Science*, **189**, 137–9.
Okamoto, H., Takahashi, K. & Yamashita, N. (1977). Ionic currents through the mammalian oocyte and their comparison with those in the tunicate and sea urchin. *J. Physiol., Lond.* **267**, 465–95.
Pearson, K. G. (1976). Nerve cells without action potentials. In *Simpler Networks and Behavior*, ed. J. C. Fentress, pp. 99–110. Sunderland, Mass: Sinauer.
Pearson, K. G. (1979). Local neurons and local interactions in the nervous systems of invertebrates. In *The Neurosciences, Fourth Study Program*, ed. F. O. Schmitt & F. G. Worden, pp. 145–60. Cambridge: MIT Press.

Pearson, K. G. & Fourtner, C. R. (1975). Non-spiking interneurons in the walking system of the cockroach. *J. Neurophysiol.* **38**, 33–52.
Pellionicz, A. & Llinás, R. (1977). A computer model of cerebellar Purkinje cells. *Neurosci.* **2**, 37–48.
Phillips, C. G., Powell, T. P. S. & Shepherd, G. M. (1963). Responses of mitral cells to stimulation of the lateral olfactory tract in the rabbit. *J. Physiol., Lond.* **168**, 65–88.
Pinching, A. J. & Powell, T. P. S. (1971). The neuropil of the glomeruli of the olfactory bulb. *J. Cell Sci.* **9**, 347–77.
Rakic, P. (1976). *Local Circuit Neurons.* Cambridge, Mass: MIT.
Rall, W. (1957). Time constant of motoneurons. *Science*, **126**, 454.
Rall, W. (1964). Theoretical significance of dendritic trees for neuronal input-output relations. In *Neural Theory and Modeling*, ed. R. F. Reiss, pp. 73–97. Stanford.
Rall, W. (1967). Distinguishing theoretical synaptic potentials computed for different soma-dendritic distributions of synaptic input. *J. Neurophysiol.* **30**, 1138–68.
Rall, W. & Shepherd, G. M. (1968). Theoretical reconstruction of field potentials and dendrodendritic synaptic interactions in olfactory bulb. *J. Neurophysiol.* **31**, 884–915.
Rall, W., Shepherd, G. M., Reese, T. S. & Brightman, M. W. (1966). Dendro-dendritic synaptic pathway for inhibition in the olfactory bulb. *Exptl. Neurol.* **14**, 44–56.
Ralston, H. J. III (1968). The fine structure of neurons in the dorsal horn of the cat spinal cord. *J. Comp. Neurol.* **132**, 275–302.
Ralston, H. J. III (1971). Evidence for presynaptic dendrites and a proposal for their mechanism of action. *Nature, Lond.* **230**, 585–7.
Ramón y Cajal, S. (1911). *Histologie du System Nerveux de l'Homme et des Vertebres.* Paris: Maloine.
Ramon y Cajal, S. (1954). Neuron theory or reticular theory? Objective evidence of the anatomical unity of nerve cells. Madrid: Consejo Superior de Investigaciones Cientificas.
Ripley, S. H., Bush, B. M. H. & Roberts, A. (1968). Crab muscle receptor which responds without impulses. *Nature, Lond.* **218**, 1170–1.
Ritchie, J. M. (1979). A pharmacological approach to the structure of sodium channels in myelinated axons. *Ann. Rev. Neurosci.* **2**, 341–62.
Roberts, A. (1971). The role of propagated skin impulses in the sensory system of young tadpoles. *Z. Vergl. Physiol.* **75**, 388–401.
Roberts, A. & Stirling, C. A. (1971). The properties and propagation of a cardiac-like impulse in the skin of young tadpoles. *Z. Vergl. Physiol.* **71**, 295–310.
Sandeman, D. C. & Mendum, C. M. (1971). The fine structure of the central synaptic contacts on an identified crustacean motoneuron. *Z. Zellforsch. Anat.* **119**, 515–25.
Scheer, B. T., Monroy, A., Santagelo, M. & Riccobono, G. (1954). Action potentials in sea urchin eggs at fertilization. *Exptl. Cell Res.* **7**, 284–7.
Schmitt, F. O., Dev, P. & Smith, B. H. (1976). Electrotonic processing of information by brain cells. *Science*, **193**, 114–20.
Schmitt, F. O. & Worden, F. G. (eds.) (1979). *The Neurosciences: Fourth Study Program.* Cambridge, Mass: MIT.
Schurmann, F. W. (1971). Synaptic contact of association fibers in the brain of the bee. *Brain Res.* **26**, 169–76.

Setalo, G. & Szekely, G. (1967). The presence of membrane specializations indicative of somato-dendritic synaptic junctions in the optic tectum of the frog. *Exptl. Brain Res.* **4**, 237–42.

Shaw, S. R. (1968). Organization of the locust retina. *Symp. Zool. Soc. Lond.* **23**, 135–63.

Shepherd, G. M. (1972). The neuron doctrine: a revision of functional concepts. *Yale J. Biol. Med.* **45**, 584–99.

Shepherd, G. M. (1977). The olfactory bulb: a simple system in the mammalian brain. In *Handbook of Physiology*, Sect. 1. *The Nervous System*, vol. 1, part 2. *Cellular Biology of Neurons*, ed. E. R. Kandel, pp. 945–68. Bethesda: American Physiological Society.

Shepherd, G. M. (1979). *The Synaptic Organization of the Brain.* Second Edition. New York: Oxford University Press.

Sherrington, C. S. (1906). *The Integrative Action of the Nervous System.* New Haven: Yale.

Sibaoka, T. (1962). Excitable cells in Mimosa. *Science*, **137**, 226.

Sibaoka, T. (1966). Action potentials in plant organs. *Symp. Soc. Exptl. Biol.* **20**, 49–73.

Sibaoka, T. (1973). Transmission of action potentials in Biophytum. *Bot. Mag. Tokyo*, **86**, 51–61.

Siegler, M. V. S. & Burrows, M. (1979). The morphology of local non-spiking interneurones in the metathoracic ganglion of the locust. *J. Comp. Neurol.* **183**, 121–48.

Sinyuhkin, A. M. & Gorchakov, V. V. (1966). Characteristics of the action potentials of the conducting system of pumpkin stems evoked by various stimuli. *Soviet Plant Physiol.* **13**, 727–33.

Slayman, C. L., Long, W. S. & Gradmann, D. (1976). 'Action potentials' in *Neurospora crassa*, a mycelial fungus. *Biochem. Biophys. Acta*, **426**, 732–44.

Sloper, J. J. (1972). Dendro-dendritic synapses in the primate motor cortex. *Brain Res.* **34**, 186–92.

Spencer, A. N. (1975). Non-nervous conduction in invertebrates and embroyos. *Am. Zool.* **14**, 917–29.

Spencer, W. A. & Kandel, E. R. (1961). Electrophysiology of hippocampal neurons. IV. Fast prepotentials. *J. Neurophysiol.* **24**, 272–85.

Stell, W. K. (1967). The structure and relationships of horizontal cells and photoreceptor-bipolar synaptic complexes in goldfish retina. *Am. J. Anat.* **121**, 401–24.

Stell, W. K. (1972). The morphological organization of vertebrate retina. In *Physiology of Photoreceptor Organs, Handbook of Sensory Physiology*, Vol. VII/2, ed. M. C. F. Fuortes, pp. 111–213. Berlin: Springer.

Stirling, C. A. (1972). The ultrastructure of giant fiber and serial synapses, in crayfish. *Z. Zellforsch*, **131**, 31–45.

Strausfeld, N. J. (1970). Golgi studies on insects. II. The optic lobes of *Diptera. Philos. Trans. R. Soc. Lond., Series B*, **258**, 135–223.

Takeuchi, A. & Takeuchi, N. (1962). Electrical changes in the pre- and postsynaptic axons of the giant synapse of Loligo. *J. Gen. Physiol.* **45**, 1181–93.

Taraskerich, P. S. & Douglas, W. W. (1977). Action potentials occur in cells of the normal anterior pituitary gland and are stimulated by the hypophysiotropic peptide thyrotropin-releasing hormone. *Proc. Nat. Acad. Sci.* **74**, 4064–7.

Tomita, T. (1970). Electrical activity of vertebrate photoreceptors. *Q. Rev. Biophys.* **3**, 179–222.

Traub, R. D. (1977). Motoneurons of different geometry and the size principle. *Biol. Cybernetics*, **25**, 163–76.

Traub, R. D. & Llinás, R. (1979). Hippocampal pyramidal cells: significance of dendritic ionic conductances for neuronal function and epileptogenesis. *J. Neurophysiol.* **42**, 476–96.

Trujillo-Cenoz, O. (1965). Some aspects of the structural organization of the arthropod eye. *Cold Spring Harbor Symp. Quant. Biol.* **30**, 371–82.

van der Loos, H. (1960). On dendro-dendritic junctions in the cerebral cortex. In *Structure and Function of the Cerebral Cortex*, ed. D. B. Tower & J. P. Schade, pp. 36–42. Amsterdam: Elsevier.

Weight, F. (1974). Synaptic potentials resulting from conductance decreases. In *Synaptic Transmission and Neuronal Interaction*, ed. M. V. L. Bennett, pp. 141–52. New York: Raven.

Werblin, F. S. & Dowling, J. E. (1969). Organization of the retina of the mud puppy, *Necturus maculosus*. II. Intracellular recording. *J. Neurophysiol.* **32**, 339–55.

Whitehead, R. & Purple, R. L. (1970). Synaptic organization in the neuropile of the lateral eye of *Limulus*. *Vision Res.* **10**, 129–33.

Williams, S. E. & Pickard, B. G. (1972). Properties of action potentials in *Drosera* tentacles. *Planta*, **103**, 222–40.

Williams, T. H. & Palay, S. L. (1969). Ultrastructure of the small neurons in the superior cervical ganglion. *Brain Res.* **15**, 17–34.

Wilson, C. J., Groves, P. M. & Fifkova, E. (1978). Monoaminergic synapses, including dendrodendritic synapses, in the rat substantia nigra. *Exp. Brain Res.* **30**, 1–14.

GORDON L. FAIN

Integration by spikeless neurones in the retina

One of the central goals of neurophysiology is to understand how single cells in the central nervous system interact to influence behaviour. Since the human brain contains 10^{10} to 10^{11} neurones, each making perhaps 10^3 to 10^4 synaptic connections with other neurones, it is clearly not within our present capabilities to discover what all of these cells are doing. On the other hand, we may be able to discover basic patterns of integration which will elucidate even the most complex cellular arrangements by studying single cells in the least complicated regions of the central nervous system. The vertebrate retina is a favourable preparation for studying neural interaction at the cellular level. It is anatomically one of the simplest regions of the central nervous system. Although it is large, overlying most of the back of the eye, it is only a few hundred microns thick and contains just six major types of neurones. The retina receives nearly all of its input from the signals of the photoreceptors, and its only output to the central nervous system is provided by the axons of one of the cell types, the ganglion cells. Finally, the retina is a relatively convenient tissue to use for physiological experiments. It can be completely removed from the eye and perfused with artificial media without seriously affecting the nature of its responses. It is thus ideal for studying receptor function or synaptic pharmacology.

During the last ten years, we have learned to make intracellular recordings from most of the cell types in vertebrate retina. These studies have shown that the great majority of retinal neurones do not fire action potentials under normal circumstances. Although at first retinal neurones were thought to be unconventional in this respect, there is now abundant evidence that spikeless cells have an important function in signal processing in both vertebrate and invertebrate nervous systems. In the following discussion, I shall review the anatomy and physiology of vertebrate retina, beginning with the photoreceptors, which contain the photosensitive pigment, and proceeding proximally

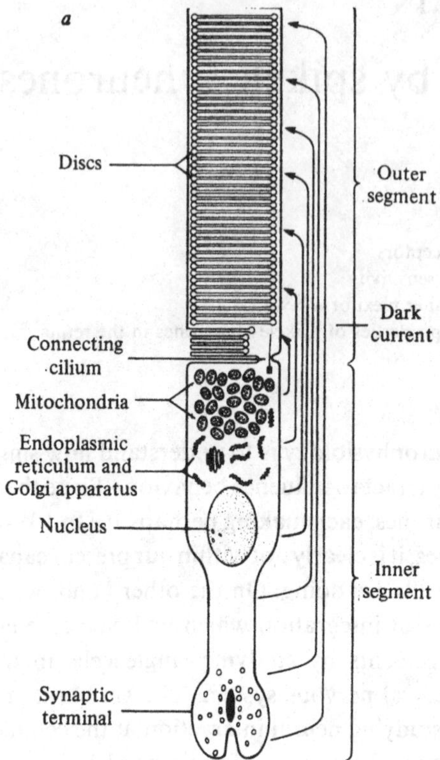

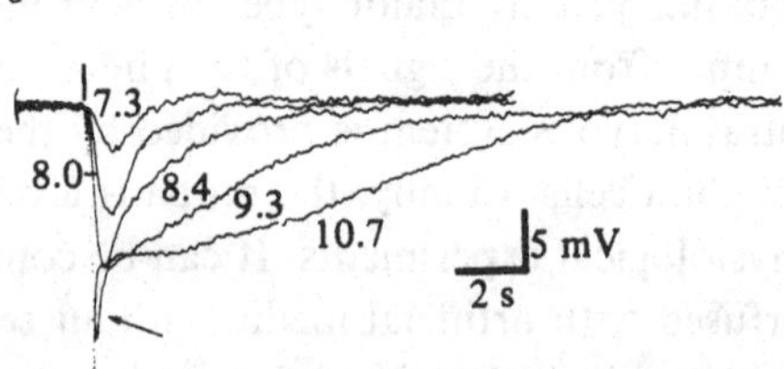

Fig. 1. ***a.* Schematic drawing of a rod, illustrating features typical of most** vertebrate species. The cell is represented as it would be seen in radial section. The distal *outer segment* encloses 1000 to 2000 membraneous discs or saccules, which contain most of the visual pigment. The proximal *inner segment* is the metabolic end of the cell, containing mitochondria, endoplasmic reticulum, the Golgi apparatus, and the nucleus. At the proximal end of the cell is the synaptic terminal, where the rod contacts second-order horizontal and bipolar cells. The arrows to the right of the rod illustrate in a schematic way the light-dependent current which continuously circulates around the rod in darkness. This current is caused by the high permeability of the outer segment to sodium ions. Light reduces this permeability, decreases the dark current, and causes negative-going changes in rod membrane potential. *b.* Intracellular

to the ganglion cells, whose axons leave the retina and terminate in the central nervous system. In this treatment, I shall emphasize the role of cells which do not generate action potentials.

Photoreceptors

Most vertebrate retinas contain two kinds of receptors, the rods and cones, which appear to work in much the same fashion (Fig. 1). In the dark, the receptor membrane is approximately equally permeable to sodium and potassium ions (Cervetto, 1973; Brown & Pinto, 1974). This is unusual for neurones or for other cells in the body, which, as a rule, have little sodium permeability in the resting (or unstimulated) condition. The sodium permeability in photoreceptors appears to be restricted to the pigment-containing region of the cell called the outer segment (Fig. 1*a*) and permits sodium ions to flow into the receptor in the dark (Hagins, Penn & Yoshikami, 1970). Light causes a chemical change in the receptor pigment, and this somehow reduces the sodium permeability and decreases the inflow of sodium ions. The resulting reduction in current flow into the receptor causes a negative-going change in its membrane potential (Fig. 1*b*). Recent evidence shows that, in rods, a single photon can produce a detectable change in the sodium current, amounting to about 10^{-12} amp (Yau, Lamb & Baylor, 1977). This means that one photon can reduce the net flow of ions crossing the rod outer segment membrane by 10^6 to 10^7 charges s^{-1}, which is close to the maximum gain which can be achieved with a conventional photomultiplier tube.

The large resting sodium permeability of the receptor generates a 'dark current' (Hagins *et al.*, 1970), which circulates continuously around the soma, flowing out of the proximal inner segment and into the distal outer segment (Fig. 1*a*). The effect of this current is to bias the receptor potential to a value which is more positive than for most neurones, so that receptors are continuously depolarised in the dark. As a result, receptors continously release synaptic transmitter on to second-order cells. This has the effect of greatly increasing the sensitivity of transmission at the receptor synapse, as I shall explain in more detail below (p. 39).

Although photoreceptors do not normally fire action potentials (Fig. 1*b*), this does not mean that the receptor membrane is passive. Though apparently

---

responses of rod in the retina of the toad, *Bufo marinus*. Bar above and to the left of responses gives duration of the light flash. Numbers next to responses give the intensity of the light, in units of log 516 nm incident quanta cm^{-2} $flash^{-1}$. Arrow, quick relaxation or 'initial transient' of responses to bright flashes. Membrane potential of cell in darkness was approximately -35 mV.

lacking TTX-sensitive sodium channels like those found in most nerve and muscle, vertebrate photoreceptors can be shown to have a variety of other voltage-sensitive conductances. For example, the current–voltage curve of rods shows a prominent outward rectification, like that observed in many nerve and muscle preparations (Jack, Noble & Tsien, 1975), which can be suppressed by TEA (tetraethylammonium chloride; Fain, Quandt & Gerschenfeld, 1977). TEA has been shown to block voltage-dependent potassium channels in a variety of nerve and muscle cells (see, for example, Armstrong, 1975). This suggests that rods have a voltage-gated potassium conductance, perhaps similar to the delayed rectifier of axons (Hodgkin & Huxley, 1952). After treatment with 6–12 mM TEA, this potassium conductance is suppressed, and rods are capable of producing action potentials (Fig.

Fig. 2. *a*. Action potentials recorded intracellularly from a toad rod perfused externally with TEA and 3.6 mM (twice normal) Ca^{2+}. Resting potential, approximately −35 mV. Spikes are spontaneous in darkness but are inhibited during exposure to a 109 ms, 516 nm full-field flash of intensity 9.8 log incident quanta cm^{-2} (note hyperpolarisation of receptor membrane potential during light response). *b*. Action potential from cell of Fig. 2*a* on expanded time scale. Spikes have a time course like calcium-dependent action potentials in other systems and are much slower in rise time and longer in duration than sodium-dependent action potentials of nerve and muscle. Reprinted with permission from Fain, Quandt & Gerschenfeld, 1977.

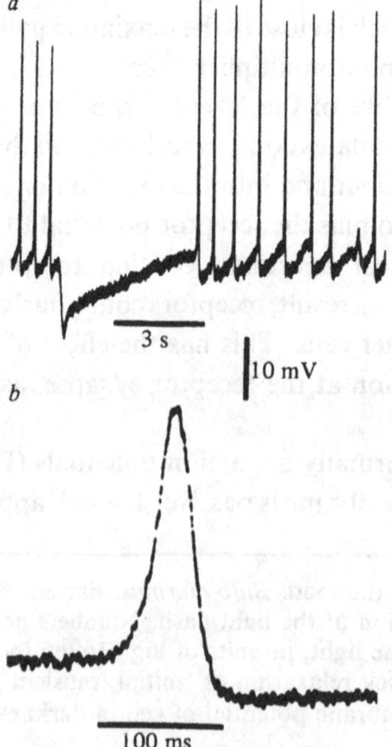

2), which in every respect resemble the calcium spikes that have been described in arthropod muscle fibers, mouse and tunicate eggs, and presynaptic terminals (see Hagiwara, 1975). Under normal circumstances, the inward calcium current does not become regenerative because it is matched by outward current through the voltage-sensitive potassium conductance. Blocking the potassium conductance with TEA destroys the balance between these two currents, and the calcium conductance becomes able to produce action potentials. The calcium conductance of rods probably serves to regulate the release of transmitter at the receptor synapse, and the potassium conductance serves as a kind of negative feedback to keep the rod membrane potential stable.

In addition to the potassium and calcium conductances, the rod plasma

Fig. 3. Effects of caesium ions on the waveform of the receptor response. Responses recorded from toad rods perfused with normal and 2 mM CsCl Ringer's solutions have been superimposed at five light intensities. Numbers to left give intensities for each pair, in units of log 501 nm quanta cm^{-2} $flash^{-1}$. Duration of flash is indicated by negative deflection on uppermost trace. At 7.6 log quanta the waveforms are nearly identical. For brighter flashes, the caesium responses are always larger. Note that caesium blocks the fast decay (initial transient – see lowermost response). Reprinted with permission from Fain *et al.*, 1978.

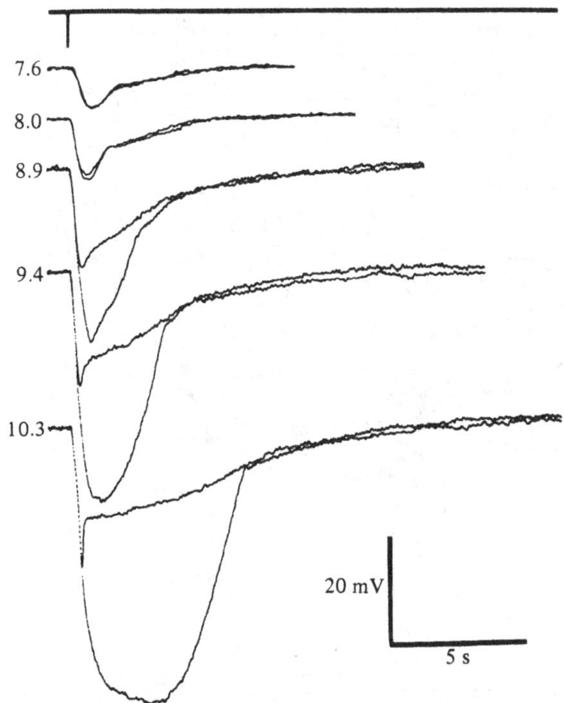

membrane contains a third voltage-sensitive conductance at least partially permeable to sodium ions, which in some respects resembles the inward rectifier of starfish eggs (Hagiwara & Takahashi, 1974) and skeletal muscle (Adrian, 1970). This conductance is activated when the rod hyperpolarises and is responsible for the fast decay or 'initial transient' of the response to bright illumination (arrow in Fig. 1*b*). Preventing the activation of this

Fig. 4. Cell types and synaptic connections in vertebrate retina. Schematic diagram summarising synaptic contacts among retinal cells, viewed as seen in radial section. In the outer plexiform layer, processes of bipolar cells (B) and horizontal cells (H) enter invaginations and make synaptic contact with the receptor terminals (RT). Flat bipolar cells (FB) make superficial contacts on the proximal membrane of some receptor terminals. Horizontal cells in some retinas contact one another (not shown) and bipolar cells outside the invaginations. In the inner plexiform layer, bipolar cell axon terminals synapse on to both ganglion (G) and amacrine (A) cell processes. Amacrine cells may synapse back on to bipolars in a feedback configuration, or may synapse on to one another in complex serial arrangements. Amacrine cells can also synapse directly on to ganglion cells. Interplexiform cells are not shown. Reprinted, with permission, from Dowling, 1970.

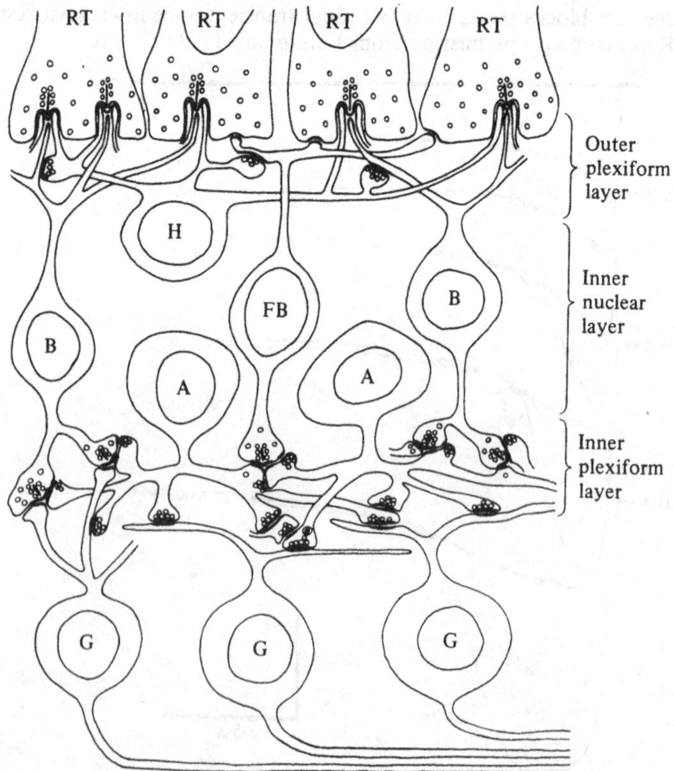

conductance by perfusing the retina with 2 mM CsCl (Fig. 3) eliminates the fast decay and causes a dramatic change in the waveform of the receptor response (Fain, Quandt, Bastian & Gerschenfeld, 1978).

The function of this 'Cs-sensitive' conductance may be to keep the rod membrane potential relatively depolarised, even in the presence of bright light. One could easily imagine that, without this conductance, the rod would hyperpolarise so much to a strong stimulus that the release of transmitter at the receptor synapse would be shut off completely. This in turn could produce a large decrease in the sensitivity of synaptic transmission in the presence of bright light. The Cs-sensitive conductance allows the rod to depolarise rapidly toward the resting membrane potential, and this may permit the receptor to transmit changes in response amplitude with nearly the same efficiency in background light as in darkness.

Synaptic contacts from receptors

Receptors terminate in a region of densely packed cell processes called the 'outer plexiform layer' (Fig. 4). Here they synapse on to the processes of two types of second-order cells, called horizontal and bipolar cells. In nearly all species examined, receptor synapses are located at least in part within an invagination of the receptor terminal (Stell, 1972). At the very top of the invagination within the receptor cytoplasm, there is a dense membranous structure called the synaptic ribbon, which may act as a sort of conveyor belt, directing the synaptic vesicles downward to release sites on the receptor plasma membrane (Gray & Pease, 1971; Raviola & Gilula, 1975). As the vesicles reach these sites, they discharge their contents into the extracellular space. In the electron microscope it is possible to visualise the detailed arrangement of processes within the invagination of the receptor terminal (Stell, 1972). Just opposite the synaptic ribbon, there are often only three post-synaptic processes; a central process, which usually comes from a bipolar cell, and two lateral processes, which usually come from horizontal cells. This arrangement is nowhere more precise than for primate cones, where terminal invaginations appear always to contain three processes, arranged with the bipolar in the centre and horizontals on either side (Kolb, 1970). In primate rods and in the rods and cones of many lower vertebrates, the invagination may contain more than three processes, and the disposition of second-order terminals may be quite complicated. However the occurrence of horizontal and bipolar cell terminals arranged in some fashion within a receptor invagination is a nearly universal feature of vertebrate retina.

In addition to the synapses at the invaginations, receptors may also make flat or basal contacts, usually at the lowest or most proximal part of their terminal membrane (Missotten, 1965; Kolb, 1970; Lasansky, 1971). These contacts sometimes occur very near to and in close association with the invaginations. Although it is often possible to see an increased membrane density at these regions of contact, which in the central nervous system is thought to be an indication of synaptic interaction, there are other features of the basal contact which still puzzle us. For example it is usually not possible to observe a clustering of synaptic vesicles in the receptor cell just opposite to the presumed area of synaptic contact. This is uncommon for synapses in the rest of the nervous system and has suggested to some investigators that these basal contacts have no physiological function. However, we now know that some bipolar cells make only this curious type of junction with the receptors, never sending processes into the invaginations (Kolb, 1970). If the basal contacts are not functional areas of synaptic transmission, it is difficult to imagine what these bipolar cells are doing.

In addition to receiving input from the photoreceptors, the horizontal and bipolar cells also interact with one another. Horizontal cells in some retinas make electric synapses onto other horizontal cells (Stell, 1972), which can have a profound effect on their physiological responses (*e.g.* Naka & Rushton, 1967). In some species horizontal cells also synapse on to bipolar cell processes within the outer plexiform layer (*e.g.* Fisher & Boycott, 1974), though contacts of this type have not been observed in very many cases.

Horizontal cell processes stretch laterally across the retina within the outer plexiform layer, and a single cell may make direct synaptic contacts with many receptors (Wässle, Boycott & Peichl, 1978; Wässle, Peichl & Boycott, 1978). Fig. 5 shows two types of horizontal cells which have been described in the cat (Fisher & Boycott, 1974; Boycott, Peichl & Wässle, 1978), and which are representative of those found in many vertebrate species. They are shown as we would seen them if we were looking down on to the surface of the retina. The type A horizontal cell in cat is axonless and has the larger dendritic field. Type A cells receive direct input only from cones, and each type A cell makes 3 to 5 contacts with between 120 and 170 receptors (Wässle, Boycott & Peichl, 1978). The cell body of the type B cell (shown to the upper right in Fig. 5) also receives direct synaptic input only from cones, but its dendritic field is about half as large as that of the A type cell. In addition, type B cells have thin axons which course in an apparently random fashion for about 400 μm through the outer plexiform layer and then suddenly balloon into a large axon terminal system, the teleodendria of which make synaptic contacts exclusively with rods. A single type B axon terminal system may contact three to four thousand receptors.

Intracellular recordings in cat and in several other species have shown that horizontal cells, like receptors, have a low resting potential (-20 to -40 mV)

Fig. 5. Morphology of horizontal cells in the cat, stained in whole mounts of retina with modifications of the Golgi method. Cells are shown as seen looking down on to the surface of the retina. There are two types of horizontal cells. The A-type has a relatively large dendritic tree and no axon. The B-type has a smaller dendritic spread (when the A and B types are compared in the same part of the retina) and an axon which may be as long as 0.5 mm. The axon is 0.5 to 1 μm in diameter for most of its length and then abruptly enlarges into an axon terminal (a.t.) with an extensive network of teleodendria. Reprinted with permission from Fisher & Boycott, 1974.

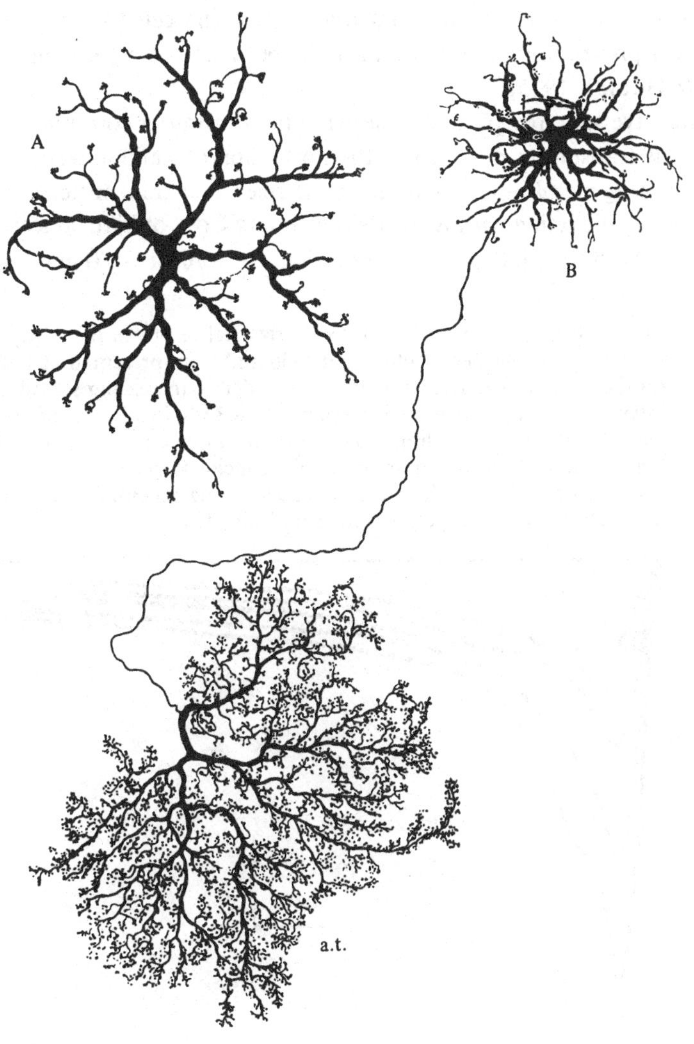

which becomes more negative when the retina is illuminated (Gouras, 1972). The response of a typical horizontal cell consists of a slow hyperpolarisation whose amplitude is graded with stimulus intensity (Fig. 6). Horizontal cells are not normally capable of producing spikes, even in those cases (as for type B cells in cat) where they have a morphologically identifiable axon. The function of the axon of type B cells appears not to be the propagation of signals between the two ends of the horizontal cell. On the contrary, the small diameter of the axon and large impedance mismatch between it and either end of the horizontal cell probably results in a large decrement in the amplitude of signals passing from one end of the cell to the other, thus effectively preventing any communication between them (Nelson, Lützow, Kolb & Gouras, 1975).

As we have seen, receptors release synaptic transmitter continuously in the dark. This transmitter depolarises the horizontal cell and is responsible for its low resting membrane potential. If the release of transmitter is blocked, by perfusing the retina with Ringer's containing Co^{2+} or high concentrations of Mg^{2+}, the horizontal cell hyperpolarises to −70 or −80 mV and is no

Fig. 6. Responses of dark-adapted horizontal cell from the mudpuppy retina. 190 ms flashes of blue light (450 nm) were presented as indicated on the lower beam at 15 s intervals in order of increasing stimulus intensity. Uppermost trace is response to 8.3 log incident quanta cm^{-2} $flash^{-1}$. In succeeding traces, stimulus intensity was increased in 0.5 log unit steps to a maximum of 10.3 log incident quanta cm^{-2} $flash^{-1}$ (lowermost trace). Dashed line indicates resting potential of cell in the dark. Reprinted with permission from Fain, 1975.

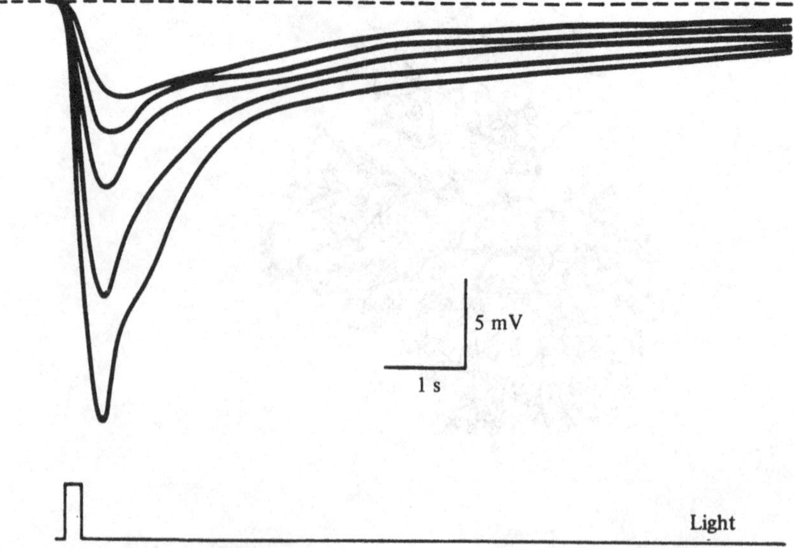

longer capable of responding to changes in illumination (see for example Kaneko & Shimazaki, 1976). Since the transmitter released by the photoreceptors depolarises the horizontal cell, light must cause a *reduction* in the rate of transmitter release. The hyperpolarisation of the photoreceptor appears to decrease the entry of Ca^{2+} into the receptor synaptic terminal, thereby lessening the rate at which synaptic vesicles release their contents into the synaptic cleft. The decrease in transmitter release produces the negative-going responses of the horizontal cell to flashes of light.

Although it may seem inefficient for receptors to release transmitter continuously in the absence of stimulation, this process probably has the effect of greatly increasing the sensitivity of transmission. To illustrate this effect, let us suppose that the input–output curve for the receptor synapse is similar to that for the giant synapse of squid. This curve, from the data of Katz & Miledi (1967), is given in Fig. 7. Here the pre-and post-synaptic voltages of the squid neurones are plotted relative to their resting membrane potentials. Fig. 7 shows that, in the giant synapse, the presynaptic terminal must be depolarised by 25–30 mV before any post-synaptic response is recorded. This initial polarisation is presumably that required to produce a significant activation of the calcium channels in the presynaptic membrane (Llinás, 1977). Once the calcium channels are open, the post-synaptic response is very sensitive to changes in presynaptic voltage. One of the functions of the

Fig. 7. Input–output curve for the giant synapse in the squid stellate ganglion. Data replotted from Fig. 9 of Katz & Miledi (1967). Voltages of pre- and post-synaptic terminals are given with respect to their resting membrane potentials. Reprinted with permission from Fain, 1977.

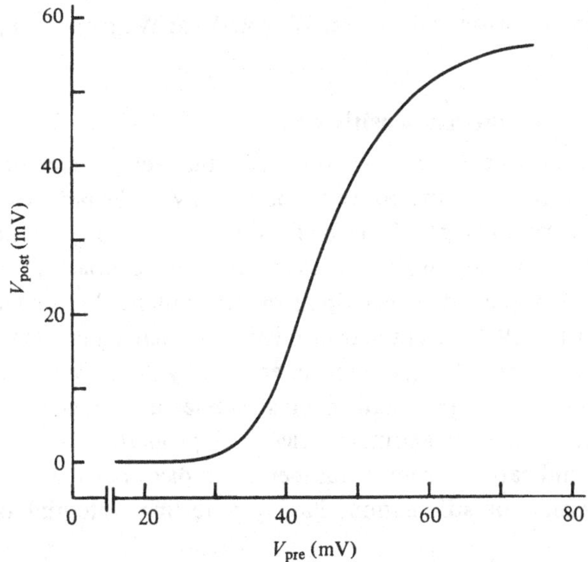

receptor dark current must be to bias the receptor membrane potential to a level where transmitter can be continuously released (Fain, 1977). As a result, it takes a signal of only a few microvolts in a photoreceptor to produce physiologically significant changes in the flow of synaptic transmitter to second-order cells (Fain, Granda & Maxwell, 1977).

Though horizontal cells do not normally fire action potentials, they should not be thought of as simple, passive neurones responding only to changes in the flow of synaptic transmitter from the photoreceptors. Since horizontal cells also release synaptic transmitter (as I shall describe below), they too must have voltage-dependent calcium channels. Since the membrane potential of horizontal cells is depolarised, it is likely that their calcium conductance is continuously activated, and that they too are continuously releasing synaptic transmitter. They must therefore also have a voltage- or Ca^{2+}-dependent outward current, to keep their Ca^{2+} conductance from becoming regenerative. In addition, there is evidence that horizontal cells have an inward rectifier (Trifonov, Byzov & Chailahian, 1974; Werblin, 1975), perhaps similar to the Cs^{++}-sensitive conductance of rods (see p. 34).

Horizontal cells in vertebrates have, in general, rather large dendritic fields and, at least in some species, are connected to one another by extensive gap junctions (see for example Fain, Gold & Dowling, 1976). As a result, they are able to sum receptor inputs over a large region of the retina. The retinal area over which light stimulation hyperpolarises (or otherwise affects) the membrane potential of the horizontal cell is called its *receptive field.* The receptive field of the horizontal cell can be measured by recording intracellularly while moving a narrow slit of light across the retina (Lamb, 1976). Receptive fields of single horizontal cells are in general quite large and, in fish, may include the whole retina (Norton, Spekreijse, Wolbarsht & Wagner, 1968).

Bipolar cells and contrast sensitivity

Bipolar cells are oriented in a radial direction: they receive receptor input in the outer plexiform layer and convey it proximally to the next level of synaptic interaction, the inner plexiform layer. Like horizontal cells and photoreceptors, bipolar cells are normally electrically inexcitable. Their responses consist entirely of graded changes in membrane potential (Werblin & Dowling, 1969; Kaneko, 1970), and the transmission of their signals from outer to inner plexiform layers is thought to occur entirely by the passive flow of current. Unfortunately, they are small in most species and difficult to penetrate, so that little is known about the electrical properties of their membrane. There are indications from a number of studies that they are depolarised in the absense of stimulation, having a resting potential of perhaps −40 mV (see especially Ashmore & Falk, 1976; Saito, Kondo & Toyoda, 1979). They are thought to release synaptic transmitter in the normal way and so probably have voltage-dependent Ca^{2+} and K^+ channels, but there is no information about the properties of these presumed conductances, nor is it known whether bipolar cells show inward rectification.

Fig. 8 shows two examples of bipolar cells from the goldfish, *Carssius auratus* (A. T. Ishida & W. K. Stell, personal comunication). The cell on the left contacts only cones and had been sectioned somewhat obliquely to illustrate the large area of its dendritic field. The cell on the right belongs to the class a1 (Stell, Ishida & Lightfoot, 1977) and contacts both rods and cones. Although there have been no attempts to construct mathematical models of bipolar cells in goldfish or in any other species, the dimensions of the cells in Fig. 8 suggest that a signal could spread passively from the distal to proximal ends of bipolar cells with little decrement. The finest processes of the cells in Fig. 8 (the axons and dendritic terminals) are approximately 0.5 μm in diameter. Assuming resistivities for the membrane of 1000 Ωcm^2 and for the cytoplasm of 100 Ωcm, the space constant of even these processes would be about 100 μm. This is greater than the distance between extremities

Fig. 8. Morphology of two types of bipolar cells in the goldfish, *Carassius auratus*, stained with the Golgi method. Retinas were sectioned radially. See text for details. Taken with permission from unpublished work of A. T. Ishida & W. K. Stell.

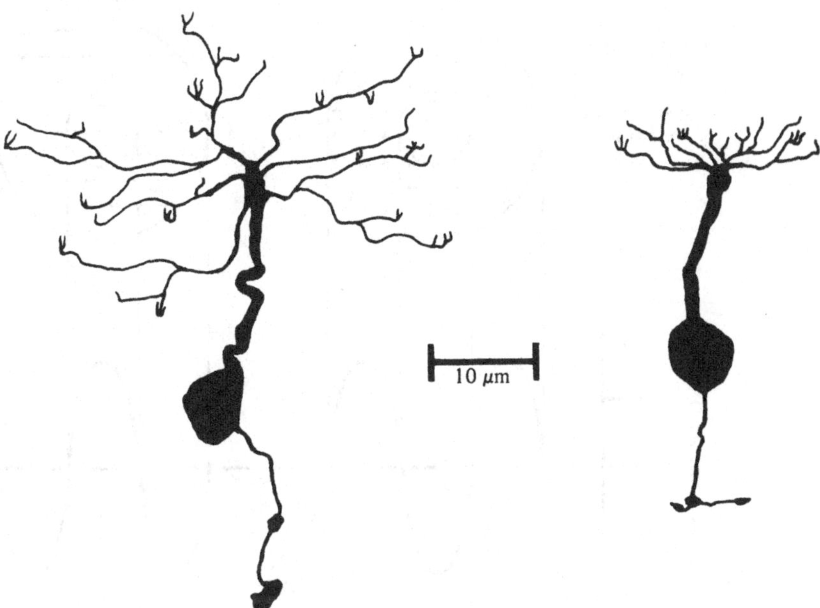

for either cell in Fig. 8. Since bipolar cells are finite rather than infinite cables, and since the dendritic trunk and soma are much larger in diameter than the axon and dendritic terminals, signal transmission would be even more efficient than this calculation suggests (see Rall, this volume).

Though signals probably pass from dendrites to axon terminals with little loss of amplitude, Fig. 8 suggests that, in some bipolar cells, voltages originating in the proximal end may have considerable difficulty passing to the dendritic terminals. The fine calibre of the axons of the cells in Fig. 8, in comparison to their somas and dendritic trunks, suggests that there is a large

Fig. 9. Receptive fields of bipolar cells. The organisation of bipolar receptive fields can be demonstrated by two simple experiments. In both, intracellular recordings are made from bipolar cells as a stimulus is presented at different places on the retinal surface. Part *a* shows the effect of small spots or rings of light on the membrane potential of the two kinds of bipolar cells. If a spot is flashed in the centre of the receptive field, it produces a response opposite in polarity from that produced by light in the annular region surrounding the centre. The antagonism of centre and surround enhances the sensitivity of the bipolar cell to contrast. The same effect can be demonstrated by moving a slit of light across the retina, as in part *b*. As the slit moves into the receptive field centre, it produces a response opposite in polarity from that produced by the flanking surround regions. Moving spots or slits of light provide a rapid method for characterising the organisation of the bipolar receptive fields (Werblin, 1970). Figure redrawn from Fain, 1979.

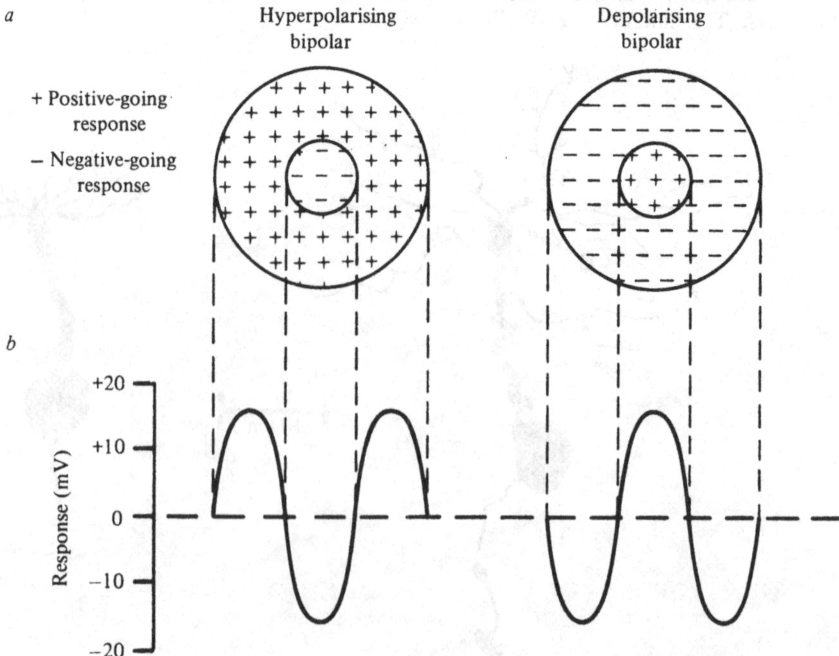

mismatch of impedance between the two ends of the cells, which may cause a significant decrease in the amplitude of signals coming from the inner plexiform layer. The large difference in the diameters of the axon and main dendritic trunk seen in the cells of Fig. 8 is a common feature of bipolar cells in vertebrate retina and may be a mechanism for preventing the flow of information in the antidromic direction.

The receptive field of bipolar cells differs from that of horizontal cells. Horizontal cell receptive fields are *homogeneous*: that is, a spot of light placed anywhere within the receptive field always produces a hyperpolarising response of approximately the same waveform. In contast, the bipolar cell receptive field is *heterogeneous*. It consists of two regions: a small central area and a large, annular field, surrounding the centre (Fig. 9) (Werblin & Dowling, 1969; Kaneko, 1970). There are two kinds of bipolar cell responses. For one type, a spot of light flashed in the centre produces a negative (or *hyperpolarising*) response; for the other, a central spot produces a positive (or *depolarising*) potential. We shall call these respectively, the hyperpolarising and depolarising bipolars. A spot of light flashed into the surround always produces a response antagonistic to that produced by the centre. For the hyperpolarising bipolar, for example, the centre produces a negative-going response and the surround, one that is positive-going.

The response of the receptive field centre is thought to be produced by direct input from receptors to bipolar cells (Kaneko, 1973; Richter & Simon, 1975). For the hyperpolarising bipolars, the centre response is thought to be generated much like the hyperpolarising response of horizontal cells (see Gerschenfeld & Piccolino, 1979). The receptors are thought to release a depolarising transmitter continuously in darkness, and light is thought to reduce this release, hyperpolarising the bipolar cell. There is evidence that the hyperpolarising light responses of horizontal and bipolar cells are accompanied by a decrease in membrane conductance, and this has led many investigators to postulate that the receptor transmitter released continuously in darkness acts upon the post-synaptic membrane of horizontal and hyperpolarising bipolar cells much like a typical excitatory transmitter, for example like acetylcholine at the motor end-plate (see Gerschenfeld & Piccolino, 1979). However, it will not be possible to test this notion rigorously until the receptor transmitter or transmitters have been conclusively identified.

For the depolarising bipolar cells, the receptor transmitter must produce a tonic hyperpolarisation. This could occur in two ways. The receptor transmitter could increase the permeability of the depolarising bipolar to chloride or potassium, much like at an inhibitory synapse in the spinal cord (Eccles, 1964). Alternatively, the transmitter could produce a tonic *decrease* in permeability, perhaps to sodium. There is as yet insufficient evidence to distinguish between these possibilities. Recent experiments in carp indicate that both mechanisms may be used, the first by the cones and the second by the rods (Saito *et al.*, 1979). We will not know whether this is a general feature of bipolar cells in vertebrate retina until similar experiments are done on other species.

The surround response of bipolars is unlikely to be generated by direct input from the receptors, since comparisons in several species suggest that the surround is larger in diameter than the dendritic tree of the bipolars (Werblin & Dowling, 1969; Kaneko, 1973; Richter & Simon, 1975). Since horizontal cells have large receptive fields and, at least in some species, synapse directly on to the bipolars, it is reasonable to suppose that *they* mediate the surround effect. However, this hypothesis is still in dispute (Piccolino & Gerschenfeld, 1977; Marchiafava, 1978; Toyoda & Tonosaki, 1978). In some species, the surround response appears to be produced (at least in part) by an antagonistic input of horizontal cells back on to receptors (Baylor, Fuortes & O'Bryan, 1971). The importance of this 'presynaptic' inhibition has also not been completely resolved (Richter & Simon, 1975).

Fig. 10. Contrast sensitivity of bipolar cells. This figure illustrates a simple experiment which demonstrates the sensitivity of bipolar cells to contrast. Part *a* shows how the experiment is done. Intracellular recording is used to measure the membrane potential of a depolarising bipolar cell as an illuminated border is moved across its receptive field. Part *b* shows the result of the experiment. Membrane potential of the bipolar cells is plotted as a function of the position of the edge on the bipolar receptive field. The bipolar cell gives little response when the receptive field is uniformly dark or uniformly illuminated (positions 1, 5). It is extremely sensitive to the position of the edge if the illumination is non-uniform (positions 2, 3, 4). See text for details.

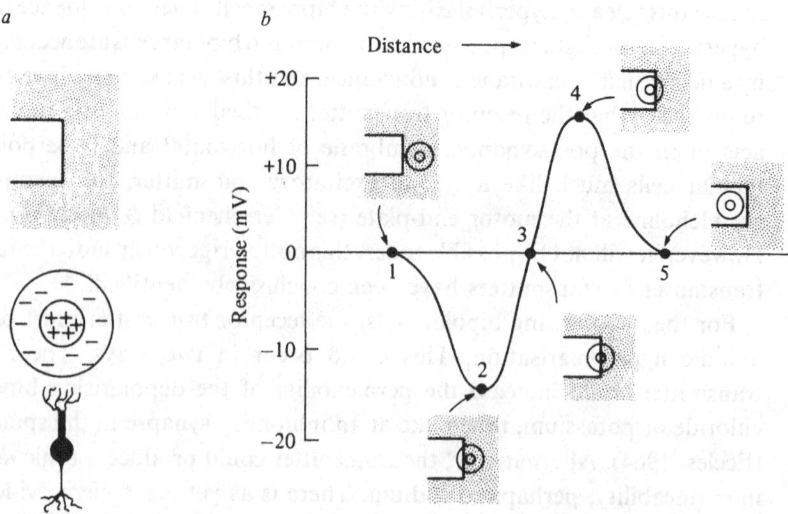

Regardless of the origin of the surround response, its effect on the bipolar cell remains the same. Since responses from centre and surround are different in sign, they are mutually antagonistic. If the retina is stimulated with a large field of light which covers both the center and surround of the bipolar cell, the response will be smaller than if the illumination is confined only to the center. As a result, the bipolar cell is not very sensitive to the mean level of illumination, but it is acutely sensitive to intensity differences, that is to contrast.

An experiment which demonstrates the sensitivity of the bipolar cell to contrast is illustrated in Fig. 10. For this experiment, we construct a stimulus consisting of a perfectly sharp border between two fields, one of which is dark and the other light (Fig. 10*a*). When this stimulus is moved across the receptive field of the bipolar cell (which we shall assume to be one of the centre-depolarising variety), it produces the reponses plotted in Fig. 10*b*. At the position labelled 1 to the far left of the curve, the stimulus has not yet reached the bipolar cell, and there is no response. As the border crosses the receptive field, the potential of the bipolar cell goes negative (or hyperpolarises). The reason for this is that the light portion of the stimulus reaches the surround of the receptive field first, and illumination in the surround hyperpolarises the membrane potential. The maximum negativity is reached near position 2, where as much as possible of the surround has been illuminated without any light reaching the centre. As light reaches the centre, the cell starts to depolarise, and a null position is reached when the border lies exactly across the centre of the receptive field (position 3). As more of the centre is illuminated, the potential depolarises and reaches its maximum positivity near position 4. Here the centre is completely illuminated, but some of the surround is still in the dark. As the rest of the surround becomes illuminated, the membrane potential returns near zero (position 5).

As this experiment shows, the bipolar cell is a rather good indicator of the position of the border. In practice, the sensitivity of the bipolar cell to contrast depends upon the dimensions of its receptive field and the strength of the surround inhibition. Often the surround is weaker than the centre, and contrast detection is not as good as in Fig. 10 (Werblin & Dowling, 1969). However, as we shall see, this device of centre-surround antagonism is repeated in the inner plexiform layer. By the time the signal reaches the brain, sensitivity to contrast is very highly developed.

Centre-surround antagonism is also used by the retina to improve colour discrimination. Animals with good colour vision, like fish and turtle, have three kinds of cone receptors, each containing a photopigment sensitive to a different region of the spectrum (Marks, 1965; Liebman & Granda, 1971). The absorption spectra of the photopigments are rather broad, so that the signals from the cones themselves are not very useful for detecting small differences in wavelength. However the pigment curves are sharpened by interactions similar to those which produce contrast sensitivity. A single bipolar cell may receive direct input from one class of cone and be inhibited (via the horizontal cells) by another (Kaneko, 1973; Yazulla, 1976). Similar interactions may occur between rods and cones (Fain, 1975). The mutual interaction between signals coming from different photopigments forms the basis of colour discrimination, and these interactions begin in the outer plexiform layer of the retina.

Visual processing in the inner plexiform layer

The bipolar cell is the only cell which is known to convey the visual signal from the outer plexiform layer to the more proximal, inner plexiform layer. Between the two synaptic layers lies a relatively thick region called the inner nuclear layer, which contains cell bodies of retinal interneurones but is relatively free of synaptic connections. In the inner plexiform layer, the bipolar cell axonal endings terminate on the processes of a third type of retinal interneurone, called the amacrine cell, as well as on to the dendrites of the output neurones, the ganglion cells (see Fig. 4). In addition to these cell types, there is a fourth class of retinal interneurone, called the interplexiform cell, whose properties are less well understood (Gallego, 1971; Dawson & Perez, 1973; Boycott, Dowling, Fisher, Kolb & Laties, 1975; Kolb & West, 1977; Dowling & Ehinger, 1978). The interplexiform cell resembles an amacrine cell and makes extensive synaptic contacts (mostly with amacrine cells) in the inner plexiform layer. However interplexiform cells also send processes back up into the outer plexiform layer, where they synapse on to horizontal and bipolar cells. Intracellular recordings have not yet been made from interplexiform cells, and little is known about their function.

In all vertebrate retinas, the inner plexiform layer is considerably thicker and more densely filled with synaptic contacts than the relatively simple outer plexiform layer. We believe that the inner plexiform layer is the place in the retina where most of the complex processing of visual signals takes place. The complexity of this layer makes it very interesting but also very difficult to study. At present, the best we can do is to describe the pattern of synaptic connections between cell types (Fig. 4), without attempting to specify in detail what any particular cell is doing. From a careful study of electron micrographs, we can say, for example, that bipolar cells synapse on to amacrine cells and ganglion cells; that amacrine cells synapse back on to bipolars, on to other amacrines, and on to ganglion cells; and that ganglion cells receive input from bipolars and amacrines (Dowling & Boycott, 1966; Dowling, 1968; Dubin,

1970; Wong-Riley, 1974; Goldman & Fisher, 1978). We can even say a few things of interest about the distribution of these synaptic connections. For example, whenever a bipolar cell synapses on to an amacrine, it is usually possible to see an amacrine synapsing back on to the bipolar, in a sort of feedback arrangement. A bipolar synapse usually has a synaptic ribbon and a characteristic configuration, with a single presynaptic process apparently transmitting information to two or more different processes at the same locus. We can also say that, in general, amacine synapses are exceedingly complicated. One cell can synapse on to a second and the second on to a third, in a sort of serial arrangement. Amacrine cells also appear to 'feedback' on to one another.

Although much effort has been devoted to the description of the anatomy and physiology of synaptic interactions among cells in the inner plexiform layer, we still know very little about what these cells do. There are two principal reasons for this. In the first place, there appear to be many different subtypes of amacrine and ganglion cells in the retina. The variety of morphological subclasses can be appreciated from Fig. 11, which is a composite drawing of amacrine and ganglion cells in dog from Ramón Y. Cajal's treatise on vertebrate retina (1893). The cells labelled with upper case letters are amacrine cells, which for the most part have their cell bodies in the inner nuclear layer and send processes down into the inner plexiform layer. The cells below (labelled with lower case italic letters) are ganglion cells.

Each subtype of amacrine and ganglion cell has a distinctive dendritic branching pattern, and most probably a characteristic synaptic input. The best evidence for this comes from work on the ground-squirrel retina (West & Dowling, 1972; West, 1976). Ground squirrels have at least five subtypes of amacrine cells and 15 subtypes of ganglion cells. Each of these subtypes has a characteristic proportion of its synaptic input coming from bipolar cells and from amacrine cells, which varies little from cell to cell within each subclassification. Since different subtypes of cells have different inputs, they most likely have different physiological properties. We are only just beginning to classify and describe the kinds of amacrine and ganglion cells in vertebrate

Fig. 11. Morphologies of representative amacrine and ganglion cells from the dog retina, stained by the Golgi method. Retina was sectioned radially. A–F, amacrine cells; *a–i*, ganglion cells. Note variety of dendritic branching patterns exhibited by different cell types. Reprinted from Ramòn Y. Cajal, 1893.

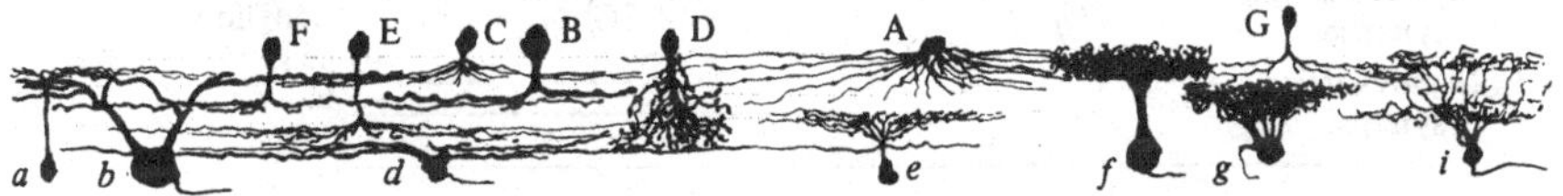

retina (see for example Boycott & Wässle, 1974; Naka, 1977; Famiglietti, Kaneko & Tachibana, 1977; Stell *et al.*, 1977; Nelson, Famiglietti & Kolb, 1978).

In the second place, we have reason to suspect that a single amacrine cell may not have just a single, clearly definable function. Amacrine cells, like horizontal cells in the outer plexiform layer, are not conventional neurones. We are used to thinking of neurones as being *polarised*, with one end (the dendritic) receiving information, and the other (the axonal) transmitting it. Amacrine cells are not organised in this way. In fact they don't even have axons (the word 'amacrine' means 'axonless'). A single ending can both receive and transmit, and synapses carrying information in opposite directions are often found in close proximity. Some amacrine cells are very large, having processes which stretch over many hundreds of square microns of the inner plexiform layer. One end of the cell may not communicate very effectively with the other, and groups of processes or perhaps even single endings may

Fig. 12. Wiring diagram of sustained cells in the inner plexiform layer. Diagram shows synaptic contacts which are thought to be principally responsible for the receptive fields of sustained amacrine and ganglion cells. In this diagram, a plus sign near a synapse means that it is sign-conserving, positive presynaptic potentials producing positive post-synaptic responses and negative, negative. A minus sign indicates a sign-reversing synapse, positive producing negative and negative, positive. In sublamina A, hyperpolarising bipolars (HB) make sign-conserving synapses on to hyperpolarising sustained amacrines (HSA) and hyperpolarising sustained ganglions (HSG), whereas HSAs make sign-reversing contacts on to HSGs. HSAs also make sign-conserving contacts on to the deplorising sustained ganglion cells (DSG). Synaptic interactions in sublamina B are thought to be similar. Figure redrawn from Fain, 1979.

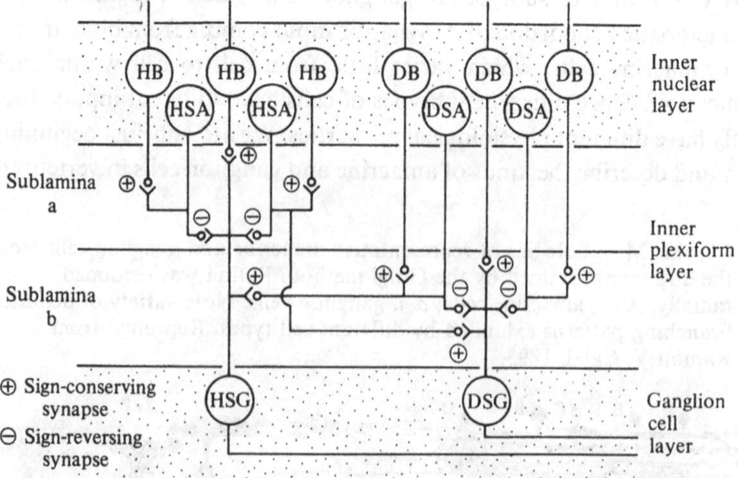

perform specific tasks, relatively free from interference from the rest of the cell. Intracellular recordings are most often made in the centre of this maelstrom at the cell body, and we do not know whether our recordings are representative of what is going on out at the terminals.

In spite of these difficulties, I shall attempt to describe our present understanding of integration in the inner plexiform layer. Needless to say, I can only hope to provide the barest outline of the variety of interactions which probably take place between different cell types, and I shall make no attempt to describe the function of individual synapses. I shall present what I believe to be the most probable description of the function of at least one part of the inner plexiform layer, based mostly on the work of K.-I. Naka (1977). However, I have little doubt that future investigations will force us to revise many of our current ideas.

Let us begin by dividing the inner plexiform layer in two (Fig. 12). The outer or more distal half we shall call 'sublamina a', and the inner, proximal half, 'sublamina b'. There is now increasing evidence that the hyperpolarising bipolars terminate primarily (if not exclusively) in sublamina a, whereas the

Fig. 13. Receptive fields of sustained cells in the inner plexiform layer. Receptive field organisation as demonstrated by moving slits, using the same method as above for bipolar cells (Fig. 9*b*). Note the large homogeneous receptive field of the hyperpolarising and depolarising sustained amacrines. The sustained amacrine cells are thought to sharpen contrast sensitivity of ganglion cells in much the same way that horizontals do for bipolars.

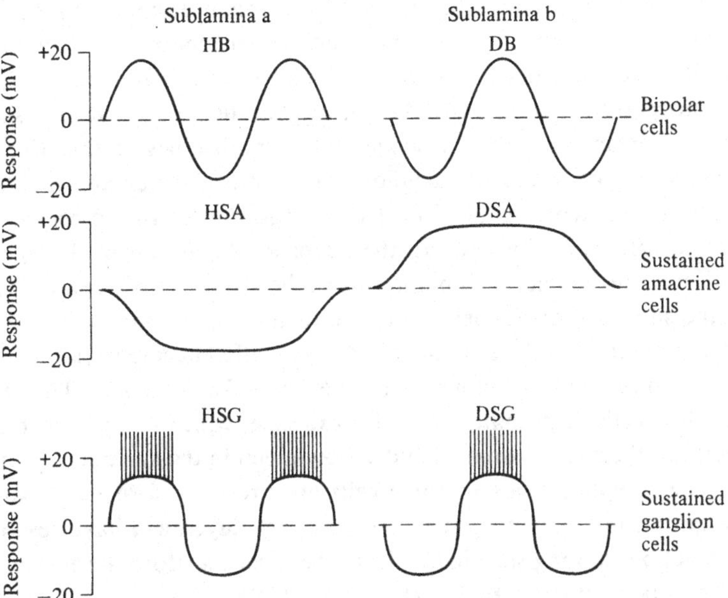

depolarising bipolars send their axon terminals into sublamina b (Famiglietti *et al.*, 1977; Nelson *et al.*, 1978). Each sublamina also contains processes from distinct subpopulations of amacrine and ganglion cells. The way these cells interact appears to be similar for the two sublaminae, though the polarity of the responses is not the same in the two layers.

In sublamina a, the hyperpolarising bipolars are thought to terminate on one class of amacrines, called 'hyperpolarising sustained amacrines'. These cells have homogeneous receptive fields, like the horizontal cells, and their responses are also similar, consisting of a sustained negativity. Also like horizontal cells, the sustained amacrines have large receptive fields, and their responses are thought to be generated by direct excitation by many bipolars over a large region of the retina. It may seem puzzling that the homogeneous receptive field of the sustained amacrines could be formed by input only from bipolars. However, as I have mentioned, the bipolar cell centre is usually stronger than the surround, so that bipolar cells give some response even to uniform illumination. Such bipolar cell responses are thought to sum together to form the response of the sustained amacrines.

Hyperpolarising bipolars are also thought to terminate on 'hyperpolarising sustained' ganglion cells, also called 'off-centre' ganglions. These cells have receptive fields much like bipolar cells, with strong antagonism between centre and surround. They too appear to receive direct excitation from bipolar cells but, in addition, receive inhibition from hyperpolarising sustained amacrines. The inhibition from the large field, sustained amacrine cells has the same effect as horizontal cell inhibition in the outer plexiform layer: it sharpens interactions between centre and surround and enhances contrast sensitivity.

In sublamina b, the depolarising bipolars terminate on 'depolarising sustained amacrines' and on to 'depolarising sustained ganglion cells', also called 'on-centre' ganglion cells. As I have said, the interactions are similar in the two sublaminae. There is evidence that both types of amacrine cells can influence both types of ganglion cells. Thus depolarising amacrines hyperpolarise on-centre ganglion cells and also depolarise off-centre cells. Hyperpolarising amacrines do just the opposite. Regardless of the sign, the effect of sustained amacrine input is always the same: for both ganglion cell types, the sustained amacrines contribute to surround antagonism.

Just as bipolar cells can respond differently to different colours, so amacrine and ganglion cells can be colour-coded (Levick, 1972; Kaneko, 1973). There are ganglion cells in primate retina, for example, which respond to blue in the centre of their receptive field and yellow-green in the surround (Gouras, 1968). The receptive fields of these cells are probably formed in part by colour-specific interactions in the outer plexiform layer, as I have described, but interactions among sustained cells in the inner plexiform layer may also make an important contribution (Kaneko, 1973).

The sustained on- and off-centre ganglion cells are the ganglion cells that mediate our sensitivity to contrast and to colour discrimination and are the simplest types of ganglion cells in the retina. Some sustained ganglion cells have huge receptive fields, indirectly receiving input from thousands of receptors. These probably aid us in perceiving very dim light. Other ganglion cells, for example the so-called 'midget' ganglions, probably have receptive field centres formed from a single cone. These are especially common in the central part (or fovea) of the primate retina (Polyak, 1941), where they probably are responsible for the high acuity at the centre of gaze.

In addition to sustained ganglion cells, the retinas of most species (including mammals) contain a varied assortment of other ganglion cell types, having much more complicated receptive fields (Levick, 1972, 1975). There are ganglion cells, for example, which are sensitive only to changes in illumination. Others are sensitive only to motion in a particular direction, responding vigorously (let us say) when a spot moves from left to right, but very little or not at all when a spot moves from right to left. One of the most surprising features of these 'direction-selective' ganglion cells is that they respond identically to white spots on black backgrounds and to black spots on white backgrounds. They usually have large, antagonistic surrounds, which are themselves not direction-selective. There is another type of ganglion cell called the 'uniformity detector', which has a large receptive field and is excited only when the whole field is uniformly illuminated (or uniformly darkened). Any discontinuity, whether it be a black or white spot, a bar, or even just an edge of light, will inhibit the cell. Another cell type, observed in amphibians, responds only when the light is turned off. The brighter the light was when

Fig. 14. Response of transient amacrine cell in mudpuppy retina to a 2 s flash of white light. Note initial spike-like response, followed by transient depolarisation at both beginning and end of stimulus.

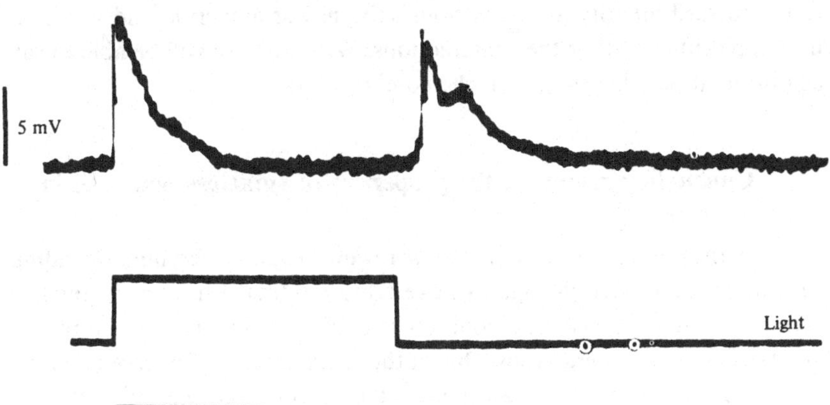

it was on, the larger the response when the light is extinguished (Lettvin, Maturana, McCulloch & Pitts, 1959).

We do not known how the receptive fields of complex ganglion cell types are formed. It is often assumed that the complex processing of the visual signal in the inner plexiform layer is not done by the sustained amacrines, but rather by another subtype or group of subtypes called 'transient amacrines' (Dowling, 1970). These cells, like sustained amacrines, have large, homogeneous receptive fields, but respond with a quickly decaying depolarisation whenever the light intensity is changed (Werblin & Dowling, 1969). As Fig. 14 shows, transient amacrine cells are capable of generating action potentials. The amplitude and waveform of these spikes vary from recording to recording, and in some cells two kinds of spikes can be recorded, some perhaps originating in the soma and others in the dendrites (Miller & Dacheux, 1976). The action potentials of amacrine cells can be blocked by relatively low concentrations of TTX (Miller & Dacheux, 1976), and they are therefore probably classical sodium spikes.

Transient amacrines appear to send processes into both sublaminae a and b (Famiglietti *et al.*, 1977), and they probably receive direct excitation from both kinds of bipolars. They respond equally well to white spots on black backgrounds and to black spots on white backgrounds. Though not sensitive to direction, they respond well to moving stimuli (Werblin, 1970). Furthermore, these cells, at least in some species, can be shown to have a direct effect on the responses of some classes of ganglion cells (Werblin, 1972). These observations, and others (*e.g.*, Caldwell, Daw & Wyatt, 1978), suggest that transient amacrines may contribute to the receptive fields of motion-sensitive and direction-selective ganglion cells, and perhaps to those of other complex types as well. However the truth is that we know practically nothing about what these cells do. Understanding integration in the inner plexiform layer is the greatest challenge the retina presents to us. A lot of effort is being directed toward identifying the various subtypes of amacrine and ganglion cells and piecing together their interactions. With luck we will be able to say much more about this subject in the coming years.

Concluding remarks: the properties of spikeless neurones in the retina

In the preceeding sections, I have reviewed our present understanding of the integration of visual signals in vertebrate retina. I have attempted to describe the kinds of synaptic contacts that cells make and to outline the patterns of synaptic interactions. One of the striking facts to emerge from the research I have reviewed is that most of the cell types in the retina do not normally generate action potentials. In the following, I shall attempt to summarise the properties of these neurones, emphasising their cellular physiology and their role in neural integration.

Spikeless cells in the retina share several common features. They generally have rather low (depolarised) membrane potentials. The reason for this appears to be to produce a tonic activation of their calcium conductance, so that neurones in the retina are able to release synaptic transmitter in the absence of stimulation. As I have said, this enables them to signal small changes in presynaptic voltage. It also allows them to communicate voltage differences of either sign. It might be thought that all neurones would benefit from having a low resting potential. However, for cells which must generate spikes of the kind normally used in signal communication, a low resting potential would tend to inactivate the Na^+ conductance responsible for the regenerative inward current of the action potential. Spike-generating neurones generally have rather high membrane potentials and must rely upon the large amplitude of their spikes to release transmitter at their presynaptic endings.

Although most neurones in the retina do not fire action potentials, their membranes are not simple, linear resistances. Cells which make chemical synapses must have some kind of voltage-dependent calcium conductance. Spikeless cells which have voltage-dependent inward calcium currents must also have voltage-sensitive or Ca^{2+}-dependent outward currents to keep their calcium currents from becoming regnerative. In addition to these, photoreceptors and perhaps horizontal cells show inward rectification. It is not known what sorts of voltage-senstive conductances are present in the other classes of retinal interneurones.

Why do most cells in the retina not generate action potentials? The answer usually given is that since these cells are small and for the most part lack long axons, they can convey signals passively with little decrement. They therefore do not need to spike. However, this begs the question. It assumes that spiking is somehow disadvantageous, though there is every indication that the regenerative mode of communication ultimately requires less energy than the electrotonic mode, since cells which rely entirely upon the latter must have a high Na^+ permeability and must therefore expend more energy maintaining their Na^+ and K^+ gradients. It is perhaps true that spikeless communication is more accurate over short distances, since it avoids the necessity of having continually to convert analogue signals to digital ones, and vice versa. It is also possible that neurones may be able to integrate their inputs more precisely if calculations are done entirely in the analogue mode. On the other hand, the electrotonic spread of potential, even over short distances, is accompanied by some loss of amplitude and distortion of waveform. It is not clear how the advantages of spikeless communication outweigh the disadvantages.

Although I have framed my conclusions specifically within the context of vertebrate retina, it should be clear that many of the properties I have described for retinal cells must apply to spikeless neurones wherever they are found. For example, there is now excellent evidence from the work of Burrows and his colleagues (see pp. 199–221 of this volume) that many of the interneurones in locust ganglia do not generate action potentials. These cells tend to have low membrane potentials, and many appear to show a tonic release of transmitter. The current–voltage curves from at least some interneurones in locust show strong outward rectification, suggesting the presence of a voltage-dependent potassium conductance (Burrows & Siegler, 1978). It seems likely that non-spiking neurones in other systems will have similar characteristics.

The study of non-spiking neurones in the retina has revealed many of the difficulties we may confront as we attempt to understand the role of these cells in other parts of the nervous system. For example, work on retina has illustrated the difficulty of making inferences about the function of neurones from their shape or synaptic input. Cells with axons (such as type B horizontal cells) can be incapable of spike generation. Cells without a morphologically distinguishable axon (*e.g.*, the transient amacrines) can produce TTX-sensitive spikes like those found in nerve and muscle. The synaptic input to a cell may be so complex (for amacrine cells in particular) that clear inferences about their function may be very difficult. Single cells may not even have single functions but may participate in several 'local circuits', each of which may do something slightly different.

Work on vertebrate retina has also taught us to be extremely cautious in making inferences about the function of systems of neurones from the responses of single cells. For example, when intracellular recordings were first made from vertebrate retina, the amacrine cell response most commonly recorded was that of the transient amacrine (see Werblin & Dowling, 1969). Many investigators assumed from these initial experiments that all amacrine cells were transient, and detailed inferences were made about the physiology of ganglion cells on the basis of their input from bipolar cells and transient amacrines (see, for example, Dubin, 1970). We now know that there are also amacrine cells giving sustained responses, and that transiently-responding cells represent some as yet unknown fraction of the amacrine cell population (Naka, 1977). An adequate description of the relation between amacrine cell input and ganglion cell output will not be possible until we have much more information about the anatomy and physiology of the inner plexiform layer. We will need to know how many subtypes of amacrines there are in the retina, and which give transient responses, which sustained, and which (if any) responses unlike any so far recorded. We will then need to be able to identify the synapses of different subtypes of amacrines on to different subtypes of ganglion cells and construct mathematical models of the cells which will enable us to calculate the effectiveness of these synapses with some accuracy. These models will need to incorporate the nonlinear properties of the membranes of the different cell types (see Jack *et al.*, 1975).

We have learned that we must be rather cautious in interpreting the function of single cells in a system as complex as the retina. The same must surely be true of systems of even greater complexity. Do we have any real evidence, for example, that granule cells in the cerebellum all behave in exactly the same fashion? Is there any way for us to know at the present time how many subclasses of stellate cells there are in cerebral cortex? Isn't it premature to talk about *the* function of periglomerular cells in the olfactory bulb? It is of course necessary to begin by making some simplifying assumptions in order to form any notion whatever of the function of a system as complicated as the vertebrate central nervous system. However, we must clearly recognize how quickly our assumptions tend to become accepted as established fact. The study of integration in the vertebrate retina has demonstrated how necessary it is for us to view our conceptualisations with suspicion. Only by continually revising our theories and rephrasing out questions can we hope to understand how the nervous system works.

I am grateful to A. T. Ishida and W. K. Stell for permission to publish Fig. 6 from their unpublished work, and to W. K. Stell, R. E. Greenblatt, B. B. Boycott, and the editors of this volume for commenting on a draft of the manuscript. Work in my laboratory is supported by grants from the National Eye Institute and Research to Prevent Blindness, Inc., New York City. I am a James S. Adams Research to Prevent Blindness, Inc., Scholar.

References

Adrian, R. H. (1970). Rectification in muscle membrane. *Prog. Biophys. Mol. Biol.* **19**, 344–69.

Armstrong, C. M. (1975). K pores of nerve and muscle membranes. *Membranes—A Series of Advances*, vol. 3, ed. G. Eisenman, pp. 325–58. New York: Marcel Dekker.

Ashmore, J. F. & Falk, G. (1976). Absolute sensitivity of rod bipolar cells in a dark-adapted retina. *Nature, Lond.* **263**, 248–9.

Baylor, D. A., Fuortes, M. G. F. & O'Bryan, P. M. (1971). Receptive fields of cones in the retina of the turtle. *J. Physiol., Lond.* **214**, 265–94.

Boycott, B. B. & Dowling, J. E. (1969). Organisation of the primate retina: light microscopy. *Phil. Trans. R. Soc. Lond. Series B*, **255**, 109–84.

Boycott, B. B., Dowling, J. E., Fisher, S. K., Kolb, H. & Laties, A. M. (1975). Interplexiform cells of the mammalian retina and their comparison with catecholamine-containing retinal cells. *Proc. R. Soc. Lond. Series B*, **191**, 353–68.

Boycott, B. B., Piechl, L. & Wässle, H. (1978). Morphological types of horizontal cell in the retina of the domestic cat. *Proc. R. Soc. Lond. Series B*, **203**, 229–45.

Boycott, B. B. & Wässle, H. (1974). The morphological types of ganglion cells of the domestic cat's retina. *J. Physiol., Lond.* **240**, 397–419.

Brown, J. E. & Pinto, L. H. (1974). Ionic mechanism for the photoreceptor potential of the retina of *Bufo marinus*. *J. Physiol., Lond.* **236**, 575–91.

Burrows, M. & Siegler, M. V. S. (1978). Graded synaptic transmission between local interneurones and motor neurones in the metathoracic ganglion of the locust. *J. Physiol., Lond.* **285**, 231–55.

Caldwell, J. H., Daw, N. W. & Wyatt, H. J. (1978). Effects of picrotoxin and strychnine on rabbit retinal ganglion cells: lateral interactions for cells with more complex receptive fields. *J. Physiol., Lond.* **276**, 277–98.

Cervetto, L. (1973). Influence of sodium, potassium and chloride ions in the intracellular responses of turtle photoreceptors. *Nature, Lond.* **241**, 401–3.

Dawson, W. W. & Perez, J. M. (1973). Unusual retinal cells in the dolphin eye. *Science*, **181**, 747–9.

Dowling, J. E. (1968). Synaptic organization of the frog retina: an electron microscopic analysis comparing the retinas of frogs and primates. *Proc. R. Soc. Lond. Series B*, **170**, 205–28.

Dowling, J. E. (1970). Organization of vertebrate retinas. *Invest. Ophthal.* **9**, 655–80.

Dowling, J. E. & Boycott, B. B. (1966). Organization of the primate retina: electron microscopy. *Proc. R. Soc. Lond. Series B*, **166**, 80–111.

Dowling, J. E. & Ehinger, B. (1978). The interplexiform cell system. I. Synapses of the dopaminergic neurons of the goldfish retina. *Proc. R. Soc. Lond. Series B*, **201**, 7–26.

Dubin, M. W. (1970). The inner plexiform layer of the vertebrate retina: a quantitative and comparative electron microscopic analysis. *J. Comp. Neurol.* **140**, 479–505.

Eccles, J. C. (1964). *The Physiology of Synapses*. New York: Springer-Verlag.

Fain, G. L. (1975). Interactions of rod and cone signals in the mudpuppy retina. *J. Physiol., Lond.* **252**, 735–69.

Fain, G. L. (1977). The threshold signal of photoreceptors. In *Vertebrate Photoreception*, ed. H. B. Barlow & P. Fatt, pp. 305–23. New York: Academic Press.

Fain, G. L. (1979). Le fonctionnement de la rétine. *La Recherche*, **10**, 355–62.

Fain, G. L., Gold, G. H. & Dowling, J. E. (1976). Receptor coupling in the toad retina. *Cold Spr. Harb. Symp. Quant. Biol.* **40**, 547–61.

Fain, G. L., Granda, A. & Maxwell, J. (1977). The voltage signal of photoreceptors at the visual threshold. *Nature, Lond.* **265**, 181–3.

Fain, G. L., Quandt, F. N., Bastian, B. L. & Gerschenfeld, H. M. (1978). Contribution of a caesium-sensitive conductance increase to the rod photoresponse, *Nature, Lond.* **272**, 467–9.

Fain, G. L., Quandt, F. N. & Gerschenfeld, H. M. (1977). Calcium-dependent regenerative responses in rods. *Nature, Lond.* **269**, 707–10.

Famiglietti, E. V., Jr., Kaneko, A. & Tachibana, M. (1977). Neuronal architecture of on and off pathways to ganglion cells in carp retina. *Science*, **198**, 1267–9.

Fisher, S. K. & Boycott, B. B. (1974). Synaptic connexions made by horizontal cells within the outer plexiform layer of the retina of the cat and the rabbit. *Proc. R. Soc. Lond. Series B*, **186**, 317–31.
Gallego, A. (1971). Horizontal and amacrine cells in the mammal's retina. *Vision Res. Suppl.* **3**, 33–50.
Gerschenfeld, H. M. & Piccolino, M. (1979). Pharmacology of the connexions of cones and of L-horizontal cells in the vertebrate retina. In *The Neurosciences: Fourth Study Program*, Cambridge, Massachusetts: MIT Press.
Goldman, K. A. & Fisher, S. K. (1978). Synaptic organization of the inner plexiform layer of the retina of *Xenopus laevis*. *Proc. R. Soc. Lond.Series B*, **201**, 57–72.
Gouras, P. (1968). Identification of cone mechanisms in monkey ganglion cells. *J. Physiol., Lond.* **199**, 533–47.
Gouras, P. (1972). S-potentials. In *Handbook of Sensory Physiology*, vol. VII/2, ed. M. G. F. Fuortes, pp. 513–29. Berlin: Springer-Verlag.
Gray, E. G. & Pease, H. L. (1971). On understanding the organization of the retinal receptor synapses. *Brain Res.* **35**, 1–15.
Hagins, W. A., Penn, R. D. & Yoshikami, S. (1970). Dark current and photocurrent in retinal rods. *Biophys. J.* **10**, 380–412.
Hagiwara, S. (1975). Ca-dependent action potential. In *Membranes—A Series of Advances*, vol. 3, ed. G. Eisenman, pp. 359–81. New York: Marcel Dekker.
Hagiwara, S. & Takahashi, K. (1974). The anonalous rectification and cation selectivity of the membrane of a starfish egg cell. *J. Mem. Biol.* **18**, 61–80.
Hodgkin, A. L. & Huxley, A. F. (1952). Currents carried by sodium and potassium ions through the membrane of the giant axon of *Loligo*. *J. Physiol., Lond.* **116**, 449–72.
Jack, J. J. B., Noble, D. & Tsien, R. W. (1975). *Electric Current Flow in Excitable Cells*. London & Oxford: Clarendon.
Kaneko, A. (1970). Physiological and morphological identification of horizontal, bipolar, and amacrine cells in the goldfish retina. *J. Physiol., Lond.* **207**, 623–33.
Kaneko, A. (1973). Receptive field organization of bipolar and amacrine cells in the goldfish retina. *J. Physiol., Lond.* **235**, 133–53.
Kaneko, A. & Shimazaki, H. (1976). Synaptic transmission from photoreceptors to the second-order neurons in the carp retina. In *Neural Principles in Vision*, ed. F. Zettler & R. Weiler, pp. 143–57. Berlin: Springer-Verlag.
Katz, B. & Miledi, R. (1967). A study of synaptic transmission in the absence of nerve impulses. *J. Physiol., Lond.* **192**, 407–36.
Kolb, H. (1970). Organization of the outer plexiform layer of the primate retina: electron microscopy of Golgi-impregnated cells. *Phil. Trans. R. Soc. Lond. Series B*, **258**, 261–8.
Kolb, H. & West, R. W. (1977). Synaptic connections of the interplexiform cell in the retina of the cat. *J. Neurocytol.* **6**, 155–70.
Lamb, T. D. (1976). Spatial properties of horizontal cell responses in the turtle retina. *J. Physiol., Lond.* **263**, 239–55.
Lasansky, A. (1971). Synaptic organization of cone cells in the turtle retina. *Phil. Trans. R. Soc. Lond. Series B*, **262**, 365–81.
Lettvin, J. Y., Maturana, H. R., McCulloch, W. S. & Pitts, W. H. (1959). What the frog's eye tells the frog's brain. *Proc. Inst. Radio Engrs.* **47**, 1940–51.

Levick, W. R. (1972). Receptive fields of retinal ganglion cells. In *Handbook of Sensory Physiology*, vol. VII/2, ed. M. G. F. Fuortes, pp. 531–66. Berlin: Springer-Verlag.

Levick, W. R. (1975). Form and function of cat retinal ganglion cells. *Nature, Lond.* **254**, 659–62.

Liebman, P. A. & Granda, M. (1971). Microspectrophotometric measurements of visual pigments in two species of turtle, *Pseudemys scripta* and *Chelonia mydas*. *Vision Res.* **11**, 105–14.

Llinás, R. R. (1977). Calcium and transmitter release in squid giant synapse. In *Society for Neuroscience Symposia*, vol. 2, ed. W. M. Cowan & J. A. Ferrendelli, pp. 139–60. Bethesda, Maryland: Society for Neuroscience.

Marchiafava, P. L. (1978). Horizontal cells influence membrane potential of bipolar cells in the retina of the turtle. *Nature, Lond.* **275**, 141–2.

Marks, W. B. (1965). Visual pigments of single goldfish cones. *J. Physiol., Lond.* **178**, 14–32.

Miller, R. F. & Dacheux, R. (1976). Dendritic and somatic spikes in mudpuppy amacrine cells: identification and TTX sensitivity. *Brain Res.* **104**, 157–62.

Missotten, L. (1965). *The Ultrastructure of the Human Retina.* Brussels: Arscia Uitgaven N. V.

Naka, K.-I. (1977). Functional organization of catfish retina. *J. Neurophysiol.* **40**, 26–43.

Naka, K.-I. & Rushton, W. A. H. (1967). The generation and spread of S-potentials in fish (*Cyprinidae*). *J. Physiol., Lond.* **192**, 437–61.

Nelson, R., Famiglietti, E. V., Jr. & Kolb, H. (1978). Intracellular staining reveals different levels of stratification for on- and off-center ganglion cells in cat retina. *J. Neurophysiol.* **41**, 472–83.

Nelson, R., v. Lützow, A., Kolb, H. & Gouras, P. (1975). Horizontal cells in cat retina with independent dendritic systems. *Science*, **189**, 137–9.

Norton, A. L., Spekreijse, H., Wolbarsht, M. L. & Wagner, H. G. (1968). Receptive field organization of the S-potential. *Science*, **160**, 1021–2.

Piccolino, M. & Gerschenfeld, H. M. (1977). Lateral interactions in the outer plexiform layer of turtle retinas after atropine block of horizontal cells. *Nature, Lond.* **268**, 259–61.

Polyak, S. L. (1941). *The Retina.* Chicago, Illinois: Chicago University Press.

Ramon Y Cajal, S. (1893). La rétine des vertébrés. *La cellule*, **9**, 17–257.

Raviola, E. & Gilula, N. B. (1975). Intramembrane organization of specialized contacts in the outer plexiform layer of the retina. A freeze-fracture study in monkeys and rabbits. *J. Cell Biol.* **65**, 192–222.

Richter, A. & Simon, E. J. (1975). Properties of centre-hyperpolarising, red-sensitive bipolar cells in the turtle retina. *J. Physiol., Lond.* **248**, 317–34.

Saito, T., Kondo, H. & Toyoda, J.-I. (1979). Ionic mechanisms of two types of on-centre bipolar cells in the carp retina. I. The responses to central illumination. *J. Gen. Physiol.* **73**, 73–90.

Simon, E. J., Lamb, T. D. & Hodgkin, A. L. (1975). Spontaneous voltage fluctuations in retinal cones and bipolar cells. *Nature, Lond.* **256**, 661–2.

Stell, W. K. (1972). The morphological organization of the vertebrate retina. In *Handbook of Sensory Physiology*, vol. VII/2, *Physiology of Photoreceptor Organs*, ed. M. G. F. Fuortes, pp. 111–213. Berlin: Springer-Verlag.

Stell, W. K., Ishida, A. T. & Lightfoot, D. O. (1977). Structural basis for on- and off-centre responses in retinal bipolar cells. *Science*, **198**, 1269–71.

Toyoda, J.-I. & Tonosaki, K. (1978). Effect of polarization of horizontal cells on the on-centre bipolar cell of carp retina. *Nature, Lond.* **276**, 399–400.

Trifonov, Yu. A., Byzov, A. L. & Chailahian, L. M. (1974). Electrical properties of subsynaptic and non-synaptic membranes of horizontal cells in fish retina. *Vision Res.* **14**, 229–41.

Wässle, H., Boycott, B. B. & Peichl, L. (1978). Receptor contacts of horizontal cells in the retina of the domestic cat. *Proc. R. Soc. Lond. Series B*, **203**, 247–67.

Wässle, H., Peichl, L. & Boycott, B. B. (1978). Topography of horizontal cells in the retina of the domestic cat. *Proc. R. Soc. Lond. Series B*, **203**, 269–91.

Werblin, F. S. (1970). Response of retinal cells to moving spots: intracellular recording in *Necturus maculosus*. *J. Neurophysiol.* **33**, 342–50.

Werblin, F. S. (1972). Lateral interactions at inner plexiform layer of vertebrate retina: anatagonistic responses to change. *Science*, **175**, 1008–10.

Werblin, F. S. (1975). Anomalous rectification in horizontal cells. *J. Physiol., Lond.* **244**, 639–57.

Werblin, F. S. & Dowling, J. E. (1969). Organization of the retina of the mudpuppy, *Necturus maculosus*: II. Intracellular recording. *J. Neurophysiol.* **32**, 339–55.

West, R. W. (1976). Light and electron microscopy of the ground squirrel retina: functional considerations. *J. comp. Neurol.* **168**, 355–78.

West, R. W. & Dowling, J. E. (1972). Synapses on to different morphological types of retinal ganglion cells. *Science*, **178**, 510–12.

Wong-Riley, M. T. T. (1974). Synaptic organization of the inner plexiform layer in the retina of the tiger salamander. *J. Neurocytol.* **3**, 1–33.

Yau, K.-W., Lamb, T. D. & Baylor, D. A. (1977). Light-induced fluctuations in membrane current of single toad rod outer segments. *Nature, Lond.* **269**, 78–80.

Yazulla, S. (1976). Cone input to bipolar cells in the turtle retina. *Vision Res.* **16**, 737–44.

STEPHEN R. SHAW

Anatomy and physiology of identified non-spiking cells in the photoreceptor–lamina complex of the compound eye of insects, especially Diptera

Introduction

It has been argued recently that a common optical design strategy may govern construction of all different types of advanced eyes, the selective advantage of which has guided evolution of these structures (Snyder, Laughlin & Stavenga, 1977). The optimum detection of environmental contrast or of movement, for instance, is likely to be a common imperative for species-survival of all visually guided animals, regardless of whether these possess the vertebrate or spider type of camera eye, or the optically quite different multi-lens eyes of the arthropod group. It is not yet known whether, similarly, there is some best way to organise the neurones behind the optics in a retina to optimise information flow, though it is already clear that the basic building blocks upon which evolution has acted are again diverse in both vertebrates and invertebrates. The response polarity of the photoreceptors to illumination is in most cases opposite in the two divisions, for example. Nevertheless, despite differing details of mechanism, many similarities of neural operation common to both groups will emerge from analysis of peripheral organisation of the insect eye, in the summary presented here: hyperpolarising and depolarising, high-fidelity slow potential cells; single photon detection and amplification; decrementally transmitted signals over at least the first two stages; presynaptic neural convergence and synaptic hypertrophy, increasing reliability of operation and tuning synaptic transmission to tiny signals; high amplification synaptic mechanisms releasing transmitter reliability for very small voltages; presynaptic electrical coupling between photoreceptors, averaging signals and optimising synaptic amplification; lateral inhibition between receptors; auto-ranging mechanisms by which synaptic transmission adjusts to new lighting conditions without losing sensitivity; no doubt others remain to be discovered.

The account here will deal principally with the known properties and interactions of the photoreceptors and post-synaptic neurones of the lamina synaptic neuropile in insects, emphasizing work on Diptera, the flies. Optically, anatomically, physiologically and even behaviourally, the Diptera have been the most intensively studied of the insects, particularly the common cyclorraphid flies like *Musca* and *Calliphora.* Complementary information from other insects, and from analogous but larger cells in the insect ocellar system and the simple eyes of barnacles, fills out some of the gaps in knowledge. Particularly useful from the physiological standpoint, all of the cell types and most of their interconnections have now been described anatomically, in the Dipteran lamina, the outermost nucleus of the compound eye. The relation of this centre to the others in the optic lobe is shown in Fig. 1*d.* The lamina consists of first, second, third-order and intrinsic neurones, and receives efferent innervation, and so occupies roughly the position that the retina holds in the vertebrate visual pathway, though the two systems are not functionally equivalent in all respects.

The anatomy of the receptor layer and the wiring scheme of the lamina, along with some functional considerations, are described first, prior to consideration of more purely physiological topics.

Structure and optics of the photoreceptor zone

The precise repeating unit structure of the insect compound eye is well known and indicated outwardly by the individual hexagonally packed facet lenses (Fig. 1*a*), each of which overlies one *ommatidium*, a group of eight photoreceptors. In each half of one eye, this group is assembled in a stereotyped ring-like pattern, in which the light absorbing rods, the *rhabdomeres*, are asymmetrically but precisely placed, allowing the individual cells to be uniquely labelled 1–8 (Fig. 1*b,c*). Receptors with larger peripherally placed rhabdomeres, R1–R6, form short axons all of which end in the first optic neuropile, the *lamina*, whilst the two central cells which lie end-to-end, R7 and R8, send thinner axons to the second, deeper nucleus, the *medulla* (Fig. 1*d*). A rhabdomere comprises thousands of aligned tubular microvilli grown out from the photoreceptor surface membrane, each about 70 nm across, forming a light conducting rod about 300 μm long in *Calliphora.* By virtue of this length, the visual pigment held in the microvillar membrane

Fig. 1. The layout of a typical housefly or bluebottle eye. *a*, scanning electron micrograph of the eye surface of a large male *Calliphora auger*, showing the prominence of the array of lenses or facets, covering most of the head. The head is about 4 mm across, and the largest facets 50 μm (inset); (*b*, *c*), transverse and cross-sections through a schematic ommatidium, the eye's repeating sub-unit which underlies each facet. Each tubular unit contains 8 photoreceptors R1–R8 arranged in a characteristic trapezoidal pattern, R8 lying in line with and below R7. The pattern in *c* occurs in the upper half of the left eye, with the two halves and the two eyes showing mirror symmetry. In the fly, but unlike many insects, the light-sensitive rhabdomeres (*rh*) of one ommatidium are physically and optically separate from one another; *d*, the neuropiles of the optic lobe behind the eye, where synaptic interactions take place. Axons from receptors R1–R6 separate and terminate within different 'optic cartridges' in the first of these, the *lamina*; R7 and R8 pass through the synapses in the second, the *medulla*. The divergence pattern of R1–R6 is given in detail in Fig. 2, and the detailed wiring between the cells in a single cartridge, in Fig. 3 and Table 1. *c*, corneal lens; *cc*, crystalline cone; *pc*, pigment cell; *rc*, receptor or retinular cell; *rh*, rhabdomere; *bm*, basement membrane.

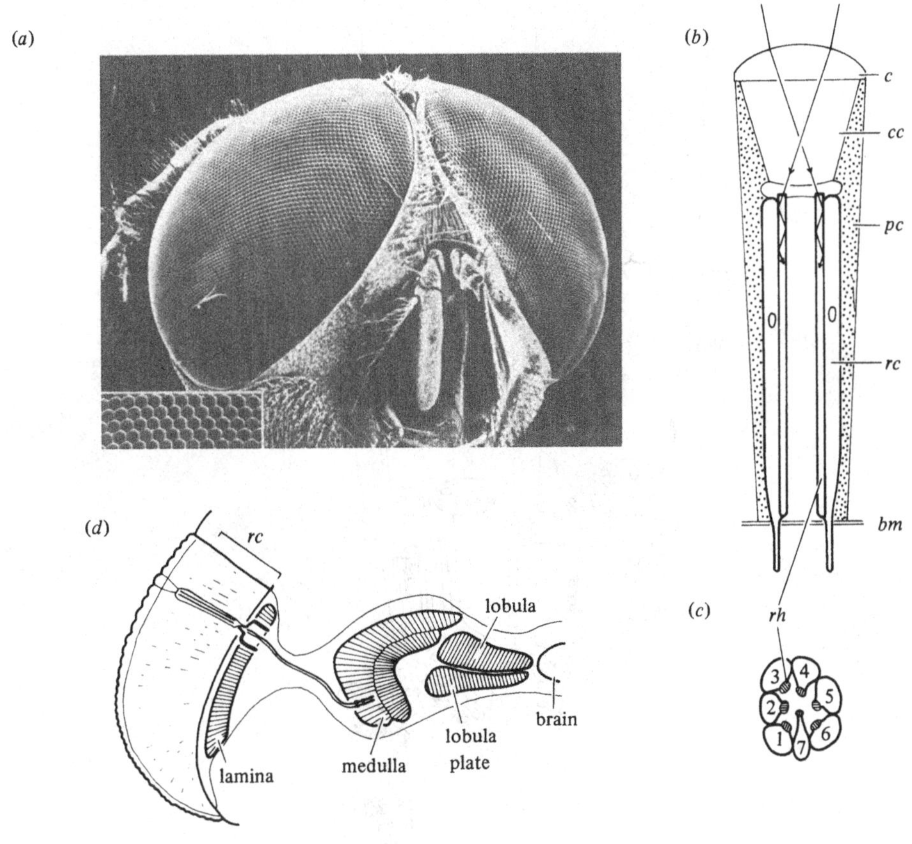

Fig. 2. The precision 'wiring' pattern of the photoreceptors on to the first synaptic region in a cyclorrhaphid fly like *Musca*, used to increase light-capture efficiency. The receptors in every ommatidium are ordered in the particular trapezoidal pattern of Fig. 1*c*. Shown at the left, six ommatidia (three shown in detail) lying in exactly this same trapezoidal formation, send one axon each into a particular synaptic column or cartridge. Note that the parent cells participating in this projection, R1–R6, occupy the same position in the ommatidia that their ommatidium holds in the larger pattern; this ensures that only the six receptors with identical lines of sight combine information at each single cartridge. To achieve the proper convergence on each column, the axons from a *single* ommatidium must diverge and interweave amongst their neighbours, as shown for the ommatidium at the right. Because a simple positive lens like that of the fly facet rotates the image through 180° on to the rhabdomeres, the required divergence geometry is into a trapezoidal arrange of cartridges that is 180° rotated from the pattern of rhabdomeres. This is achieved by a 180° twist imparted during morphogenesis on the axon bundles, as shown. R7–R8 grow directly through the lamina without making synapses. The outline of a cartridge with its two large monopolar cells L1 and L2 is shown at the bottom; for detailed structure, see Fig. 3.

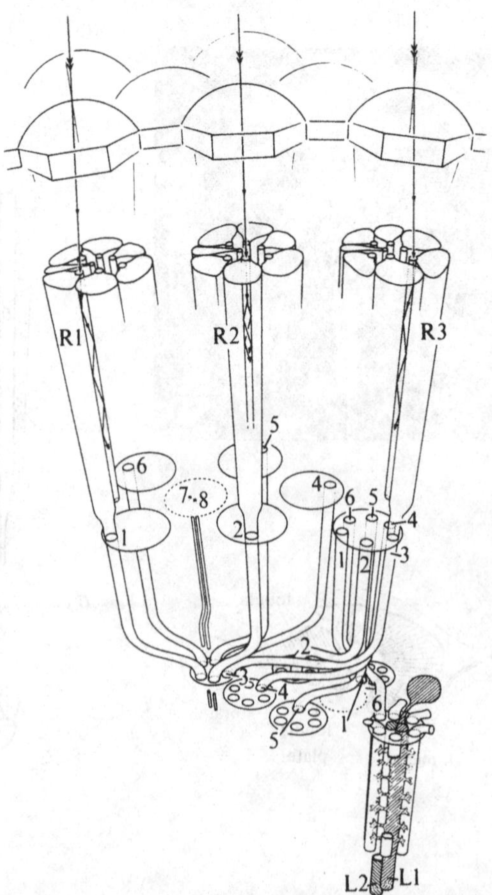

absorbs most of the light focussed upon it by its lens. Ommatidia are separated and to some extent optically insulated from each other by pigmented non-neural cells (Fig. 1*b*).

The optical arrangement behind a lens is shown diagrammatically in the upper part of Fig. 2, based on the analysis of Kirschfeld (1967) and Franceschini (1975). The rhabdomeres lie separated from each other in the focal plane behind the corneal lens, the front surface of which acts as a simple positive lens to 'invert' its image – that is, it *rotates* a local portion of object space through 180° and projects this image on to the rhabdomeres. The dimensions are such that neighbouring rhabdomeres in one ommatidium have visual axes which diverge by about 2°, in *Musca* (Kirschfeld, 1967), a value close to that by which the central anatomical axes of neighbouring whole ommatidia are inclined to each other, because of the eye's curvature. The consequence is that seven different rhabdomeres behind seven separate facets, arranged in the familiar trapezoidal pattern, look outwards from the eye along a common line of sight, shown in Fig. 4 (actually there is a slight convergence: Pick, 1977). Kirschfeld (1967) pointed out that the function of this arrangement might be to sum the light collected through several facets, in effect increasing the optical aperture of the collector and thus the light-catching power, but retaining the high visual acuity characteristic of the open rhabdomere system. Acuity is maintained by virtue of a narrow visual field conferred by having narrow, isolated rhabdomeres, but is worsened if the width of this light-collecting structure is broadened, as when rhabdomeres are pressed together to form a single collector in many insects. This optical summation scheme only works in the fly if the signals from photoreceptors with a common axis can somehow be summed together, and this is achieved in the higher flies by an intricate criss-crossing of axons behind the receptor layer, which brings only the appropriate axons into a single *optic cartridge* in the lamina synapse (Fig. 2, bottom), where they synapse with second-order fibres (Trujillo-Cenóz & Melamed, 1966; Braitenberg, 1967).

Each cartridge is the column-shaped, repeating structural unit of the lamina, arranged in a regular lattice one-for-one behind the array of ommatidial columns. One of the more surprising things about this complicated axonal projection, which arises from a simpler pattern during morphogenesis (Trujillo-Cenóz & Melamed, 1973), is the lack of developmental errors in the wiring pattern, nearly all the axons reaching the predicted cartridges (Horridge & Meinertzhagen, 1970). Unlike R1–R6, the axons from the two central cells R7 and R8 from the seventh facet of each group, bypass each cartridge to synapse in the medulla. Rotation of the image by the optics of each facet results in the R1–R6 axons all emerging on the opposite side of their ommatidium from the cartridges to which they must respectively connect, but

Table 1. *Dipteran lamina: connections to and from the fourteen neuronal classes.* Based mainly on Strausfeld & Campos-Ortega (1977)[a]

Cell class	Lamina connections: Lamina input(s) from: Narrow field sources	Wide field sources	Lamina output(s) to: Narrow field cells	Wide field cells	Inferred function	Field size	Medulla endings: Lateral spread of terminal arborisation (SC 72)
R1–R6	R1–R6[b] L2 C2 EGCs?	Tan 1 (~ 40c) field potentials?	R1–R6[b] L1–L3 T1(βs)	Amacrine αs (~ 17c)	Photoreceptor	Narrow	∅
R7	∅	∅	∅	∅	Photoreceptor	Narrow	Narrow
R8	∅	∅	∅	∅	Photoreceptor	Narrow	Narrow
L1	R1–R6 L2 C2 C3	L4 (own + 2c) Tan 1 (~ 40c)	∅	∅	Pure output cell	Narrow (JZ 73)	Narrow
L2	R1–R6 C2 C3	L4 (own + 2c) Tan 1 (~ 40c)	R1–R6 L1	∅	Output[c]– feedback	Narrow (JZ 73)	Narrow
L3	R1–R6 C2	Tan 1 (~ 40c)	∅	∅	Output	Narrow?	Narrow
L4	Other L4	Amacrines (~ 17c)	L1 own + 2c L2 own + 2c L4 at 2c	∅	? Output[c]/lateral interaction	Narrow?	Narrow
L5	∅	Amacrines (~ 17c) Tan 2 (~ 40c)	∅	∅	Output	?	Narrow
T1	R1–R6 (at βs)	Amacrine αs (~ 17c)	EGCs	Amacrine αs	? Output[c]	Narrow (JZ 73)	Narrow
T1a?[d]	?	?	?	?	?	?	Narrow[d]
Amacrine	R1–R6 T1 ($\alpha \rightleftharpoons \beta$)	Other Amacrines	T1 ($\alpha \rightleftharpoons \beta$) L5 (in ~ 17c) EGCs	L4 Amacrines	Intrinsic lateral interaction cell	Wide? (≥ 17c)	∅
C2	∅	∅	R1–R6 L1–L3	∅	Centrifugal	Narrow?	Wider than one column
C3	∅	∅	L1, L2	∅	Centrifugal	Narrow	Narrow
Tan 1	∅	∅	R1–R6 in ~ 40c L1–L3 in ~ 40c	∅	Centrifugal	Wide	'La-tan' spans many columns[e]
Tan 2	∅	∅	L5 (~ 40c)	∅	Centrifugal	Wide	

[a] *Abbreviations*: ∅, no connections: c, cartridges; αs, βs, the α and β processes of amacrine and T1 cells respectively; EGCs, lamina epitheli glial cells; JZ 73, Jarvilehto & Zettler, 1973; SC 72, SC 77, Strausfeld & Campos-Ortega, 1972, 1977.

[b] R1–R6 are interconnected by gap junctions, presumed electrical connections.

[c] There is no internal evidence on which to decide whether the axon of L4, T1, or even L2 is centrifugal (T1: Boschek, 1971) or centripe (T1: Trujillo-Cenóz, 1972, JZ 73, SC 77). All of these cells make both input and output connections in the lamina, and the polarity the medulla arborisations is unknown. All the other cell types with axons have either pure inputs or pure outputs in the lamina, whic allows the axon's conduction direction to be specified with some certainty. Response latency dictates that L2 and probably T1 are afferen

[d] Amacrine α processes were originally thought to derive from the centrifugal axons of T1a medulla cells (SC 72), the destination of whic is therefore now unaccounted for. In medulla arborisation, T1a closely resembles C3 (SC 72, Fig. 4).

[e] Tan 1 and Tan 2 have not been distinguished in the medulla by any author. La-tan of SC 72 is presumably the origin of one; perha T1a is the other.

this is corrected by the precise 180° twist which each bundle makes during development (Fig. 2, lower right). The projection pattern of axons at the equator of the eye is slightly more complicated with extra axons interpolated in each cartridge, but serves the same function of optical summation (Kirschfeld, 1967).

The more primitive Biblionid flies appear to have a different optical summation pattern, using receptors in a next-but-one facet array rather than next-neighbours, which requires the axons to diverge even further laterally to reach the correct synaptic station (Zeil, 1979).

The structure of the lamina optic cartridge

Not only does the sensory projection described by Fig. 2 remain practically invariant across the eye, but the complement of cells obviously repeats itself from one optic cartridge to the next for the most prominent neurones, in those flies examined. Regularly repeated columns appear to make up the lamina in all the insects studied but are particularly clear in flies because the six regular profiles of R1–R6 give a circular outline to each cartridge, thrown further into relief by the epithelial glial cells, three of which invest each cartridge (Figs. 3, 11; Boschek, 1971). Our knowledge of the synaptic structure and connections has developed over the last fifteen years from several sources, but except for a brief review (Strausfeld & Campos-Ortega, 1977), has not been systematised recently in an accessible form. Table 1 is an attempt to condense the currently known information and Fig. 3 represents a more digestible, idealised view of a single cartridge, incorporating the known directionality of the synapses. In each cartridge group, there are fourteen neural types though two of these, the long axons of R7 and R8, run straight through the lamina near R6 without making synapses. This contrasts with one of the long fibres in dragonfly lamina, which does make synaptic connections (Armett-Kibel, Meinertzhagen & Dowling, 1977), and this may happen also in the bee (Ribi, 1975).

The reader is referred directly to the legend to Fig. 3 for a synopsis of the main features of lamina connectivity, some of which are discussed in finer detail next.

Lamina output cells, L1–L3

The only direct sensory input to each cartridge comes from the six R1–R6 photoreceptors of each optical projection. The most common synaptic contact is from R 1–R6 to three monopolar cells L1–L3, which have cell bodies just above the lamina. L1 and L2 form the prominent central pair of axons

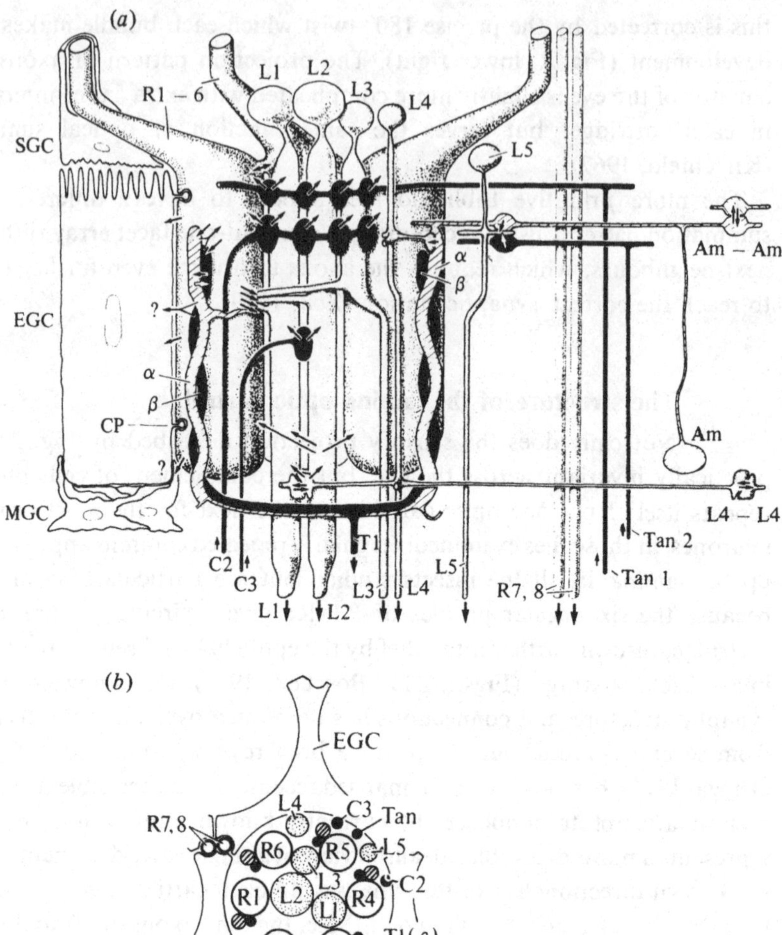

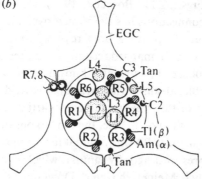

Fig. 3. (*a*) Diagram of the neuronal types and their synaptic interconnections, in a single optic cartridge in the lamina synaptic nucleus of the fly's eye, based on Table 1. Direct sensory input to each cartridge (downwards arrows) comes solely from the six photoreceptors with short axons, R1–R6, one each from a different ommatidium (see Fig. 2), and three of which are illustrated. R1–R6 form numerous sm all tetrad (1 → 4) synapses at their points of contact with the grouped lateral spines of the second-order monopolar cells L1–L3 (which form the main lamina output), plus an α-amacrine process. In addition to an internal loop within the lamina from L2 back to R1–R6 and L1, four other *efferent feedback pathways* from the medulla reach back into each cartridge, three of which involve receptor terminals and/or L1–L3: C2, C3 (narrow field) and Tan 1, Tan 2 (wide field). Two *laterial interaction pathways* across

in all cartridges, and put out lateral spines upon which many of the synapses are made, all along the cartridge. L3 contributes only in the top third of the cartridge, then runs outside the column (Fig. 3). During development, both receptor axons and monopolars put out many slender filopodia, establishing contact with each other (Trujillo-Cenóz & Melamed, 1973; Meinertzhagen, 1973), and whilst those of R1–R6 are later withdrawn, the spines from monopolars are probably these filopods which have persisted through to the adult. In section, the synapses themselves nearly always appear to be multiple with a varying number of post-synaptic elements for each presynaptic site, but Burkhardt & Braitenberg (1976) suggest that this results from variation in the plane of section, and that the structure is quite uniform. There are really four processes, forming a post-synaptic *tetrad*, one each from L1 and L2 as the two more central elements, the more lateral elements coming from L3 and from an amacrine cell (Fig. 3). The main distinctive ultrastructural feature seen at each tetrad is an osmophilic presynaptic table often appearing T-shaped in section, which all authors have used as the main morphological criterion for the presence of a functional synapse. Small electron-lucent vesicles appear more ordered in the vicinity of these small tables, and something like 500000 others densely pack the rest of the terminal (Burkhardt & Braitenberg, 1976).

The reduplication of spine contacts between Rs and Ls has encouraged speculation that this hypertrophy may indicate a synapse with improved signal–noise characteristics (Laughlin, 1973; Strausfeld & Campos-Ortega, 1977). This cannot be assessed simply from the number of synaptic subdivisions (Shaw, 1979), but a rough comparative check on the suggestion can be made if it is conservatively assumed that the presynaptic tables delineate active zones of vesicle release (Wood, Pfenninger & Cohen, 1977), like those at the neuromuscular junction. From a preliminary serial section analysis of

---

the lamina are established reciprocally along collaterals of the third order monopolar L4, and by the widely ramifying and interconnecting amacrine cells (*Am*); the two networks also interact at the α processes. Midget monopolar L5 is also a third-order neurone, and T1 is now thought to be another second-order output neurone. The central photoreceptor pair R7/8 bypass each cartridge to synapse in the medulla, but might still influence the lamina through one of the efferent pathways. Apart from Am, Tan 1 and Tan 2, which are aperiodic serving up to 40 columns, all the neurones shown repeat in each cartridge. Two of the types of presumed glial cells sometimes appear as post-synaptic elements (left, see text). The uppermost, double-headed arrow indicates the site of gap junctions coupling R1–R6. (*b*) The approximate topography of the elements in a cross-section of the upper cartridge. SGC, EGC, MGC: satellite, epithelial, marginal glial cells; CP, capitate projections; α, β, the intertwined, reciprocally synapsing pairs of fibres from amacrine and T1 cells.

cartridges in *Musca* (A. Fröhlich & I. A. Meinertzhagen, personal communication), there could be as many as 400 presynaptic tables per receptor, rather than the 40–50 estimated earlier (Trujillo-Cenóz, 1965; Burkhardt & Braitenberg, 1976). Each table, about 0.1×0.3 μm across, could accommodate some 10–15 vesicles along its edges, suggesting that for all six receptors, there could be in the neighbourhood of 30000 quantal release sites per cartridge on to the four post-synaptic cells. Whilst these involve six different photoreceptor terminals, evidence is discussed later that all six are strongly electrically coupled together. The value obtained compares in magnitude with more directly based estimates of 15000–20000 active sites for one end-plate to one muscle fibre at the amphibian neuromuscular junction, where the fractional

Fig. 4. Frontal view of the compound eyes of a male fly, *Calliphora stygia*, photographed down a microscope whilst illuminated by a narrow 1° beam of light, coaxial with the viewing axis. The two sets of six trapezoidally arranged spots are reflections in the two eyes from sets of six rhabdomeres in six ommatidia in each eye, the only ones which are looking exactly back along the observer's line of sight. This is the same projection drawn in Fig. 2, the cells of which converge their axons into a single cartridge to sum the light responses through the six lenses, as discussed first by Kirschfeld (1967). The reflections are made visible because reflecting granules muster round the rhabdomeres of the cells illuminated. They do not show up in the ommatidium containing cells R7–R8, which have a higher threshold. Except for the two central reflection patterns, the rest of the eye has been underexposed to show up its outline. The bright spots at the top are reflections from the three ocelli.

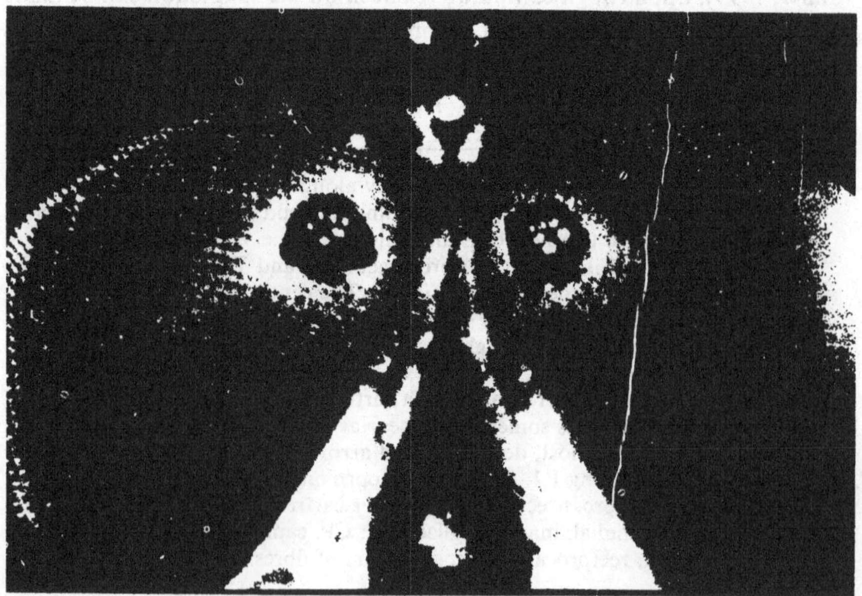

release per nerve impulse is known to be about 200 (see Heuser & Reese, 1979), with accordingly high transmission reliability. The fractional release factor is not known in the lamina cartridge, but recent physiological work on the related ocellar synapse indicates that even a quantised post-synaptic potential, driven by a presynaptic single photon response of only about 1 mV, as in the eye, represents the release of many transmitter quanta (Wilson, 1978*c*). These estimates need refining further, but lend some credit to the idea that elaboration of many monopolar spines with active synaptic sites greatly enlarges the synaptic area, and so establishes more reliable transmission of small signals at this important main sensory relay.

Basket cell, T1

The relations involving the distinctive 'mimetic pairs' of α and β climbing fibres discovered by Trujillo-Cenóz (1965), has the most complicated history of all the cartridge connections, and is still not fully resolved. As shown in Fig. 3, a pair of intertwining fibres ascends the cartridge next to each of the six photoreceptor terminals, though Boschek (1971) shows seven pairs. These were originally thought to branch from two separate efferent neurones from the medulla, but Golgi-electron-microscope studies now indicate that the α processes come from several separate cells intrinsic to the lamina, the amacrines (Campos-Ortega & Strausfeld, 1973). All the β processes in one cartridge spring from a single neurone named T1 by Strausfeld, which has a cell body in the outer medulla, and a nearby dendritic bush which associates closely with the terminal of a lamina monopolar, probably L2 (Trujillo-Cenóz, & Melamed, 1970; Campos-Ortega & Strausfeld, 1973). In the housefly *Musca*, α and β can be distinguished ultrastructurally, but this seems difficult in the greenbottle *Lucilia*. In *Musca*, β has many fewer presynaptic bars or tables than α, fewer synaptic vesicles, is often less electron-dense, and unlike α is visited by curious ingrowths from the epithelial glial cells, called 'gnarls' (Fig. 11; Campos Ortega & Strausfeld, 1973; Burkhardt & Braitenberg, 1976). In *Musca*, all authors have found either α and β to be *post*-synaptic to R1–R6 (Boschek, 1971; Campos-Ortega & Strausfeld, 1973; Burkhardt & Braitenberg, 1976), Boschek finding this one of the most common synapses in the cartridge. In major disagreement, Trujillo-Cenóz (1965) thought both α and β were *pre*synaptic to R1–R6, but did not distinguish one from the other. This was confirmed later for α, when distinction could be made by Golgi impregnation of β, of course obliterating βs ultrastructure (Trujillo-Cenóz & Melamed, 1970), and Campos-Ortega & Strausfeld (1973) concur. It therefore remains possible that both *Lucilia* and *Musca* are similar in that the T1 β processes are exclusively post-synaptic to R1–R6, and that only the amacrine α

processes differ in the two species; it is to be hoped that future work will more conclusively resolve this uncertainty in *Lucilia*.

α and β synapse reciprocally with each other in *Musca* (Boschek, 1971; Campos-Ortega & Strausfeld, 1973) but most commonly in the direction $\alpha \rightarrow \beta$. Burkhardt & Braitenberg (1976), by contrast, would classify most or even all such occurrences not as reciprocal, but as cases where α and more occasionally β synapses on to a thin epithelial glial process which insinuates between them over much of their length, and Boschek too found direct synapses of this type very common. However, the first two papers cited do show what appear to be direct $\alpha \rightleftharpoons \beta$ synapses. Lastly, Strausfeld & Campos-Ortega (1973) recorded one instance of a L4 $\rightarrow \beta$ synapse but do not refer to this in any later papers.

From this complex distillation it seems likely that, at least in *Musca*, β processes of T1 are directly post-synaptic to all six photoreceptors in one cartridge, and separately to the six α processes which arise from several different amacrine cells, dictating that T1 is another lamina output cell, like L1–L3. If the physiological action of these two synaptic inputs were opposite, the presumed wide fields of the amacrines in contrast to the narrow field of one cartridge should confer centre-surround antagonism on T1s receptive field. Only one recording has been certainly made from T1, supported by dye injection, and indicates that T1 is a hyperpolarising, slow potential cell with a centre response like L1–L2, but with a receptive field even narrower than these monopolars (Järvilehto & Zettler, 1973). This is consistent with the anatomical summary above and suggests that the amacrine input may provide an antagonistic, depolarising surround to the centre hyperpolarising response. If the synaptic delay of 1 ms quoted by Järvilehto & Zettler (1973) is correct, a centrifugal origin of the centre response is definitely excluded: conduction of a signal to the medulla and back along at least two fibres and across at least two synapses would be expected to take at least 5 ms. A reciprocal inhibitory synapse from β back to α could enhance the transient response of T1 to centred illumination, but T1 responses show no obvious differences from L1–L2 in this respect in the records of Järvilehto & Zettler.

Third-order monopolar cells, L4 and L5

Comparison of the remaining monopolar types L4 and L5 to the functionally homogeneous group L1–L3, reveals that cell shape is by itself no indicator of a common cell function, a common assumption in studies of the vertebrate retina. L4 and L5 receive no direct input from receptors, and are both third-order neurones driven by the widely ramifying network of lamina intrinsic cells, the amacrines. L5 has no outputs in the lamina, and receives input only from the amacrines and from the medulla efferent cell Tan 2 (Fig. 3). Since tangential neurones and amacrines send processes into as many as 40 and 20 cartridges respectively, the anatomical evidence at first sight suggests that L5 activity should reflect a summation pool of at least twenty ommatidia or even more, for amacrine cells both overlap and interact further with each other. A contrary view emerges below.

Fig. 5. *a*, Receptive field measured from a lamina 'sustaining unit' by Arnett (1972), defined approximately by the iso-effect contours for central excitation (+), and two flanking patches of inhibition (−). This type of unit is attributed in the text to monopolar cell L4, the tripartite branching and interconnection pattern of which is known, and illustrated below; *b*, *c*, two variant schemes compatible with the anatomy of the basal interconnections of L4, both of which would produce the two required narrow inhibitory zones of flanking inhibition around a narrow excitatory centre. The centre response originates from sustained excitation of L4 by the Amacrine α-processes in its own cartridge. In the simplest case, (*b*, *d*), the two incoming L4 collaterals each make reciprocal inhibitory synapses with the third L4 process belonging to that cartridge, but not with each other, producing the pattern of inhibition shown; (*c*, *e*), Braitenberg & Debbage (1974) claim evidence of reciprocal synapse between *all* pairs of L4 collaterals, and if the extra pair required are incorporated as excitatory synapses, the resulting receptive field becomes extended vertically. If the extra excitatory coupling is weaker than direct excitation via a central α input, *e* tends towards *d*. If all L4 interactions are inhibitory, an annular inhibitory surround results, contrary to *a*. L4 is the only cell known in the lamina with the branching pattern needed to explain the sustaining unit response.

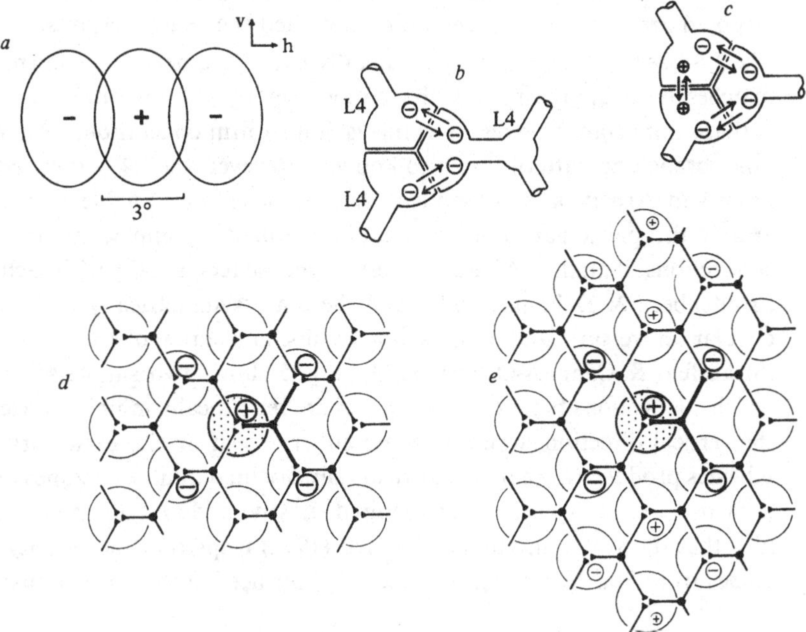

The most complicated monopolar, L4, establishes a regular pattern of reciprocal synaptic contacts through three basal collaterals, reaching its homologues laterally in two neighbouring cartridges on either side of itself. These processes also synapse back on to L1 and L2 (Figs. 3, 5; Strausfeld & Braitenberg, 1970; Strausfeld & Campos-Ortega, 1973; Braitenberg & Debbage, 1974). The 1973 paper also detected occasional other outputs from L4, particularly to some R1–R6 terminals, but this is not mentioned again in later papers from the same authors (Strausfeld, 1976*a*, Strausfeld & Campos-Ortega, 1977). The mutual interaction between the three L4 collaterals at the base of each cartridge, whatever its sign, obviously cannot provide the input to the L4 plexus, which must therefore come through the only other known input, via the web of amacrines in the outer cartridge. Again, simplistically, L4 should be a wide-field neurone driven from many ommatidia. In making such deductions, however, it has been necessary to assume conventionally that a cell with a quite wide branching pattern like the amacrine behaves more or less as an integral unit, rather than having branches which behave independently. In reality this may be incorrect, and the physiological evidence points in this direction (see below).

There are no conclusive reports yet of either L4 or L5 activity in the lamina, but some strong suggestive evidence for L4. Arnett (1972) regularly made extracellular recordings from impulse-generating, sustaining units in the optic chiasm below the lamina which were presumed to be afferent, since they were unaffected after efferent activity had been interrupted surgically. These cells possessed a narrow excitatory field centre flanked by two narrow, horizontally disposed areas in which illumination inhibited the centre response (Fig. 5). The L4 basal plexus is the only laterally extending network known which matches even approximately the pattern set by the inhibition, and it is perhaps important to stress that this is quite a firm conclusion, since all the other branching patterns are also known. However, the L4 pattern was not known to Arnett, and subsequent authors have been unable to reconcile exactly the particular zones of inhibition found by him with the known connections: if all the L4 basal processes synapse reciprocally with each other (Braitenberg & Debbage, 1974) and these are symmetrically inhibitory, a circular centre-surround organisation results, unlike that observed by Arnett (Strausfeld & Campos-Ortega, 1973). Fig. 5 shows two simple schemes in which connections between L4 branches are reciprocally inhibitory except in the vertical direction, where they are either absent or else excitatory. Both schemes produce two narrow, horizontally oriented inhibitory zones around a narrow excitatory centre, as required. Lastly, additional support for the idea that the sustaining units of Arnett (1972) respresent the activity of L4 comes from Hardie (1978), who recorded intracellularly and dye-marked a single cell which depolarised to illumination with a train of small impulses, like Arnett's sustaining units. The cell apparently possessed basal collaterals like L4, although these are not clear in the reproduction (Hardie, 1978, Fig. B3). The evidence available therefore suggests that L4 is an impulse generating cell, unlike L1–L2, which is excited in a sustained fashion by the amacrine input confined to its own cartridge, and inhibited by the L4 collaterals on its lateral flanks.

This has the important implication that the α processes of the amacrine cell must function relatively independently of one another, at least with small field stimulation, as discussed in the next section. If this is the correct interpretation, it is likely that L5 also has a restricted field centre, in contast to the suggestion made initially above. We now reach the interesting position that all of the six anatomically known afferent outputs from the cartridge have at least tentatively been assigned a correlated physiological response except L5, and that one type of known physiological response remains unassigned. It therefore seems possible that the transient on-off units recorded extracellularly by Arnett (1972), often simultaneously with his sustaining units, and also shown to be afferent, come from L5, the axon of which runs close to that of L4 both in the cartridge and near the medulla terminal (Strausfeld & Campos-Ortega, 1972). These on-off units had no centre-surround organisation, but had relatively narrow receptive field centres, in line with the suggested isolation between amacrine α processes. It is appropriate from this incomplete evidence to go no further than to note that if the L5 response has been identified correctly, the much greater transience of its response than that assigned to L4, may be explained by its other input, a delayed feedback from the efferent Tan 2. The delay around the loop is unknown, but must be several milliseconds at least, which could explain the rapid curtailment of the initial burst of spikes to create the on-response.

Amacrine cells

The amacrine cells are aperiodic neurones whose irregularity belies the precise organisation of the rest of the lamina (Fig. 6). They form the main lateral network in the lamina apart from that coming from L4, and thus might be expected to be involved in any lateral interactions amongst cartridges. Since they often stain incompletely with silver techniques, are irregularly distributed and have thin tortuous processes, there has been some confusion about their connections, which are complicated (see T1, above). The cell bodies lie just below the lamina in a fairly regular array, about one for every 5–6 cartridges (Strausfeld & Campos-Ortega, 1977), but the actual shape and extent of the lamina arborisation seems to be more variable, with radially extending α processes reaching into 6–17 cartridges (Figs. 3, 6; Campos-Ortega & Strausfeld, 1973), though most of these cartridges are contiguous. Usually one amacrine provides only two or three of the adjacent α processes in a particular catridge, and 2–3 amacrines visit each cartridge, giving different members a potential unique in the lamina, that of selectively connecting with particular pairs of receptor terminals, for instance to preserve the dichroism common to pairs of R1–R6. They default in this however, one amacrine sending α processes to different R1–R6 in successive cartridges, in the few cells examined (Campos-Ortega & Strausfeld, 1973). This and the lack of selective outputs for the amacrines suggests strongly that they all connect into one uniform network without subdivisions, at the reciprocal synaptic contacts between amacrines identified on the outer lamina surface (see Fig. 3).

The presumed main input to the amacrine network from R1–R6 is present at the many tetrad synapses, where a spine from an α process forms one of the lateral elements alongside L1–L3 (Burkhardt & Braitenberg, 1976), and also at direct, superficial contacts (Boschek, 1971). If this is not allowed and

Fig. 6. Perspective drawing of a lamina amacrine cell in the fly *Calliphora* following a relatively rare, apparently complete silver impregnation. The fibre ascending from the soma just below the lamina branches several times, then sends radially oriented, fatter, α-processes down again along the receptor terminals to synapse in several cartridges (not shown: see Fig. 3). There is usually more than one α-process from a particular amacrine in each cartridge visited, and what appear to be two pairs of such fibres have been indicated (arrows). The amacrine cells are relatively irregular, aperiodic, lamina neurones, but visit mostly cartridges in one locality, as shown in plan view at the right for this cell. It projects into 17 cartridges (dots) in the hexagonal matrix of cartridges, the three axes of which are also indicated. After Strausfeld (1976*a*).

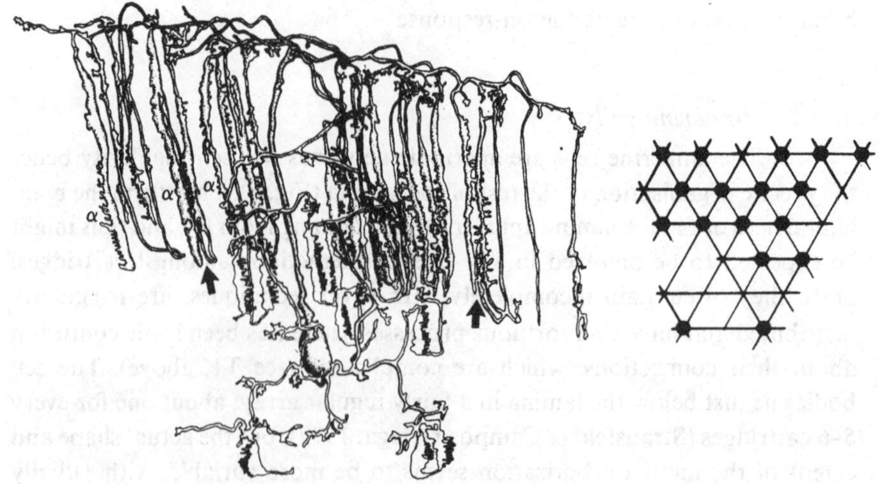

the amacrine spine is presumed entirely efferent, it is then difficult to account for the sensory input to the amacrine net, since the only other input to it apart from other amacrines, comes from the infrequent synapses in the T1 β element, discussed above. The same argument holds incidentally for the other three processes of L1–L3 at the tetrad, all of which must be presumed to be post-synaptic. However, this does not discount the possibility that the post-synaptic spines can act in both directions, and some evidence pointing in this direction is given later.

In contrast to the multiple α inputs from receptors, the outputs from the amacrine net act upon L4 and L5 at rather sparse contacts in the outer lamina, and reciprocally upon other amacrine collaterals. Campos-Ortega & Strausfeld (1973) drew a positional analogy between amacrines and the network of horizontal cells in the vertebrate retina, as mediators of lateral interaction. Whilst this may apply to T1, it does not apply to the largest output cells with the strongest synaptic input drive, L1–L3, which have no direct connections with amacrines. Recordings from L1–L2 and corresponding cells in dragonfly provide the current main evidence of inhibitory lateral interaction in the lamina (Zettler & Järvilehto, 1972; Laughlin, 1974*b*). A solution to the dilemma of what causes lateral interaction might seem in sight because the amacrines directly drive the laterally connecting L4 network; this does synapse back on to L1–L2 (Strausfeld & Campos-Ortega, 1977), and even R1–R6 (Campos-Ortega & Strausfeld, 1973). But there are very few synaptic bars in each small L4 ending (about 4; Burkhardt & Braitenberg, 1976), making it difficult to see how even full activation of the L4 system could generate conductance changes in each L1–L2 to rival at all the much larger effects expected from several hundred direct synapses from R1–R6. This might be a less crucial objection if the output characteristic from L4 were more sensitive than that from other synapses, but the available evidence is instead that the R → L tetrad has these properties (see Fig. 10). It therefore appears unlikely that the only known laterally-extending neural net in the lamina, that of the amacrines plus L4s, could significantly affect the main lamina monopolar output during direct photoreceptor activation. Although this does not apply to the small monopolars L4 and L5, some alternative mechanisms of lateral inhibition must exist to explain the L1–L2 results, and are discussed in a following section.

No records are known to have been taken from amacrines, with the possible exception of some unidentified wide-field cells described by Mimura (1976) which hyperpolarised like L1–L2 to illumination, and one of which had a wide-angle, star-shaped receptive field, hard to explain otherwise. This would be expected if the branches of an amacrine were to some degree in electrical continuity with each other, although the receptive field illustrated by Mimura extends over a much wider range than expected from the anatomy of a single amacrine. The possibility that the separate processes of one cell can maintain some degree of electrical autonomy, has been suggested specifically for such amacrines (Pearson, 1979), and remains possible since the membrane resistance of the connecting fibres is unknown. Assuming that amacrines are slow potential generators, some signal decrement and local weighting of responses might be expected (see, for example, Graubard & Calvin, 1979), but not complete isolation, since cartridges are at the most 12 μm apart and visited almost sequentially by α processes (Fig. 6). The finding that there are synapses between the amacrine connecting fibres also indicates that some conduction must take place along them. The on–off activity ascribed above tentatively to L5, which would in that case be driven by amacrine input, does indicate a receptive field somewhat wider than expected for a single cartridge. The sustaining fibres attributed to L4 have narrower central fields, which may reflect the encroachment of the relatively large test object used by Arnett (1972), into the inhibitory surround.

An unresolved problem with the amacrine cell which needs re-investigation, is the earlier finding of a predominance of a synapses from the α processes back on to R1–R6 terminals in *Lucilia*, which do not occur in *Musca* (see section on T1 above).

Efferents from the medulla: C2, C3, Tan 1, Tan 2

The proliferation of efferent pathways is perhaps the most striking feature of the insect lamina by contrast with the vertebrate retina, though it is common in other vertebrate sensory systems such as lateral line and hearing. The lamina efferents in the fly feed back mainly to the receptor terminals and output monopolars, arising from cells in the superficial layers of the medulla: centrifugals C2, C3, and tangential centrifugals Tan 1, Tan 2. C2 and C3 are periodic and distribute only to a single cartridge, and may be narrow field cells as gauged from their relatively narrow input dendritic fields in the medulla, though this is little guarantee since the presynaptic elements are unknown. The two tangentials are aperiodic, wide field cells which serve as many as 40 cartridges, and at least one of them has a wide collecting arborisation in the medulla (Table 1). Unlike the other three, Tan 2 synapses only with the midget monopolar L5.

The function of the efferents is unknown, with the exception of the earlier suggestion here that Tan 2 may represent a delayed inhibitory line back to L5, and account for the extreme phasic on-off response attributed to L5. The main monopolars L1–L2 and the corresponding cells in dragonfly respond much more phasically than receptors (e.g. Laughlin & Hardie, 1978), for which some of the remaining efferents could be responsible. The efferent routes are also the only ones known by which the R7/8 photoreceptor subsystem could influence the lamina. In Orthoptera, auditory and tactile impulse-generating cells have been localised to the lamina (Northrop & Guignon, 1970) and must represent efferent activity, perhaps with some arousal function. Surgical removal of efferent activity in fly left the response to light of sustaining units unaffected (probably L4, q.v.), but noticeably elevated the threshold of the other, transient unit (possibly L5) recorded by Arnett (1972).

Synaptic and membrane associations between neurones and 'glia'.

The tetrads from R1–R6 are the most abundant synapses in the lamina, but, surprisingly, the next most common type finds processes from the large epithelial glial cells (EGCs) at the post-synaptic site. These occur at between 10 and 100 sites from each amacrine α process at the 'gnarl' invaginations, and less frequently from the T1 fibres (Boschek, 1971; Burkhardt & Braitenberg, 1976). This suggests the possibility of a potent, delayed effect upon the EGCs indirectly from R1–R6 over a wide field, presuming that the many presynaptic bars do indicate functional synapses with competent post-synaptic membranes. This perhaps invites comparison with the horizontal cells in vertebrate retina, originally classed by most workers as glial cells but now considered as neurones. But unlike the horizontal cells, the EGCs make no known synaptic junctions with neurones or with each other.

The function of the synapses on to the EGCs is not clear, but one possibility is that they serve to trigger a recycling process involving the EGCs with receptor terminals, at the capitate projections, outlined below. Another is their possible involvement in generating the large extracellular depolarisation recorded in the lamina by most workers, particularly convincingly when large tipped, and therefore certainly extracellular, probes have been used (Shaw, 1968; Mote, 1970*a*, *b*). The wide-field nature and slow time-course of these potentials may reflect a triggered release of potassium by EGCs, which then spreads laterally in the lamina compartment, depolarising the neurones. It seems necessary to assume something like this, since the time course of the extracellular field development is slower than either the receptor or monopolar cell waveforms in response to light (Shaw, unpublished), although current flow directly from these neurones is a major contributor to the extracellular fields (review: Shaw, 1979).

By far the most striking type of intercellular specialisation in the dipteran lamina is a peculiar one between R1–R6 terminals and the large epithelial glial cells which invest each cartridge (Figs. 3, 11). Discovered by Trujillo-Cenóz (1965) who christened them *capitate projections*, these appear as thin stalks carrying larger heads about 0.1 μm across, and many hundreds of them invaginate each axon terminal. No useful function has been offered for them to date, but a suggestion here is that they are neither neural junctions nor even stable structures at all, but part of a recycling system by which glial contents and membranes are retrieved by endocytosis into the receptors, finally to break down and be digested. The stalks are quite short and most projections are seen on the rim of each axon, but in some micrographs occasional apparently detached heads in the process of lysis can be seen, in the middle of the axon where no stalks extend (e.g. Pedler & Goodland, 1965, Figs. 22, 25). On the hypothesis outlined, the relative rarity of detached processes would suggest that the final digestion process is comparatively rapid. Such an uptake system might be connected with the recovery of liberated receptor neurotransmitter. In darkness, the epithelial glia will concentrate ^{3}H-GABA (Campos-Ortega, 1974), but no systematic studies of possible lamina transmitters have been published. Obvious capitate projections have not been reported from the medulla endings of R7 and R8, although inclusions of a similar size appear to be undergoing lysis in some micrographs (Figs. 27, 28, Melamed & Trujillo-Cenóz, 1968). There are no capitate projections in R7–R8 as these pass through the lamina and in agreement with the notion that the capitate system is related to synaptic function, they develop contemporaneously with synaptic bars in the late pupa (Trujillo-Cenóz & Melamed, 1973). Capitate projections have not been reported in the presumed functionally similar laminae of other arthropods, but comparative studies of the recycling of photoreceptive membranes has revealed a variety of endocytotic disposal methods in different groups all having microvillar type receptors (Blest, 1980), so diversity of mechanism would not be without precedent.

The epithelial glia are also the participants in suggestively similar-shaped specialisations in between the six paired α and β climbing fibres (Figs. 3, 11), called 'gnarl' complexes. The epithelial cells send many club-like invaginations into each T1 β process, at points opposite to presynaptic densities in the apposed amacrine α process (Burkhardt & Braitenberg, 1976). This may indicate an amacrine $\rightarrow$ glia synapse, but leaves the role of the gnarl invaginations into T1 unexplained. Again, these gnarls may be labile structures involved in recycling material from glia. The α–β system has not yet been recognised ultrastructurally ouside Diptera.

The final known instance of glial involvement at synapses appears to be common to both insects in which the lamina ultrastructure has been studied in some detail, fly and dragonfly. It has been found that one or two of the post-synaptic locations at the tetradic or triadic synapses involving the photoreceptors is occupied by glial extensions. In the fly, the spines of monopolar cell L3 are present at each tetrad only in the upper cartridge, and are replaced by processes from marginal glia in the lower part (Burkhardt & Braitenberg, 1976). In the dragonfly, the corresponding synapses do not have glial involvement, but this is found occasionally in feedback L → R synaptic dyads, and as the usual occurrence at the two lateral elements in triads from the long visual fibre R7 on to one monopolar cell (Armett-Kibel, Meinertzhagen & Dowling, 1977). The latter authors do not think glial participation is functional at these locations, and it seems most probable that glial processes are capable of insinuating into any vacant positions in the neuropile formed by atrophied, inappropriate or simply absent neural connections.

We are left with the uneasy dilemma of either having regarded as glia some cells which may act in ways functionally resembling neurones, or else having to deny that the mere presence of a presynaptic T structure is a necessary proof of a competent synaptic connection. This last premise underlies most of the connectivity analysis but may not be exclusively correct for every T structure. About the most that can be said at present is that this same assumption of functional connectivity has been made in several other parts of the insect nervous system (Osborne, 1967; Lamparter, Steiger, Sandri & Akert, 1969; Boeckh, Sandri & Akert, 1970; Hengstenberg, 1971; Pfenninger, 1973; Wood, Pfenninger & Cohen, 1977). Wood *et al.* have shown that at least some of these sites appear to have active release zones for vesicles, like those identified at the neuromuscular junction.

Physiology of the photoreceptors and lamina

The photoreceptor response

This has been treated extensively in other reviews (Fuortes & O'Bryan, 1972; Hausen, 1977; Shaw, 1979). Illumination of the microvillar border (rhabdomere) of a retinular cell in an ommatidium produces an increase in membrane conductance which is graded with light intensity, and which results in graded membrane depolarisations (Fig. 7). These range in amplitude from about 1 mV to a maximum, saturation value of 60–80 mV in R1–R6 in fly, depolarisaton being continuous throughout the stimulus. As the light level is lowered, random fluctuations in voltage become more obvious, and finally, single fluctuations or 'bumps' separate out. These are 1–2 mV in height and 0.1 s long, and in locust and fly the evidence is good

Fig. 7. *Right*: receptor potentials recorded from a photoreceptor terminal in the lamina of *Calliphora*, at the different relative intensities of stimulus shown against each trace. Flash duration is shown on the lower traces. The initial response peak and following trough at the four highest intensities are much less pronounced than in records from the cell soma. A positive extracellular field develops in series with the recording site in lamina, and would have caused some fraction of the positive response shown. *Left*: the same terminal was injected with the dye Procion Navy Blue, and the eye sectioned after the electrode had been frozen in place. Seen here in a drawing of a longitudinal section, the dye appears in a single short retinular axon. A piece of the electrode (*e*) shows that it recorded from the proximal part of the lamina zone (*la*), 30–40 μm deep. *rc*, receptor cell bodies; *tr*, tracheae; *ch*, local chiasmata of receptor axons (*r.ax*); *L*, cell bodies of lamina monopolar cell. After Järvilehto & Zettler, 1970.

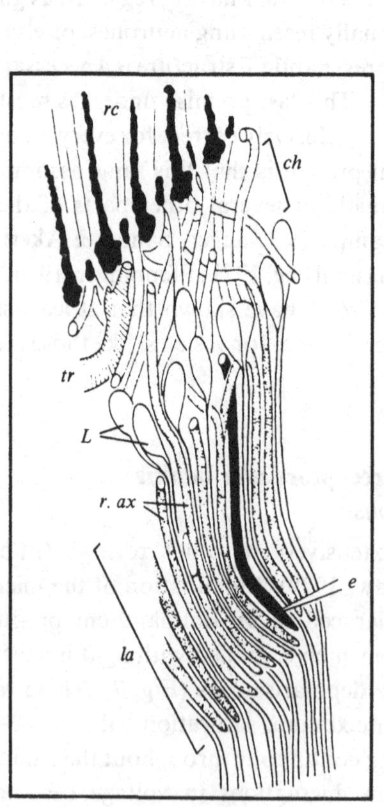

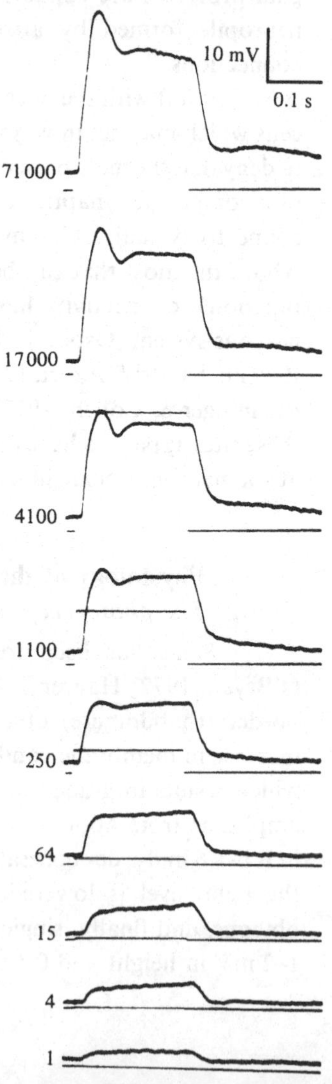

that they represent single photon responses, with a very high capture efficiency of about 60% (Scholes, 1964; Kirschfeld, 1966; Shaw, 1968; Lillywhite, 1977). Noise analysis under voltage clamp has indicated that each photon causes the coordinated transient opening of several hundred membrane channels to produce one bump, but fewer at higher intensities, where many simultaneous absorptions are taking place in each membrane region at once (*Limulus*: Wong, 1978). To uncover the intermediate mechanism which couples one photoisomerised rhodopsin molecule to many membrane channels, themselves presumably some distance apart on the microvillar membrane, is a major challenge at present. An intermediate transmitter molecule might be involved (Cone, 1973), or some cooperative interaction between rhodopsin chromophores. It is conceivable that the rhodopsin protein itself forms the channel, since the related bacteriorhodopsin functions as a transmembrane carrier under certain conditions (Henderson, 1977). The analogy is not exact, bacteriorhodopsin pumping protons actively, rather than acting as a passive sodium channel gated by light, as the physiology indicates for photoreceptors.

The range of light intensities which take the initial graded response of the photoreceptor from the level of a few bumps per second up to peak amplitude is abut 10^5, allowing the cell to operate over a favourably large range. This peak is not maintained with the brighter stimuli, but is cut back over about 0.2 s to a lower, sustained plateau, also graded with intensity, and from which further incremental responses can be elicited by even stronger stimuli. This process of light adaptation (Glantz, 1968; Laughlin & Hardie, 1978), allows the photoreceptor to stretch its response mechanism further to cover an even wider range within about 0.2 s. The process involves a rise in internal free Ca^{2+} concentration following illumination (Brown & Lisman, 1975; Lisman, 1976; Bader, Baumann & Bertrand, 1976). This range-stretching process in the receptors is only part of the adaptational mechanism, and further tuning takes place post-synaptically.

Conduction of the visual signal by the photoreceptor axons

The depolarisation in the microvillus-bearing part of the cell must reach the presynaptic terminal of the cell to promote synaptic transmission. A variety of evidence now indicates that the depolarisation simply spreads passively down the axon cable into the terminal (review: Shaw, 1979). The loss factor is relatively slight, perhaps a factor of two in the fly retinular cell, determined by recording from the terminals (Fig. 7; Zettler & Järvilehto, 1971). There appears to be very little attenuation in the orthodromic direction along photoreceptor somata in the drone bee (Carreras, 1978). None of the different photoreceptors examined in insect or crustacean eyes or ocelli produce regenerative trains of impulses. In the two cases known where a single spike is triggered on the rising phase of the slow potential, it has been thought to be a non-functional, probably local membrane event (Shaw, 1979). The most direct and compelling evidence that the receptor potential itself effects synaptic transmission comes from barnacle ocelli. Responses resembling those evoked by light can be elicited from post-synaptic cells by polarising photoreceptors which normally drive them, with currents through an intracellular microelectrode. This avoids the light stimulus altogether, and is effective over a certain range of membrane potential (Fig. 8; Shaw, 1972,

Fig. 8. Voltage dependence of synaptic transmission from photoreceptors in the barnacle ocellar visual system, monitored from the activity of spiking post-synaptic cells (left, middle traces), during intracellular polarisation of a dark-adapted photoreceptor (lowest traces). In *a* the cell body of the photoreceptor is held depolarised from resting potential (lower interrupted line), by steady current applied through a double-barrel pipette (upper trace). A pulse hyperpolarises the cell back towards this level, causing a burst of impulses in the visual cell whose axon response is recorded on the middle trace, via a suction electrode on the oesophageal connective. The response was similar at resting potential in this case, but, *b*, was suppressed at holding levels more negative than this, showing that synaptic transmission operates preferentially over a range of membrane potential depolarised from rest, utilised naturally when shadows fall on the ocellus during depolarisation by daylight. This is shown more quantitatively in *c*, where the ordinate counts the average number of impulses in the burst produced by a hyperpolarising pulse of about 20 mV amplitude, delivered at the membrane potential levels shown on the abscissa. Several bursts are averaged for each point. Mean background firing at each level is shown by hatching. The final resting potential upon withdrawal of the pipette was −46 mV. From Shaw, 1972.

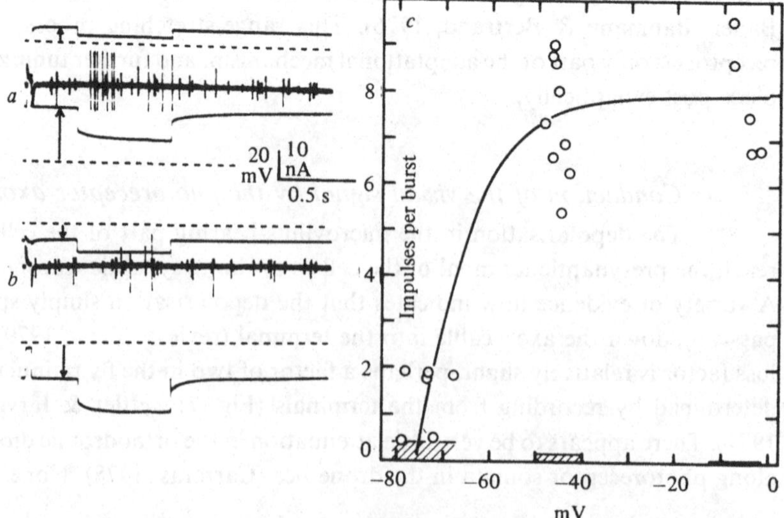

Gwilliam & Millechia, 1975; Stuart & Oertel, 1978). In a particularly sensitive preparation, a receptor response estimated at only 0.3 mV was adequate to cause an obvious post-synaptic discharge (Shaw, 1972), and sub-millivolt values are also obtained in compound eyes by comparing the low intensity ranges of post-synaptic cells (Fig. 10; Laughlin, 1973), or the behaviour (Scholes & Reichardt, 1969), with the range of receptor responses. But there is probably no threshold for transmitter release, with release taking place even at resting potential (see Fain, previous chapter).

Decremental versus impulse conduction: pros and cons

The obvious advantage of passive conduction is that the graded response of the receptor is preserved. This provides an accurate, though non-linear, transformation of the input signal. Such accuracy could in principle be retained (though not surpassed) by adding on a digital stage, for instance with a high resolution voltage-to-frequency converter driven from the slow potential output of the receptor. The biological equivalent of this has parameters less than ideal, for impulse generating mechanisms known in other neurones use voltage not just as the parameter coded, but also as the trigger mechanism in the digitising process of spike initiation itself. Neurones must therefore be protected by having a sufficiently high voltage threshold, from discharging inadvertently to, for instance, depolarisation caused by some slight change in ion balance created metabolically. Now single photon bumps have a wide size distribution about the mean in a particular receptor so that, to retain optimal efficiency by transmitting every capture, a 10 mV threshold would require every bump to exceed this level. The mean quantal unit response then has to be very large to ensure this, which produces two detrimental consequences. First, the number of levels of illumination which can be signalled without ambiguity from each receptor within the response ceiling is reduced: each level has associated with it some amplitude uncertainty or deviation, which is larger the larger the mean unit response, and the size of this deviation sets the boundaries between levels (Rose, 1974; Snyder *et al.*, 1977). The information sending capacity is reduced. Second and perhaps more important, the time resolution of the system would suffer dramatically, since the threshold for spike initiation could only be reached along the time course of the receptor potential itself. Even in the very fast photoreceptor of the drone bee, which has an initial spike response, the time taken to the spike is at least 10 ms extra beyond the first appearance of the receptor potential (Baumann, 1968; Shaw, 1969), and longer at lower intensities; in *Limulus*, it is at least 50 ms (Fuortes & Hodgkin, 1964). A related drawback in the temporal domain occurs because relatively few receptors converge on to each post-synaptic cell, by contrast with the vertebrate eye, so that signal averaging over many inputs is not possible. To obtain a reliable size estimate of an input signal, a single spike is no use, so that the receptor output would have to be time-averaged over several spikes to become reliable. Integration times of the order of 100 ms become necessary even at high light intensity. By comparison, the actual response lag measured during visual tracking by a fast flying insect, the hoverfly, is as little as 20 ms for expression of the complete tracking behaviour (Land, 1977). This gives some indication of the superior temporal performance of the 'slow' potential systems used in the first-order and second-order visual cells in Diptera, over an impluse-conducting system, and is presumably a major reason why slow potential conduction has been favoured in the evolution of advanced visual systems.

In fact, insects have reduced the temporal delay to the bare minimum, about 2 ms, by having synaptic transmission work effectively from the first appearance of the receptor potential, in contrast to other non-sensory synaptic systems. The 2 ms is the mean conduction delay recorded between the first detection of the light response in the soma and the terminal of the fly photoreceptor (Scholes, 1969), and is the result of the slow wave having to discharge the axon membrane capacitance ahead of it as it spreads along the axon (Hodgkin & Rushton, 1946; Jack, 1979). This gives rise to an interesting quirk of slow potential propagation, that whilst impulse propagation over such a distance in such an axon would be faster, when combined with the delay time to first spike, the overall delay is several-fold higher in the impulse system (Fig. 9). Of course, for very long axons a crossover point is reached where the propagation delay becomes limiting, but in Diptera, even the long, thin axons of R7 and R8 which bypass the lamina carry decremental slow potentials (Hardie, 1977).

In contrast to impulse conduction, attentuation of the signal is of course inevitable with slow potential propagation, but whether this is ever a drawback in the real system depends first upon what the main sources of masking noise are, and second, upon whether the synaptic transmission process can respond to the attenuated signal. On the first count, the two dominant sources of membrane potential fluctuation (noise) in insect photoreceptors are the inevitable shot noise due to random arrival of photons, superimposed on which is transducer noise, a term coined by Laughlin (1976) to describe the long known, wide range of bump amplitudes recorded from each cell, but each due to a single photon absorption. The two sources of uncertainty are of similar magnitude in locust eye (Lillywhite & Laughlin, 1979), and since they appear at the origin along with the signal, are attenuated equally with it, by passive conduction. The signal-to-noise (S/N) ratio is therefore unimpaired as regards these two sources of uncertainty, with passive conduction. On the second count, we have seen already that very small presynaptic voltages can activate post-synaptic cells, indicating that transmitter can be released from the axon terminals close to resting potential.

Of course, where other types of noise contribution come into prominence, a threshold-protected digital transmission system with its relative noise immunity might be favoured. For instance, in very fine, high resistance dendritic terminals, spontaneous opening and closing of membrane channels would cause appreciable membrane voltage fluctuations, in the absense of an input. This may be nullified in an impulsive system having a high spike

Fig. 9. Explanation of the apparent paradox that in a neural network with relatively short axons, transmission using slow potential signals can act faster than in one using impulses. *a*, In the impulsive system, spike initiation threshold (*th*) is reached along the rising slope of the receptor potential, and is needed to protect against accidental firing. This creates a delay in transmission, *l*, of at least 10 ms in the bee and over 50 ms in *Limulus* eye, before the receptor system (*R*) can begin to signal to the post-synaptic cell *L*, involving a further 1 ms synaptic delay, *d*. Conduction delay of spikes in the axon is assumed to be negligible in this case. If in addition reliable intensity information is required, the burst of spikes must be averaged over a similar or larger time epoch, creating an additional delay in the visual system; *b*, A slow potential system like that of the photoreceptor cells in the lamina of most insects. There is only a short conduction delay *c* of 1–2 ms caused by the cable capacitance before the signal appears in the axon terminals (dotted curve), plus the standard synaptic delay as in the impulsive system. This occurs because the synapse releases transmitter for very small presynaptic signals, as soon as the receptor potential becomes evident. In a slow potential system of this type, with axons a few hundred microns long using high gain synapses at the end, information can travel through the first relay several times faster than in the threshold-limited type of impulse system.

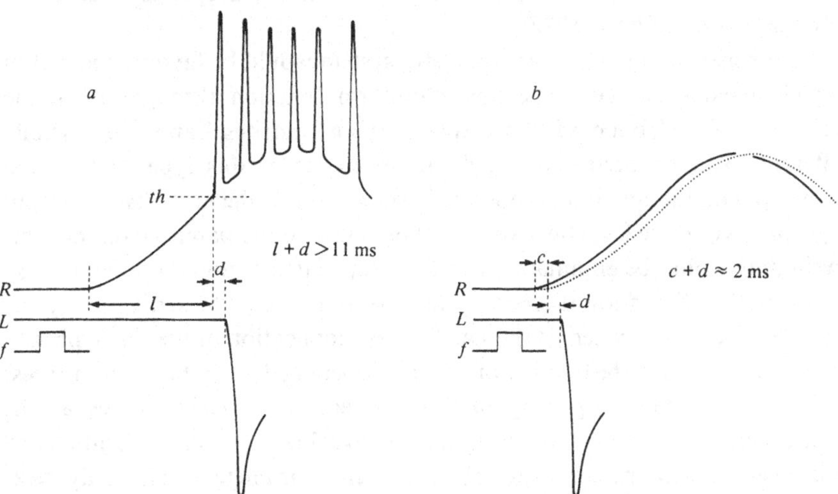

initiation threshold, in effect a form of non-linear amplification favourably weighting large inputs. However, Hille (1970) points out that in an extreme situation of this sort, voltage-sensitive sodium channels can also be a drawback, since spontaneous opening of a single channel might accidentally fire the whole dendrite and then the cell. It is perhaps noteworthy in this context that insect photoreceptors operate with a relatively low membrane resistance, with values around 10^3 ohm cm^2 (Shaw, 1969; Carreras, 1978), despite the decremental loss factor this introduces. Even the much longer barnacle ocellar receptors operate effectively with values around 10^5 ohm cm^2 (Shaw, 1972; Hudspeth, Poo & Stuart, 1977), amongst the highest values reported for biological membranes (see also Bush, this volume). This suggests that modest signal losses during decremental spread are not so disadvantageous as to be worth preventing. More important than the absolute size of signal should be the S/N ratio at the synapse, which is improved by post-synaptic convergence or presynaptic coupling between receptors, known to occur in insects (see below).

Membrane resistance also affects receptor frequency response. A distinct limitation of a decremental signalling mechanism is the sharp increase in attenuation which occurs at high frequencies of periodic stimulation, in the range up to several hundred Hertz in which impulsive systems are not degraded. The membrane time-constant determines the high frequency characteristic, and better relative performance in the higher range is therefore obtained with a lower membrane resistance (see Jack, 1979). However, the visual transduction mechanism itself behaves as a low-pass filter with a relatively sharp cutoff, which limits the usefully transmitted band to about 100 Hz in the fly (Järvilehto & Zettler, 1971; Leutscher-Hazelhoff, 1975; French & Järvilehto, 1978*a*), and much less in other species (Fuortes & Hodgkin, 1964; Shaw, 1972).

Another situation where an impulsive system would be favoured is that in which a particular cell type has a uniform function throughout all the cartridges visited, but has fine, widely ramifying dendrites. Tan 1 and 2, wholly efferent to many cartridges, might be examples of this type. If the final dendritic conduction process operated decrementally, the most distant output synapses would be less effective than those close to the parent axon, but this weighting would be eliminated if spikes propagated through the final stage.

From this discussion it emerges that the main factor to arrange properly in a slow potential system is the satisfactory amplication of the single photon response. Provided the number of channels opening during this is far in excess of the spontaneous opening rate, subsquent decremental losses at the frequencies of interest become quite manageable in axons of quite small dimensions, with rather ordinary membrane parameters. The only clear advantages of an impulse carrying system arise where the axon must be both long and thin or carry high frequency information, or where high noise immunity is needed. In most respects, it is inferior in performance to a simple decremental system of the sort found in an insect eye.

Synaptic transmission from photoreceptors to monopolar cells

The larger second-order monopolar cells in the compound eye, and the similarly placed L-cells in the ocellus are all hyperpolarised by illumination (Fig. 12). The response is not only inverted compared to the receptors, but is also larger at low stimulus intensity and this, combined with the clear synaptic latency, indicates that the synapse is chemically mediated. There is some evidence that acetylcholine may be the transmitter both in the ocellus (Klingman & Chappell, 1978) and in the eye, where it increases the membrane potassium conductance (Zimmerman, 1978).

The synaptic amplification factor going from six receptors to one of the main monopolars probably somewhat exceeds the value of 14 estimated for the initial response by Laughlin (1973) in the dragonfly, because this figure was measured from receptor cell body responses before decremental losses had occured, and because the synaptic latency affects the estimate. A figure of up to eight is given for the fly and seven for the bee (Järvilehto & Zettler, 1973; Menzel, 1974), though the value for the bee appears larger than it should because the responses were first normalised. These high values for the gain are not maintained during longer light exposures, because the monopolar potential rapidly reduces in size (Fig. 12), indicating the rapid onset of light adaptation. As it is in the receptors, the post-synaptic response at low intensity is clearly quantised in locust eye (Shaw, 1968), and ocellus (Wilson, 1978*a*, *c*). Preliminary experiments on the locust lamina have established that the frequency of these hyperpolarising potentials is higher than that of bumps in the preceding photoreceptors stimulated similarly, by a factor expected for a 6-to-1 convergence, if each receptor bump evokes a direct post-synaptic equivalent (Shaw, 1968). The alternative explanation that each hyperpolarisation might instead represent the release of a single transmitter quantum now seems unlikely, from the results of artificially depolarising locust ocellar photoreceptors with a high K^+ perfusate. The follower cells become smoothly and progressively hyperpolarised with no sign of the large quantal responses seen during illumination, as expected if each transmitter package produces a very much smaller unit response than that due to each photon, which in turn therefore must be composed of many packages (Wilson, 1978*c*). If single hyperpolarising bumps are therefore single photon arrivals counted faithfully by the post-synaptic cell, comparison of the pre- and post-synaptic bump amplitude gives directly the 1-to-1 synaptic gain at low intensity, a factor of 5 or less in the locust eye.

Small changes in synaptic output produced by corresponding changes at the input can be represented by a different measure of synaptic amplification, which has been called the sensitivity or *dynamic gain*, i.e. the instantaneous slope of the pre- versus post-synaptic relationship, dV_{post}/dV_{pre} (Falk & Fatt, 1972; Ashmore & Falk, 1976). The maximum value of the dynamic gain usefully identifies the most effective operating range of presynaptic voltage at a particular synapse. Results averaged from dragonfly retina (Laughlin, 1973), illustrated in Fig. 10, give a maximum dynamic gain of about 34 at presynaptic polarisations of only some 0.7 mV, indicating that the synapse is optimised to transmit very small presynaptic signals from the receptors. Photon bumps are difficult to observe in isolation in dragonfly, but some records of Laughlin (1974*a*) show fluctuations presumably of this origin, about 1 mV amplitude as in other insect receptors, which thus would be

Fig. 10. Synaptic transfer relationship, for photoreceptors transmitting through a chemically-mediated high gain synapse, to lamina monopolar cells. Data points are from averaged curves from both sets of cells, taken from dragonfly eye by Laughlin (1973). The initial transfer characteristic is exponential and very steep with a slope of about 0.8 mV/decade, indicating an unusually sensitive transmitter release process, tuned to transmit optimally at presynaptic polarisations of < 1 mV. The roll-off with larger signals is expected in part from non-linear post-synaptic voltage summation, but may also reflect some negative feedback process acting upon the terminals.

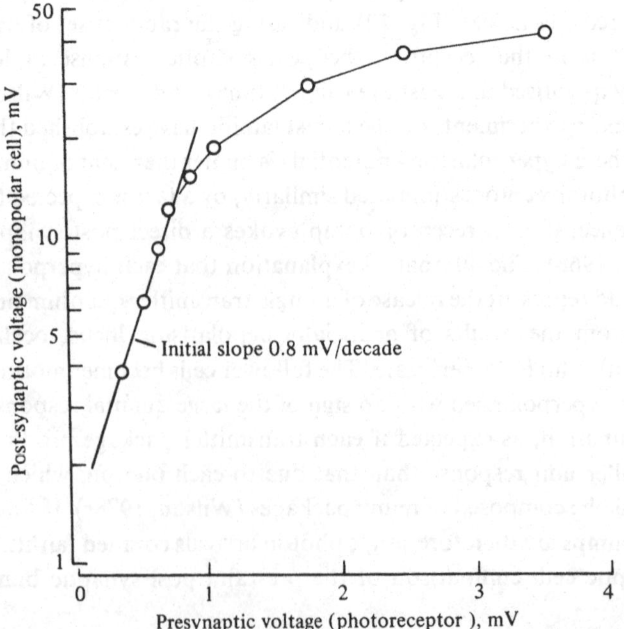

optimally transmitted by the synapse. The dynamic gain has also been estimated at the rod to bipolar synapse in the dogfish retina by Ashmore & Falk (1976, 1979), though rather less directly, since they were unable to measure the presynaptic voltages in their preparation. They arrive at a value of at least 50, for a system in which a large but unknown number of rods, probably electrically intercoupled (references: Owen & Copenhagen, 1977), converge on to the bipolar cell under study. By contrast, values for the impulse-transmitting, non-sensory synapses analysed are considerably less, ~ 0.3 at a lamprey central synapse (Martin & Ringham, 1975), ~ 3 at the squid giant synapse (Katz & Miledi, 1967), in both cases optimised at presynaptic polarisations of 60–80 mV, with practically no observable transmission taking place at < 30 mV. Even non-spiking neurones in *Aplysia*, which operate detectably at much lower voltage levels, have very low dynamic gains of ~ 0.3 (Graubard & Calvin, 1979). The evidence therefore suggests that in both the invertebrate and vertebrate photoreceptor systems analysed, the synapse on to the second-order cells is specialised not only in transmitting very small voltages with no obvious threshold, but in achieving this by virtue of a particularly sensitive transmitter release characteristic, is illustrated either by the dynamic gain values or by the slope of the line in Fig. 10, of < 1 mV/decade. As suggested elsewhere (Shaw, 1979), it appears that some electroreceptor synapses fall into the same category, with similarly steep slopes.

Llinás (1979) has suggested, conversely, that such so-called high gain synapses are really just a variant of the better known version, exemplified by the squid giant synapse, suggesting as evidence that even in the squid, quite small presynaptic signals can be made to show detectable post-synaptic activity with careful recording. Actually this particular demonstration does not bear at all upon the issue, the resolution of which depends on being able to model the synapse to see which parameters are useful for distinction between different supposed species of synapse. Whilst little is known about the events underlying sensory synaptic transmission, and few data are available by comparison with the squid, the result in Fig. 10, with an initial exponential section followed by a roll-off, is of the form expected for the behaviour of a conventional synapse in the analysis of Falk & Fatt (1972), which can therefore be employed to clarify the situation. Following them in assuming an exponential release of transmitter proportional to presynaptic voltage, V_{pre}, and post-synaptic voltage V_{post} produced by a linear conductance response to this transmitter, and additionally supposing that n presynaptic fibres converge on to each post-synaptic cell, it follows that

$$V_{\mathrm{post}} = V_{\mathrm{r}} \Big/ \left(1 + \frac{c}{n} e^{-bV_{\mathrm{pre}}}\right) \tag{1}$$

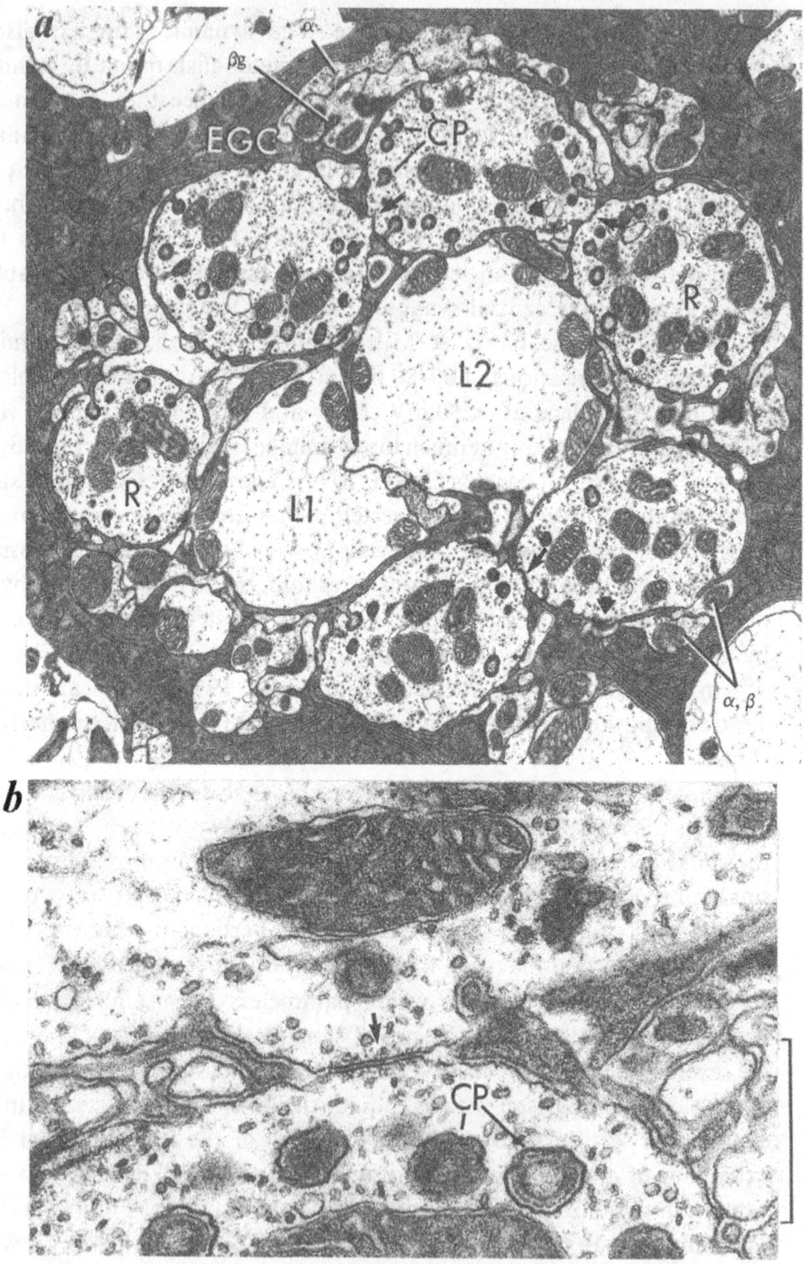

Fig. 11. *Above:* Electron micrograph of an optic cartridge in *Calliphora*, sectioned just distal to the main synaptic zone. The six large profiles in a ring (R–R) are the terminals of photoreceptors R1–R6, which surround the two central monopolar cells L1–L2. R1–R6 are all connected laterally with each other at this level by symmetrical close appositions thought to

where V_r is the reversal potential of the post-synaptic mechanism, b is the transmitter release constant which defines the synaptic characteristic, and c is another constant which affects the matching between the pre-and post-synaptic responses. As shown by Falk & Fatt (1972), the numerical value of b directly sets the exponential 'slope' on semilog coordinates as in Fig. 10. The evidence above that this slope is steeper by at least a factor of 10 at synapses involving insect photoreceptors, some electroreceptors, and probably rods, indicates that these synapses are quantitatively different from the other squid type of synapse, where the slope factor is about 10 mV/decade.

The high value of b is the factor primarily responsible for shifting the useful range of transmitter release to lower values of presynaptic voltage, so to speak tuning the synapse to work most sensitively over a narrower range of small signals. If b and c in eqn (1) are held constant at values which would apply approximately in Fig. 10, the effect of increasing n, the number of cells converging, is to shift the output curve somewhat further towards smaller signals, and particularly to extend the dynamic range downwards for the smallest signals. Increasing the effective synaptic area acts in an identical way, so that the hypertrophy of the receptor–monopolar synapse in insects and 6-to-1 convergence of receptors can both be viewed as additional modifications to enhance selectively transmission of small signals in the size range of single photon bumps, in addition to enhancing the transmission reliability as argued here earlier. These two features of convergence and reduplication of synapses also mark the projection from rods and cones to bipolar cells in the periphery of the vertebrate retina.

An obvious property of the photoreceptor high-gain synapse is that the post-synaptic response has a narrowed dynamic range, because the output saturates at low intensity. The range is 10^1–10^2 for insect monopolar cells, in contrast to receptors, about 10^5. However, not all post-synaptic cells show this truncation. The impulse-generating cells found by Arnett (1972) have output curves similar to receptors, which may indicate that these are driven

---

be gap junctions, three of which appear in this section (arrows); one is shown at higher magnification in the lower frame. Numerous processes (capitate projections, CP) invaginate each R1–R6 from the surrounding epithelial glial cells (EGC), often appearing as disembodied heads in section, and suggested in the text to be labile structures, part of some membrane renewal process. Of the several other profiles visible, only the twin climbing fibres, α from an amacrine and β from basket cell T1, can be readily identified in a single section. The uppermost β process can be distinguished from the nearby α by a gnarl invagination (βg) from EGC. Occasional T-shaped presynaptic bars may be seen within R1–R6 (arrowheads). Scales: *a*, 2 μm; *b*, 0.5 μm. Courtesy Dr W. A. Ribi; see Ribi (1978).

through relatively low-gain synapses. Ashmore & Falk (1979) make a similar distinction between the low-gain synapse from rods to horizontal cells, in contrast to that to bipolar cells.

Synaptic gain also seems to depend upon the rate of rise of the presynaptic potential (Järvilehto & Zettler, 1971), and the gain measured with sinusoidal or white noise stimuli increases continuously from low frequency up to the limit of measurements, 100 Hz (Järvilehto & Zettler, 1971, 1973; French & Järvilehto, 1978*b*). This feature may compensate somewhat for the losses in high frequency performance imposed by decremental signal transmission, electrical coupling, and by transduction itself, all of which have low-pass filter characteristics.

A final comment on the amplification properties of cells in the peripheral retina may be made in relation to the transduction process itself. Most insect eyes contain a few cells in each ommatidium which have both shorter and narrower rhabdomeres, which because of this and their spectral properties, catch less light than their larger neighbours operating in the middle wavelength range. Laughlin (1976) has found that such ultraviolet-light-sensitive cells in dragonfly produce much larger voltage signals per photon than other cells, and suggests that this allows an intrinsically poor photon-catcher to operate still at the same voltage level as the larger cells, at its synaptic outlet. Hardie (1979) has found the same result for the two small central cells in the fly, R7 and R8, which send signals through the lamina to the medulla.

Although the amplification at the receptor–monopolar synapse is larger, the synaptic delay is comparable to that at other synapses. It is difficult to specify the latency precisely because the waveforms develop slowly, particularly in the receptors. In the fly's eye, Järvilehto & Zettler (1971) measured the average difference to the point of first appearance of the waves in the receptor axons and monopolars to be about 1 ms, with which a more recent estimate obtained by detecting feedback to the terminals agrees (< 1.5 ms; see Fig. 13). Larger values in the locust eye produce a similar result when corrected for a 2–3 ms conduction delay in the axons, described earlier, and Laughlin (1973) found a delay of about 2 ms in the same type of experiment in dragonfly, where the axons are somewhat shorter. A delay of 6–10 ms in the barnacle ocellar system to electrical stimulation of the very long ocellar nerve may have the same origin (Ozawa, Hagiwara & Nicolaysen, 1977), and even longer conduction delays have been recorded in the 1-cm long lateral ocellar nerve (Shaw, 1972). A latency of 1–2 ms has been measured directly at motor output synapses in the locust thoracic ganglian, at which a motoneurone is driven by small presynaptic potentials in a non-spiking interneurone (Burrows, 1979). In the insect eye, the synaptic delay remains fairly constant over a wide range of stimulus intensity (Järvilehto & Zettler, 1971; Laughlin, 1973; Shaw, 1979).

Signal transmission by monopolar cells L1 and L2

The conduction mechanism of L1–L2 is still controversial, and is reviewed in more detail elsewhere (Shaw, 1979). Briefly, Zettler & Järvilehto (1971, 1973) found very little if any decrement of the monopolar cell response, in identified penetrations taken various distances downstream from the lamina synapse, and this led them to suggest that the hyperpolarising response of L1–L2 is propagated regeneratively along the 2 μm diameter axons to the medulla, 400–700 μm away. It is not clear at present, however, that such a mechanism can be reconciled with the graded nature of the hyperpolarising transient and plateau responses to different light intensities, known from all published accounts of monopolar cells and similar neurones in ocelli. Neither is it clear why hyperpolarising current injected into monopolar cells fails to evoke a regenerative response. Electrotonic conduction by itself would also entail very little decremental loss of signal if the membrane resistance were reasonably high, around 20000 ohm cm^2, but Zettler & Järvilehto (1973) reject this alternative because their measurements suggest much lower values, determined from the input resistance. More recently, Wilson (1978*b*) has shown that the equivalently positioned cells in locust ocellus which respond similarly to L1–L2, must have a relatively high membrane resistance in the conducting segment of the axon, but that this can appear anomalously low because of shunting by the low-resistance synaptic membrane.

The transmission mechanism in the ocellar L-cells has been examined definitively with separate electrodes in different parts of the same axon. This reveals decremental transmission both of the hyperpolarising signal in the normal centripetal direction, and also of a centrifugal, depolarising impulse at light-off, triggered further along the cell (Wilson, 1978*b*). These findings provide a suggestive argument by analogy with the transmission process in L1–L2, but this of course needs to be tested directly, perhaps with recordings from the enlarged medulla terminals of L1–L2.

Lateral inhibition, light adaptation and field potentials in the lamina

The familiar pioneering work on *Limulus* eye showed powerful, reciprocal lateral inhibition amongst the impulse-generating eccentric cells, each excited electrically by the photoreceptors of its own ommatidium. This demonstrated a means by which the representation of the external spatial environment could be enhanced or sharpened by the neural system, so that edge effects become more prominent. It also showed one way in which the output from the eccentric cell could be suppressed by general background illumination, a form of local gain control needed in light adaptation, to forestall output saturation (review: Hartline & Ratliff, 1972).

It was therefore natural to look for a functionally analogous lateral inhibitory system amongst the slow potential generating cells in the insect eye, and this was reported first by Zettler & Järvilehto (1972) and Zettler & Autrum (1975), who determined that the receptive fields of the L1–L2 monopolar cells were narrower by a factor of about two, than the receptive fields of the receptor terminals in the same preparation. A problem arises in interpreting this result, because very large positive extracellular fields develop in the lamina upon illumination of the eye (e.g. Shaw, 1968; Mote, 1970; Laughlin, 1974*a*) and therefore add an artefact to the intracellular recordings, which will act to broaden the receptor terminal response spuriously, but to narrow that of L1–L2. Most authors have ignored this field potential, have thought it due to direct electrotonic spread from the receptors, or have believed that it was somehow picked up selectively by damaged receptors (Zettler & Järvilehto, 1971; Laughlin, 1974*a*), but in locust it now appears to be caused by extracellular current flow mostly from receptors across part of the blood–brain barrier, and is therefore present at all recording sites as a voltage in series with the true transmembrane potential (Shaw, 1975, 1979). Though its arrangement is probably different in detail, a barrier system appears to be present in Diptera also (Heisenberg, 1971; Zimmerman, 1978), and the extracellular potentials recorded in the lamina can be very large (Mote, 1970). It is still not clear what contribution this series extracellular artefact makes to the demonstration of inhibition in the fly lamina even in the later experiments with annular stimuli (Zettler & Weiler, 1976). More extensive tests with dragonflies, obtained more properly by plotting full response– intensity curves at several angular positions, showed inhibition acting during the L cell plateau phase and to a lesser extent during the transient (Laughlin, 1974*b*). An earlier series using the original, less satisfactory technique of Zettler & Järvilehto (1972) failed to show any difference between receptors and monopolars, for reasons that are not entirely clear. But again, neither technique takes account of the lamina extracellular field evoked by light (Laughlin, 1974*a*), which by adding to the recorded response will produce an artifically narrow monopolar receptive field, in both types of experiment.

An attempt to clear this up in the fly and reveal any inhibition present is shown in Fig. 12, in which the lateral influence is isolated more completely with a single facet stimulus, and the real transmembrane voltage is revealed by subtracting off the contaminating field, recorded at a second electrode. It then becomes possible to observe a true antagonistic process which slowly depolarises the monopolar cell membrane, when a bright flash is beamed into a single facet several ommatidia away. During this depolarisation, the plateau response is practically abolished to a test flash delivered through one of the six projection facets which drive the monopolar cell directly (see inset, Fig. 12), leaving only a highly phasic on-response, characteristic of the light adapted state in both the fly and the dragonfly (Laughlin & Hardie, 1978). This shows that there is a strong, lateral, depolarising influence acting directly upon the monopolar cell itself, in addition to any presynaptic effect upon the preceding receptor terminals. This effect has a relatively short latency but has a time constant of 30–40 ms, slower by an order of magnitude than the rise time of the monopolar cell on-response, though similar to that of the

Fig. 12. Differential transmembrane recordings taken from a lamina monopolar cell in *Calliphora*, according to the protocol at the right. The voltage in the extracellular space near the cartridge is also recorded with a coarse-tipped microelectrode about 15 μm behind the intracellular barrel, and electrically subtracted before averaging, to provide the transmembrane voltage $V_{(i-e)}$. The monopolar cell is stimulated directly through the R2 input of its projection with a single fibre optic carrying a green monochromatic test flash S1, producing the hyperpolarising response at the bottom left. A longer flash S3 is delivered from a second fibre optic to a facet 5 rows dorsal to the first, but outside the direct projection of the recorded cell. This produces an initial hyperpolarisation probably due to internally scattered light, followed by a pronounced transmembrane depolarisation which indicates a slow, laterally originating process antagonistic to the centre response (upper trace). The test response is reduced during the antagonistic depolarisation, so that only an initial transient remains. The middle trace shows the complementary test in which a similar flash S2 is piped as background adaptation directly into the test facet. It also adapts the test response, and produces a smaller repolarising wave of similar time course to the top trace. Each record is the average of eight repetitions.

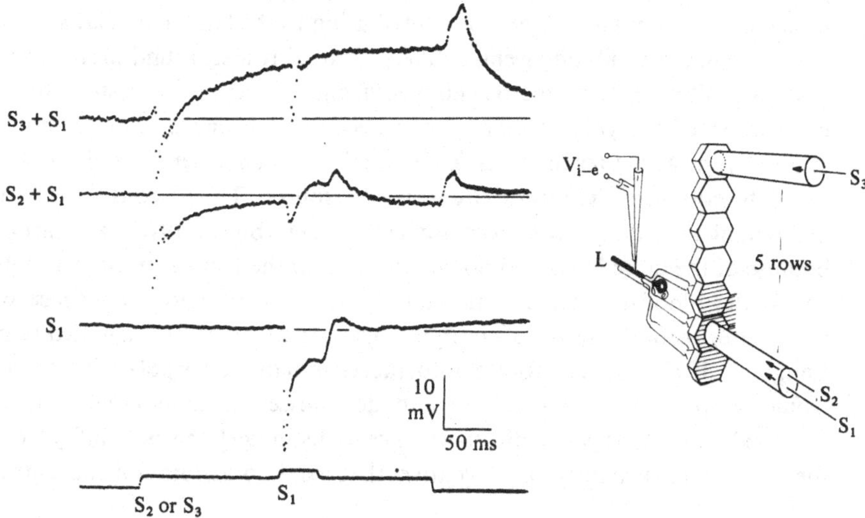

repolarisation of the monopolar membrane when both the test and adapting flash are combined at the test facet (middle trace, Fig. 12), suggesting that the two effects perhaps have a common origin. The laterally directed flash produced the larger depolarisation, and when bright enough to evoke a substantial effect, also produced an initial hyperpolarising response even from several facets away, though with a latency which exceeded the start of the antagonistic effect. This delayed response, also noted by Zettler & Weiler (1976), appears to be a direct response to light scattered internally in the eye, for although optical isolation between neighbouring facets can be as high as 10^5, the flash is bright and the receptor-monopolar system extremely sensitive. The delay in appearance occurs because receptor and therefore monopolar cell latency is strongly dependent upon stimulus intensity. Because this scatter response is still present even with lateral illumination and difficult to get rid of, it is still impossible to say how much of the adaptation shown by the reduced response to the test flash is of lateral origin, and how much originates in the same cartridge.

The origin of the depolarising, laterally triggered response is of some interest in view of the detailed knowledge of the lamina wiring discussed earlier. The collateral system of L4 is the only known neural candidate which connects laterally to L1–L2, but is grossly outnumbered by about 100:1 compared to receptors in the frequency of its contacts with these monopolars, and therefore would be unable to produce an overriding effect like that of Fig. 12. Moreover, if the L4 response has been correctly equated with Arnett's (1972) 'sustaining' fibres, the somewhat phasic initial discharge pattern of these cells does not obviously fit the slow growth of the depolarisation. However, if the maintained output from the receptors is strongly attentuated presynaptic to L1–L2 by inhibition at the receptor terminals, it remains possible that L4 might make some contribution to the lateral depolarisation.

If direct neural inhibition seems unlikely to explain the surround antagonism, a second alternative might be adapted from an earlier suggestion for a mechanism of presynaptic inhibition in receptor terminals. In this, photocurrent from active terminals is forced back through other laterally placed axons to complete the return circuit to the source. This is consistent with differential recordings from receptor cell bodies (Shaw, 1975), and arises because of the large local field potential set up in the lamina by current flow across the blood–eye barrier. If there exists some background release of transmitter from the terminals even in darkness, as has been suggested (e.g. Laughlin, 1973), current forced into the terminals, hyperpolarising them, would suppress this dark release and depolarise the monopolar cell, as observed. There is as yet no direct evidence to distinguish this possibility from the only obvious competing alternative, that local potassium-ion concentration slowly builds up in the lamina because the lamina 'compartment' is isolated from its surroundings by a diffusion barrier (presciently suggested by Laughlin, 1974*a*), and that this depolarises the neurones locally. Zimmerman (1978) suggests from resistance measurements that the lamina is an isolated compartment in the fly, and it is striking that large extracellular potentials are recorded only in this region, in contrast to the locust eye where these change in a gradual manner between the receptor layer and the lamina, and in which the barrier lies between the two (Shaw, 1975, 1977, 1978 and unpublished). Freeze-fracture preparations of the fly reveal complex arrays of intramembrane particles among the satellite and marginal glia which frame the lamina, and which may constitute junctions responsible for sealing off its space (C. Chi & S. D. Carlson, personal communication), and extensive septate junctions are common amongst the satellite glia and axons (W. A. Ribi, personal communication). Potassium release would be expected from the active response of the large monopolar cells to light, in which a potassium conductance is thought to be activated (Zimmerman, 1978), and might be augmented by currents leaving the receptor terminals, which in the absence of local active responses are likely to be predominantly potassium-permeable. Finally, the surprisingly large number of synapses on to the epithelial glial cells mentioned earlier might conceivably act in triggering the release of potassium into the local cartridge space. There is no anatomical indication of any lateral barrier to diffusion within the lamina itself, but rapid uptake by glia could limit lateral diffusion of potassium ion and localise its effect, for which there is a precedent in the bee retina (Coles & Tsacopoulos, 1979). The lateral spread of the field potential measured with single facet stimuli in the locust lamina is roughly exponential with a half-width of 10–12 ommatidial diameters, about half that in the photoreceptor zone (S. R. Shaw, unpublished). There are no corresponding results from the fly lamina, though the spread in the photoreceptor zone of *Drosophila* has a similar half-width to that in the locust, about 20 ommatidia (Heisenberg, 1971).

The effect of surround antagonism at monopolar level has been analysed in detail by Laughlin (1975) and Laughlin & Hardie (1978), using wide field targets to adapt both the centre and surround of the monopolar cell. The general problem is to explain how the visual system manages to see at all in the presence of a strong background illumination overload, and the rough answer is that the design gets rid of the background to concentrate only on the incremental test signal. Laughlin and Hardie have found that in addition to the long-known shift of the operating range of the photoreceptors, induced by strong background illumination, an extra light adaptational, desensitising effect is introduced at the synapse, acting even at low light levels. An important finding is that amplification at the synapse seems to remain relatively constant at different adaptational levels. This would not be expected from a conventional adapting post-synaptic conductance change, acting at the level of the monopolar cell membrane, which would attenuate both the wanted test signal as well as the unwanted background. Thus the contribution of lateral dendrites of L4, and of medulla feedback cells C2, C3 and Tan 1 to adaptation is likely to be minimal, an argument again reinforced by their relatively small number of synaptic specialisations, compared to the numerical dominance of the receptor input (Strausfeld, 1976*b*, Tables 7–1, 7–2). Instead, the system somehow acts to subtract out most of the steady background intensity information present in the receptor response, to concentrate on purely transient information, leaving the synaptic gain unchanged and therefore properly poised to deal sensitively with transient signals superimposed upon the particular background level.

Since the available post-synaptic permutations are unlikely to explain the subtraction feature, this argues for a presynaptic effect upon the receptor terminals themselves, but the recordings available are insufficient to establish this. The receptor terminal response (Fig. 7) is a different shape from that in the soma, with a smaller initial peak and following inflection, but qualitatively at least, this is the sort of low-pass filter distortion expected for a signal spreading passively down a thin axon cable. It is additionally complicated because of the series-recorded artefact arising extracellularly, which is expected to add to the recording. However, the response at a terminal can be recorded back in the cell body to illumination of the other inputs to the appropriate cartridge, with a lesser problem of contamination, and preliminary tests show that there is marked non-linear response summation if two of these facets are stimulated together (S. R. Shaw, unpublished). This result is expected if one input turns on a feedback mechanism which causes sustained, partial suppression of the response to the second input, and is the result anticipated for the lamina field potential scheme proposed earlier (Laughlin, 1974*a*; Shaw, 1975), generated by current flow across part of the blood–brain barrier (Shaw, 1975, 1979). A steady voltage established relatively slowly around the synaptic terminals will oppose, i.e. subtract from, the overall voltage across that part of the terminal, in turn responsible for transmitter release. It remains to be shown whether the size, timing and distribution of the field potential are adequate to explain quantitatively the more phasic response of the monopolar cell compared to the photoreceptors (cf. Figs. 7, 12), or whether some extra process is needed.

Finally, experiments with single facet stimulation have revealed one unexpected piece of evidence suggesting the importance of electrical field interaction in the lamina of the fly. Responses can be regularly observed in receptor cell bodies, originating from the terminals of adjacent axons in the same cartridge, presumably via the junctions shown in Fig. 11, provided the penetrated cell is not itself illuminated. These resemble an attenuated, filtered receptor potential, with the exception that a distinct notch is often evident early on the rising edge of the receptor potential, momentarily repolarising the terminal membrane (Fig. 13). Its usual latency of 1.5 ms is in the same range as the forward latency from receptors to L1–L2 (Järvilehto & Zettler, 1971), suggesting that feedback is electrical, but there is sufficient uncertainty as to the exact forward delay that a disynaptic pathway with no signficant conduction time between synapses, remains just possible. This definitely rules out any of the feedback pathways via the medulla (Table 1), and requires that the effect originate in the lamina. A positive transient of similar time course can be recorded in the extracellular space near the cartridge, and this with other properties of the notch response suggests that it originates by field

Fig. 13. Differential transmembrane recording from the cell body of a *Calliphora* photoreceptor, identified as R3 from the pattern of electrical interaction via the lamina which it shows to stimulation of single facets (Shaw, in preparation). R3s' axon is coupled to neighbour R2 from the next facet by the junctions in the cartridge shown in Fig. 11, and when this R2 input is tested with a single fibre optic (inset), the R2 terminal response is picked up back in the retina, after conduction decrementally across the junction and antidromically along R3. The response is a minaturised, slowed version of the normal receptor potential recorded in R3 to direct stimulation (50 mV in this case; not shown), with the usual depolarising afterpotential. In addition, not seen with direct stimulation, the early part of the R2 response is cut back sharply by a transient hyperpolarisation arising in the terminal itself, which develops very rapidly within 1.5–2.0 ms of the first sign of the receptor potential (arrows), in different preparations. This negative feedback probably arises by field potential coupling between the axon terminals and large monopolar cells in the cartridge, which momentarily curtails transmitter release and produces the pronounced transient response of the monopolar cell (see text). The response is averaged 32 times, with a time interval between sample bins of 1 ms. The duration of the monochromatic green light flash (487 nm peak) is shown by the bar at the bottom.

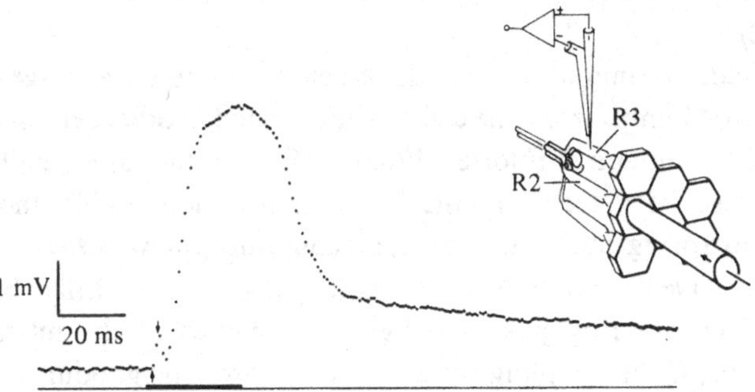

potential coupling between the monopolars and receptor terminals. The only other admissable alternative, the reciprocal synapse from L2 back to R1–R6 (Fig. 3; Strausfeld & Campos-Ortega, 1977), cannot be reconciled with all the characteristics of the notch response. The timing of the effect suggests that it should be responsible for a rapid suppression of transmitter release locally at the terminal, shortly after transmission begins, and thus may be a major cause of the rapid repolarisation of the monopolar cell following the initial peak, creating the highly phasic response. An interesting though speculative thought is that if electrical feedback has been identified correctly and is functionally useful, evolution should have favoured a tetradic synapse where all the four post-synaptic elements have the same sign of response, or the opposite post-synaptic currents generated would cancel and reduce the effect. This would imply that not only L1–L3 but also the amacrine cell have hyperpolarising responses, which accords with another speculation made earlier in this chapter.

Presynaptic electrical coupling between receptors

In most insect orders, the six short photoreceptors which end synaptically in each cartridge come from a single ommatidium (Meinertzhagen, 1976), and by virtue of their common lens and fused central rhabdom share the same receptive field (Shaw, 1969). In addition to this 'optical coupling' via a common waveguide, direct tests with pairs of electrodes have revealed electrical coupling within ommatidia of several species (Smith & Baumann, 1969; Shaw, 1969; Muller, 1973). This seems to involve all the cells in one ommatidium, except in the crayfish where only those cells with the same microvillar orientation were coupled to each other (Muller, 1973). The strength of the coupling varies widely within ommatidia of the same species, but it now seems doubtful whether all this variation is due to the distance apart of two cells in the ommatidium as originally supposed (Smith & Baumann, 1969; Shaw, 1969), since in barnacle ocelli a wide variation in coupling strength was found between the same cell pair in different animals (Shaw, 1972).

In the *Limulus* ommatidium, in which cell pairs are on average strongly coupled, photon bumps from one cell can be detected in other cells in the same group (Smith, Baumann & Fuortes, 1965). In favourable cases small coupled bumps can also be detected in the locust (Lillywhite, 1978), the weaker interaction mirroring the weaker average coupling (Shaw, 1969). Recently, similar results have been obtained in mantids (Rossel, 1979). Lillywhite could account for the differing polarised light sensitivities of the intrinsic and coupled bumps, if the coupling were lateral to direct neighbour cells in the ommatidium. Dye injections into single cells in the locust show that in many cases the small fluorescent molecule Lucifer Yellow spreads into up to three other cells also, predominantly by lateral spread, although there is no direct evidence in this case linking the electrical and the dye-coupling phenomena (S. R. Shaw, unpublished). The locus of coupling appears to be the cell soma, and the most likely site lies where the microvilli from one cell touch those of a lateral neighbour. In *Limulus*, what appear to be gap junctions between microvilli of adjacent cells have been reported by Lasansky (1967). The rest of the photoreceptor soma is remarkable for the lack of obvious intercellular junctions in the adult, except for belt desmosomes which run in insects along the whole length of the ommatidium and probably help to hold the cells together. Chi, Carlson & St. Marie (1979) have recently suggested that there may be other intercellular membrane specialisations when the fixation technique is modified, but it is not yet clear whether the images seen in freeze-fracture of the eye are really all from true junctional particles, or from non-junctional, intrinsic membrane molecules (cf. Lane, 1979).

The spectral sensitivities of insect photoreceptors inferred from intracellular recordings often show a very prominent secondary spectral peak or shoulder at a different wavelength from the main peak, and in some cases this has been attributed tentatively to electrical coupling between the cell bodies of spectrally different cells (review: Menzel, 1975). This argument is plausible but is not the only explanation, since careful analysis has now revealed at least one stable accessory pigment, within R1–R6 photoreceptors in the fly retina, responsible for extending the absorption spectrum without involving coupling (review: Kirschfeld, 1979). The accessory pigment transfers the ultraviolet energy it absorbs to the main visual pigment and thus broadens the overall absorption spectrum, increasing the total light-gathering power of the receptors.

There exists one case of interreceptor coupling for which there is now both anatomical and physiological evidence. The six axons converging on each cartridge in the fly come from six different, adjacent ommatidia and touch only at cartridge level (Fig. 2). Gap junctions, elsewhere associated with electrical coupling (Bennett, 1978), have recently been found connecting up the six presynaptic terminals (Fig. 11; Ribi, 1978). The recordings of Scholes (1969), taken from sites in the lamina, possibly these terminals, demonstrated strong presynaptic interaction, but may be subject to strong contamination by field potentials, which is particularly bad in the lamina (Mote, 1970; review, Shaw, 1979). Fortunately, coupling via the lamina can be observed also back in the retina, using single facet stimulation, and avoiding the contamination problem (Shaw, unpublished). Coupling is strong to next neighbour axons in the ring, matching the occurrence of the gap junctions

(Fig. 11) and its calculated effect is to unite the axon terminals so strongly that they have practically indistinguishable photon bump responses. Ribi (1978) also mentions finding gap junctions between receptor terminals from the same ommatidium in the bee cartridge, so that presynaptic coupling may be more common than supposed, and explain some of the coupling observed between receptor cell bodies. The strength of the observed coupling (Shaw, 1969) and the results with dyes suggest that there must be soma coupling too, and this has to be the explanation of the strong coupling between barnacle photoreceptors, where the separation and extended geometry of the axons rules out a large coupling contribution from the distant terminals (Shaw, 1972).

Finally, some degree of less specific electrical coupling between nearby receptors has been proposed to arise from the position of the blood–brain barrier in the locust eye, which creates large field potentials in the lamina (Shaw, 1975, 1979), as discussed earlier. Menzel & Blakers (1976) have since adopted this same scheme of interaction to account for what they regard as *negative* (inhibitory) electrical coupling within bee ommatidia, wherein negative-going reponses can be recorded with intracellular techniques at certain wavelengths. However, as pointed out originally (Shaw, 1975), current passively leaving the receptor soma during this type of interaction *depolarises* this part of the cell. The misinterpretation arises because the 'intracellular' response recorded with a single active probe by Menzel & Blakers was not corrected for the large extracellular response in series with it, and thus is not the real transmembrane potential of that part of the cell. Where corrections have been made by simultaneous intra- and extracellular recordings, the depolarising nature of interaction at cell soma level has been revealed (Shaw, 1975). Moreover, the importance of this interaction must lie not in its modest effect at cell body level, but in the larger expected polarisation of the axon terminals, which is opposite to that recorded in the cell soma (i.e. hyperpolarising).

What is the function of electrical coupling?

Electrical coupling occurs extensively between photoreceptors in both vertebrates and invertebrates, and seems to be the rule rather than the exception in visual systems. As a mechanism it obviously smooths out noisy responses due to the random arrival of photons in individual cells, but at the seeming expense of degrading retinal acuity, which depends upon keeping the photoreceptor outputs separate. However, the cells which are coupled in insects are those which share the same field of view in the same ommatidium or cartridge, whilst those in vertebrates are known from the peripheral retina, where there is anyway considerable output convergence on to bipolar and horizontal cells. As pointed out by Ribi (1978), the problem in the insect cartridge is that the same response averaging performed by coupling is achieved in any case by post-synaptic summation of the responses of the same receptors that are coupled, on to post-synaptic cells. If this summation process is highly reliable as argued earlier from the large synaptic area, no extra benefit is derived from adding a further averaging mechanism in the form of presynaptic electrical coupling, which therefore must have some other rationale. Ribi (1978) suggests that in restricting response fluctuation to a smaller voltage range by coupling, synaptic linearity would be better approximated, which is compatible with the notion presented earlier that the dynamic gain is tuned to a particular low amplitude range of signals. Another possibility suggested for the vertebrate system (T. D. Lamb, personal communication), is that the individual single photon responses in an isolated cell may be large enough to saturate the following cells if the synaptic gain is high enough, so that it is advantageous to average the responses over several cells by coupling, reducing the mean size of quantal fluctuation at any one synapse. The quantal responses observed postsynaptically in locust monopolar cells can exceed 10 mV (Shaw, 1968), about one third of the total output voltage range of the cell.

Another function where coupling is important lies in the optimisation of synaptic transmission, in enhancing the synaptic gain by minimising the known natural variation of photoreceptor responsiveness. Laughlin (1973) and Rossel (1979) have shown that there is some variability in absolute sensitivity of photoreceptors within a sample, even from the same eye region. Since the monopolar response-intensity curve is much steeper than that for the receptors, it covers a much narrower intensity span (Laughlin, 1973), not much larger than the total spread of sensitivity variation measured in receptors, about a factor of ten in the light adapted state. There is thus a strong probability within each cartridge group that some receptor responses will be much larger and drive the synapse maximally, whilst other inputs are barely driven at all. This is confirmed by calculations based on eqn (1), for instance for an extreme pair of receptors with light sensitivities differing by a factor of ten. Synaptic output is governed almost entirely by the more sensitive of the two until the post-synaptic output comes within a few percent of its ceiling. Coupling the receptors together strongly so that they come to have a common response-light intensity curve tunes the post-synaptic voltage output to somewhat smaller presynaptic voltages, but its main effect is to broaden the dynamic range at the lower end of the scale, so that the smallest receptor signals are more effectively transmitted.

Once electrical equivalence has been established between them by the electrical coupling linkages, any feedback control over the receptor terminals can be made identical for each, where otherwise separate control lines would be required. Three feedback routes on to the receptors have been described anatomically in the fly, from Tan 1, C2 and L2 (Table 1 and Fig. 3), and whilst their functions are unknown, it is notable that each is a single cell which sends branches to all six receptors in one cartridge; no individual control channels are present.

The function of the lamina in Diptera

This account has focussed heavily upon the anatomical and physiological features of the individual cells and the junctions between them, to the detriment of considering the retina–lamina complex from a functional point of view, as part of the larger visual system which governs perception and behaviour. Despite a widely held view that a study of its parts will not identify the function of the whole in a complex system, in this case the dissection approach does reveal something about the place of the lamina in the scheme of things; it certainly shows what this nucleus does not do. This optimism must first be qualified, though, by acknowledging that where recurrent feedback routes connect between layers in the visual hierarchy, there can strictly be no such thing as the function of one level in isolation of the others. We have virtually no idea of the function of the lamina efferents, which must to some degree modify the output from the lamina.

On the negative side, it now seems clear that the lamina in Diptera is not concerned in any important way with the processing of colour, polarised light or movement information. Although some differences have been found between the spectral sensitivities of receptors R1–R6 (Järvilehto & Moring, 1976), these are of a type which might be expected from natural variation in the proportions of the two photopigments now believed to reside in each rhabdomere, one a green-sensitive rhodopsin and the other an ultraviolet-sensitive antenna pigment (Kirschfeld, Franceshini & Minke, 1977). As reviewed here, all six receptor endings in each cartridge converge on to all four output cells L1–L3 and T1, and are additionally coupled together electrically. There are no outputs from the cartridge which could collect information selectively from any individual short photoreceptor, and even the amacrines which could in principle do this fail to connect appropriately and anyhow also lack individual outputs to the medulla. It therefore must be concluded that R1–R6 information is pooled in the lamina and that all individuality of these six cell types is lost. The cell types with differing spectral responses, R7 and R8, make no synapses within the lamina, so that interactions which give rise to the spectral discrimination shown by Diptera (review: Heisenberg & Buchner, 1977) must arise within deeper centres.

Identical arguments can be made for polarised light analysis, possible by virtue of the alignment of dichroic photopigment molecules in the parallel microvilli of each photoreceptor (see Goldsmith & Bernard, 1974). This gives rise to the well-known dichroic responses shown by fly R1–R6 photoreceptors, but again these are obliterated by convergence at the lamina (Scholes, 1969). The two remaining independent dichroic channels, again R7 and R8, plus a non-dichroic intensity channel supplied by R1–R6, are sufficient in theory to build a polarised-light detection system (Kirschfeld, 1973; Bernard & Wehner, 1977). Finally the detection of movement stimuli requires detectors to correlate information from separate directions in space, which cannot be done within a cartridge with its single field of view (Fig. 2). Connections between cartridges are maintained only by L4 and L5, neither of which seems suitable for the basis of a comprehensive movement detection system, nor do the recorded physiological responses which seem likely to arise from them show sensitivitity to movement, but only to the spatial location of the stimulus (Arnett, 1972).

The strictures may well not apply entirely to other insect groups or to crustaceans, many of which slow stratification of receptor endings in the lamina, which may well partition off spectral or polarised light cell classes as a first step in the analysis of these modalities (e.g. Ribi, 1975; Nassel & Waterman, 1977), and which have synaptic input from the long visual fibre analogues of R7/8.

The main function which has emerged for the receptor–lamina complex in Diptera is that of a first relay at which light intensity information is gathered from sets of six similar receptors and conditioned to amplify and make it more reliable within the bandwidth of the transduction process, especially for very small, transient, low-light signals. At the same synapse, a powerful gain control mechanism supplements that in the receptors themselves to match quickly the system's current operating range to the prevailing mean lighting conditions, whilst preserving the eye's dynamic sensitivity (Laughlin & Hardie, 1978). It seems likely that this same system, needing as it must to integrate light intensity laterally to find the weighted local mean, is also responsible for the lateral inhibitory interactions described in the lamina, expected to enhance border contrast and accentuate the differences in output between neighbouring cartridges. It is not possible in a biological, evolved system to pick out with certainty which of two useful features is the more adaptively important, but some indication in the lamina can be had from the range of effect produced. Light adaptation control altered the output sensitivity over a range of $\geqslant 10^4$ and even then was not tested to its limits by Laughlin & Hardie (1978), whilst the changes in receptive field width reported by all authors barely exceed a factor of two.

The future for the fly

Several advantages are apparent already in the choice of the fly as a model for investigating peripheral visual function in a neural system of some sophistication. Foremost is the presence of a known and limited complement of neural types within the lamina, the connections between all of which are known. The presence of twelve connectivity classes of cell in the lamina might seem to suggest that this part of the insect system is much more complicated than the whole vertebrate retina put together, which has only about six main neurone types (Fain, this volume). However, these six are basically *shape* classes not *connectivity* classes, and connectivity is the determinant of retinal function. There are at least seven connectivity classes of bipolar cell in fish (Scholes, 1975), and almost anyone's classification of vertebrate ganglion cells runs into double figures (e.g. 15 subdivisions, Cleland & Levick, 1974; West, 1976); presumably the amacrine cell types are similarly prolific. It is therefore likely that if a numerical list of functional or connectivity types can be taken as an index of complexity, that the vertebrate retina considerably outstrips the receptor–lamina complex of the fly. In no vertebrate eye has anything like as complete an index as Table 1 been compiled, and to do so remains a daunting task because of the absence of clear repeating subdivisions like the optic cartridges in insects. These recognisable subunits in the eye also make stimulation much easier to control, since the facet rows map out the neural subdivisions on to the eye surface. Light stimuli can be successfully contained almost entirely to single ommatidia, and even to single cells with reasonable isolation (Kirschfeld & Lutz, 1974).

Physiological responses have been recorded from about half the neurones in the lamina, and there seems no reason why this proportion should not rise. The major uncertainty at present lies in the function of the efferent connections, and here recordings from identified parent cell bodies in the medulla may hold the key. The lamina in the fly provides a test situation in which to compare the relative virtues of decremental and impulsive transmission, by comparing L1–L2 with the two spiking cells of Arnett (1972), possibly L4 and L5, which signal from the same sources along the same route. The lamina also provides an accessible neuropile in which to study the role of field potentials in neural processing, since the fields are especially large and can be recorded widely, because of local extracellular diffusion barriers. Finally, the lamina amacrine cell provides perhaps the unique opportunity to test directly the interesting notion that separate dendrites can form independent republics within the same cell. The α processes are just large enough to record from in some places, and different dendritic inputs can be sampled with stimuli delivered to single receptors in different ommatidia. Local dendritic outputs can also be monitored via the activity of L4, to test the extent to which input interaction occurs between the different α processes of one cell (Fig. 5). Hopefully, before long some of these questions should have been answered.

I am grateful to Drs I. Meinertzhagen, W. Ribi and S. Carlson for discussing recent work on the synaptology of the lamina.

References

Armett-Kibel, C., Meinertzhagen, I. A. & Dowling, J. E. (1977). Cellular and synaptic organisation in the lamina of the dragonfly *Sympetrum rubicundulum*. *Proc. R. Soc. Lond. Ser. B*, **196**, 385–413.

Arnett, D. W. (1972). Spatial and temporal integration properties of units in the first optic ganglion of dipterans. *J. Neurophysiol.* **35**, 429–44.

Ashmore, J. F. & Falk, G. (1976). Absolute sensitivity of rod bipolar cells in a dark-adapted retina. *Nature, Lond.* **263**, 248–9.

Ashmore, J. F. & Falk, G. (1979). Transmission of visual signals to bipolar cells near absolute threshold. *Vision Res.* **19**, 419–23.

Bader, C. R., Baumann, F. & Bertrand, D. (1976). Role of intracellular calcium and sodium in light adaptation in the retina of the honeybee drone (*Apis mellifera*, L.). *J. Gen. Physiol.* **67**, 475–91.

Baumann, F. (1968). Slow and spike potentials recorded from retinula cells of the honey bee drone in response to light. *J. Gen. Physiol.* **52**, 855–75.

Bennett, M. V. L. (1978). Junctional permeability. In *Receptors and Recognition*, ser. B, ed. J. Feldman, N. B. Gilula & J. D. Pitts, **2**, 23–36. London: Chapman & Hall.

Bernard, G. D. & Wehner, R. (1977). Functional similarities between polarization vision and colour vision. *Vision Res.* **17**, 1019–29.

Blest, A. D. (1980). Photoreceptor membrane turnover in arthropods: comparative studies of breakdown processes and their implications. In *The Effects of Constant Light on Visual Processes*, ed. T. P. Williams & B. N. Baker, pp. 217–45. New York: Plenum Publishing Co.

Boeckh, J., Sandri, C. & Akert, K. (1970). Sensorische Eingänge und synaptische Verbindungen im Zentralnervensystem von Insekten. *Z. Zellforsch.* **103**, 429–46.

Boschek, C. B. (1971). On the fine structure of the peripheral retina and lamina ganglionaris of the fly, *Musca domestica*. *Z. Zellforsch.* **118**, 369–409.

Braitenberg, V. (1967). Patterns of projection in the visual system of the fly. I. Retina-lamina projections. *Exp. Brain Res.* **3**, 271–98.

Braitenberg, V. & Debbage, P. (1974). A regular net of reciprocal synapses in the visual system of the fly *Musca domestica*. *J. Comp. Physiol.* **90**, 25–31.

Brown, J. E. & Lisman, J. E. (1975). Intracellular Ca modulates sensitivity and time scale in *Limulus* ventral photoreceptors. *Nature, Lond.* **258**, 252–4.

Burkhardt, W. & Braitenberg, V. (1976). Some peculiar synaptic complexes in the first visual ganglion of the fly, *Musca domestica*. *Cell Tiss. Res.* **173**, 287–308.

Burrows, M. (1979). Synaptic potentials effect the release of transmitter from locust nonspiking interneurones. *Science*, **204**, 81–3.

Campos-Ortega, J. A. (1974). Autoradiographic localization of

^{3}H-γ-aminobutyric acid uptake in the lamina ganglionaris of *Musca* and *Drosophila*. *Z. Zellforsch.* **147**, 415–31.

Campos-Ortega, J. A. & Strausfeld, N. J. (1973). Synaptic connections of intrinsic cells and basket arborizations in the external plexiform layer of the fly's eye. *Brain Res.* **59**, 119–36.

Carreras, J. (1978). Propagation du potentiel recepteur dans les cellules retiniennes du faux-bourdon (*Apis mellifera*). Ph.D. Dissertation, University of Geneva.

Chi, C., Carlson, S. D. & St. Marie, R. L. (1979). Membrane specialisations in the peripheral retina of the housefly, *Musca domestica* L. *Cell Tiss. Res.* **198**, 501–20.

Cleland, B. G. & Levick, W. R. 1974). Properties of rarely encountered types of ganglion cells in the cat's retina and an overall classification. *J. Physiol. Lond.* **240**, 457–92.

Coles, J. A. & Tsacopoulos, M. (1979). Potassium activity in photoreceptors, glial cells and extracellular space in the drone retina: changes during photostimulation. *J. Physiol. Lond.* **290**, 525–49.

Cone, R. A. (1973). The internal transmitter model for visual excitation: Some quantitative implications. In *Biochemistry and Physiology of Visual Pigments*. ed. H. Langer pp. 275–82. Berlin: Springer-Verlag.

Falk, G. & Fatt, P. (1972). Physical changes induced by light in the rod outer segments of vertebrates. In *Handbook of Sensory Physiology*, ed. H. J. A. Dartnall, *VII/1*, 200–44. New York: Springer-Verlag.

Franceschini, N. (1975). Sampling of the visual environment by the compound eye of the fly: fundamentals and applications. In *Photoreceptor Optics*, ed. A. W. Snyder & R. Menzel, pp. 98–125. New York, Heidelberg, Berlin: Springer-Verlag.

French, A. S. & Järvilehto, M. (1978*a*). The dynamic behaviour of photoreceptor cells in the fly in response to random (white noise) stimulation at a range of temperatures. *J. Physiol. Lond.* **273**, 311–22.

French, A. S. & Järvilehto, M. (1978*b*). The transmission of information by first and second-order neurons in the fly visual system. *J. Comp. Physiol.* **126**, 87–96.

Fuortes, M. G. F. & Hodgkin, A. L. (1964). Changes in time scale and sensitivity in the ommaditia of Limulus. *J. Physiol.* **172**, 239–63.

Fuortes, M. G. F. & O'Bryan, P. M. (1972). In *Handbook of Sensory Physiology, VII/2*, ed. M. G. F. Fuortes, pp. 279–338. Berlin: Springer-Verlag.

Gilula, N. B. (1978). Structure of intercellular junctions. In *Receptors* and *Recognition, Ser. B*, ed. J. Feldman, N. B. Gilula & J. D. Pitts, **2**, 1–22. London: Chapman & Hall.

Glantz, R. M. (1968). Light adaptation in the photoreceptor of the crayfish, *Procambarus clarkii. Vision Res.* **8**, 1407–22.

Goldsmith, T. H. & Bernard, G. D. (1974). The visual system of insects. In *The Physiology of Insecta* (2nd edn), ed. M. Rockstein, pp. 165–272. New York: Academic Press.

Graubard, K. & Calvin, W. H. (1979). Presynaptic dendrites: implications of spikeless synaptic transmission and dendritic geometry. In *The Neurosciences: Fourth Study Program*, ed. F. O. Schmitt & F. G. Worden, pp. 317–31. Cambridge, Massachusetts: MIT Press.

Gwilliam, G. F. & Millechia, R. J. (1975). Barnacle photoreceptors: their physiology and role in the control of behaviour. *Prog. Neurobiol.* **4**, 211–39.

Hardie, R. C. (1977). Electrophysiological properties of R7 and R8 in Dipteran retina. *Z. Naturforsch.* **32c**, 887–9.
Hardie, R. C. (1978). Peripheral visual function in the fly. Ph.D. dissertation, Australian National University, Canberra.
Hardie, R. C. (1979). Electrophysiological analysis of fly retina. I. Comparative properties of R1–6 and R7 and R8. *J. Comp. Physiol.* **129**, 19–33.
Hausen, K. (1977). Signal processing in insect eye. In *Dahlem Workshop on Function and Formation of Neural Systems*, ed. G. Stent, pp. 81–111. Berlin: Abakon-Verlagsgesellschaft.
Hartline, H. K. & Ratliff, F. (1972). Inhibitory interaction in the retina of *Limulus*. In *Handbook of Sensory Physiology*, ed. M. G. F. Fuortes, *VII/2*, pp. 381–447. Berlin, Heidelberg, New York: Springer-Verlag.
Heisenberg, M. (1971). Separation of receptor and lamina potentials in the electroretinogram of normal and mutant *Drosophila. J. Exp. Biol.* **55**, 85–100.
Heisenberg, M. & Buchner, E. (1977). Role of retinula cell types in visual behaviour of *Drosophila melanogaster. J. Comp. Physiol.* **117**, 127–63.
Henderson, R. (1977). The purple membrane from *Halobacterium halobium. Ann. Rev. Biophys. Bioeng.* **6**, 87–109.
Hengstenberg, R. (1971). Das Augenmuskelsystem der Stubenfliege *Musca domestica*. I. Analyse der 'clock spikes' und ihrer Quellen. *Kybernetik*, **9**, 56–77.
Heuser, J. E. & Reese, T. (1979). Synaptic vesicle exocytosis captured by quick-freezing. In *The Neurosciences: Fourth Study Program*, ed. F. O. Schmitt & F. G. Worden, pp. 573–600. Cambridge, Massachusetts: MIT Press.
Hille, B. (1970). Ionic channels in nerve membranes. *Progr. Biophys. Molec. Biol.* **21**, 1–32.
Hodgkin, A. L. & Rushton, W. A. H. (1946). The electrical constants of a crustacean nerve fibre. *Proc. R. Soc. Lond. Ser. B*, **133**, 444–79.
Horridge, G. A. & Meinertzhagen, I. A. (1970). The accuracy of the patterns of connexions of the first- and second-order neurons of the visual system of *Calliphora. Proc. R. Soc. Lond. Ser. B*, **175**, 69–82.
Hudspeth, A. J., Poo, M. M. & Stuart, A. E. (1977). Passive signal progagation and membrane properties in median photoreceptors of the giant barnacle. *J. Physiol., Lond.* **272**, 25–43.
Jack, J. (1979). An introduction to linear cable theory. In *The Neurosciences: Fourth Study Program*, ed. F. O. Schmitt & F. G. Worden, pp. 423–437. Cambridge, Massachusetts: MIT Press.
Järvilehto, M. & Moring, J. (1976). Spectral and polarisation sensitivity of identified retinal cells of the fly. In *Neural Principles in Vision*, ed. F. Zettler & R. Weiler, pp. 214–226. Berlin, Heidelberg, New York: Springer-Verlag.
Järvilehto, M. & Zettler, F. (1970). Micro-localisation of lamina-located visual cell activities in the compound eye of the blowfly *Calliphora. Z. Vergl. Physiol.* **69**, 134–8.
Järvilehto, M. & Zettler, F. (1971). Localised intracellular potentials from pre- and postsynaptic components in the external plexiform layer of an insect retina. *Z. Vergl. Physiol.* **75**, 422–40.
Järvilehto, M. & Zettler, F. (1973). Electrophysiological-histological studies on some functional properties of visual cells and second-order neurons of an insect retina. *Z. Zellforsch.* **136**, 291–306.

Katz, B. & Miledi, R. (1967). A study of synaptic transmission in the absence of nerve impulses. *J. Physiol., Lond.* **192**, 407–36.

Kirschfeld, K. (1966). Discrete and graded receptor potentials in the compound eye of the fly *Musca.* In *The Functional Organisation of the Compound Eye*, ed. C. G. Bernhard, pp. 291–307. Oxford: Pergamon Press.

Kirschfeld, K. (1967). Die projektion der optischen Umwelt auf das Raster der Rhabdomere im Komplexauge von *Musca. Exp. Brain Res.* **3**, 248–70.

Kirschfeld, K. (1973). Vision of polarised light. In *Proc. Internat. Biophys. Congr. Moscow*, **4**, 289–96.

Kirschfeld, K. (1979). The function of photostable pigments in fly photoreceptors. *Biophys. Struct. Mech.* **5**, 117–28.

Kirschfeld, K., Franceschini, N. & Minke, B. (1977). Evidence for a sensitising pigment in fly photoreceptors. *Nature, Lond.* **269**, 386–90.

Kirschfeld, K. & Lutz, B. (1974). Lateral inhibition in the compound eye of the fly *Musca. Z. Naturforsch.* **29c**, 95–7.

Klingman, A. & Chappell, R. L. (1978). Feedback synaptic interaction in the dragonfly ocellar retina. *J. Gen. Physiol.* **71**, 157–75.

Lamparter, H. E., Steiger, U., Sandri, C. & Akert, K. (1969). Zum Feinbau der Synapsen im Zentralnervensystem der Insekten. *Z. Zellforsch* **99**, 435–42.

Land, M. F. (1977). Visually guided movements in invertebrates. In *Dahlem Workshop on function and formation of neural systems*, ed. G. Stent, pp. 161–77. Berlin: Abakon-Verlagsgesellschaft.

Lane, N. J. (1979). Intramembranous particles in the form of ridges, bracelets or assemblies in arthropod tissue. *Tissue & Cell*, **11**, 1–18.

Lasansky, A. (1967). Cell junctions in ommatidia of *Limulus. J. Cell Biol.* **33**, 365–83.

Laughlin, S. B. (1973). Neural integration in the first optic neuropile of dragonflies. I. Signal amplification in dark-adapted second order neurons. *J. Comp. Physiol.* **84**, 335–55.

Laughlin, S. B. (1974*a*). Neural integration in the first optic neuropile of dragonflies. II. Receptor signal interactions in the lamina. *J. Comp. Physiol.* **92**, 357–75.

Laughlin, S. B. (1974*b*). Neural integration in the first optic neurophile. III. The transfer of angular information. *J. Comp. Physiol.* **92**, 377–96.

Laughlin, S. B. (1975). Receptor and interneuron light-adaptation in the dragonfly visual system. *Z. Naturforsch.* **30c**, 306–8.

Laughlin, S. B. (1976). The sensitivities of dragonfly photoreceptors and the voltage gain of transduction. *J. Comp. Physiol.* **111**, 221–47.

Laughlin, S. B. & Hardie, R. C. (1978). Common strategies for light adaptation in the peripheral visual systems of fly and dragonfly. *J. Comp. Physiol.* **128**, 319–40.

Leutscher-Hazelhoff, J. T. (1975). Linear and non-linear performance of transducer and pupil in *Calliphora* retinula cells. *J. Physiol. Lond.* **246**, 333–50.

Lillywhite, P. G. (1977). Single photon signals and transduction in an insect eye. *J. Comp. Physiol.* **122**, 189–200.

Lillywhite, P. G. (1978). Coupling between locust photoreceptors revealed by a study of quantum bumps. *J. Comp. Physiol.* **125**, 13–27.

Lillywhite, P. G. & Laughlin, S. B. (1979). Transducer noise in a photoreceptor. *Nature, Lond.* **277**, 569–72.

Lisman, J. E. (1976). Effects of removing extracellular Ca^{2+} on excitation and adaptation in *Limulus* ventral photoreceptors. *Biophys. J.* **16**, 1331–5.

Llinás, R. (1979). The role of calcium in neuronal function. In *The Neurosciences: Fourth Study Program*, ed. F. O. Schmitt & F. G. Worden, pp. 555–71. Cambridge, Massachusetts: MIT Press.

Martin, A. R. & Ringham, G. L. (1975). Synaptic transfer at a vertebrate central nervous system synapse. *J. Physiol., Lond.* **251**, 409–26.

Meinertzhagen, I. A. (1973). Development of the compound eye and optic lobe of insects. In *Developmental Neurobiology of Arthropods*, ed. D. Young, pp. 51–104. Cambridge University Press.

Meinertzhagen, I. A. (1976). The organization of the perpendicular fibre pathways in the insect optic lobe. *Phil. Trans. R. Soc. Lond.* **274**, 555–93.

Melamed, J. & Trujillo-Cenóz, O. (1968). The fine structure of the central cells in the ommatidia of Dipterans. *J. Ultrastr. Res.* **21**, 313–34.

Menzel, R. (1974). Spectral sensitivity of monopolar cells in the bee lamina. *J. Comp. Physiol.* **93**, 337–46.

Menzel, R. (1975). Colour receptors in insects. In *The Compound Eye and Vision of Insects*, ed. G. A. Horridge, pp. 121–53. Oxford University Press.

Menzel, R. & Blakers, M. (1976). Colour receptors in the bee eye – morphology and spectral sensitivity. *J. Comp. Physiol.* **108**, 11–33.

Mimura, K. (1976). Some spatial properties in the first optic ganglion of the fly. *J. Comp. Physiol.* **105**, 65–82.

Mote, M. I. (1970*a*). Focal recording of responses evoked by light in the lamina ganglionaris of the fly *Sarcophaga bullata. J. Exp. Zool.* **175**, 149–58.

Mote, M. I. (1970*b*). Electrical correlates of neural superposition in the eye of the fly *Sarcophaga bullata. J. Exp. Zool.* **175**, 159–68.

Muller, K. J. (1973). Photoreceptors in the crayfish compound eye: Electrical interaction between cells as related to polarised-light sensitivity. *J. Physiol., Lond.* **232**, 573–95.

Nässel, D. R. & Waterman, T. H. (1977). Golgi EM evidence for visual information channelling in the crayfish lamina ganglionaris. *Brain Res.* **130**, 556–64.

Northrop, R. B. & Guignon, E. F. (1970). Information processing in the optic lobes of the lubber grasshopper. *J. Insect Physiol.* **16**, 691–714.

Osborne, M. P. (1967). The fine structure of neuromuscular junctions in the segmental muscles of the blowfly larva. *J. Insect Physiol.* **13**, 827–33.

Owen, W. G. & Copenhagen, D. R. (1977). Characteristics of the electrical coupling between rods in the turtle retina. In *Vertebrate Photoreception*, ed. H. B. Barlow & P. Fatt, pp. 169–92. London: Academic Press.

Ozawa, S., Hagiwara, S. & Nicolaysen, K. (1977). Neuronal organisation of the shadow reflex in a giant barnacle. *J. Neurophysiol.* **40**, 982–95.

Pearson, K. G. (1979). Local neurons and local interactions in the nervous systems of invertebrates. In *The Neurosciences: Fourth Study Program*, ed. F. O. Schmitt & F. G. Worden, pp. 145–57. Cambridge, Massachusetts: MIT Press.

Pedler, C. & Goodland, H. (1965). The compound eye and the first optic ganglion of the fly. *J. R. Microscop. Soc.* **84**, 161–79.

Pfenninger, K. H. (1973). Synaptic morphology and cytochemistry. *Prog. Histochem. Cytochem.* **5**, 1–86.

Pick, B. (1977). Specific misalignments of rhabdomere visual axes in the neural superposition eye of dipteran flies. *Biol. Cyb.* **26**, 215–24.

Ribi, W. A. (1975). The neurons of the first optic ganglion of the bee, *Apis mellifera*. *Adv. Anat. Embryol. Cell Biol.* **50**, 5–43.

Ribi, W. A. (1978). Gap junctions coupling photoreceptor axons in the first optic ganglion of the fly. *Cell Tiss. Res.* **195**, 299–308.

Rose, A. (1974). *Vision, human and electronic*. New York: Plenum Press.

Rossel, S. (1979). Regional differences in photoreceptor performances in the eye of the praying mantis. *J. Comp. Physiol.* **131**, 95–112.

Scholes, J. H. (1964). Discrete subthreshold potentials from the dimly lit insect eye. *Nature, Lond.* **202**, 572–3.

Scholes, J. H. (1969). The electrical response of the retinal receptors and the lamina in the visual system of the fly *Musca*. *Kybernetik*, **6**, 149–62.

Scholes, J. H. (1975). Colour receptors and their synaptic connexions in the retina of a cyprinid fish. *Phil. Trans. R. Soc. Lond.* **270**, 61–118.

Scholes, J. H. & Reichardt, W. (1969). The quantal content of optomotor stimuli and the electrical responses of receptors in the compound eye of the fly *Musca*. *Kybernetik*, **6**, 74–80.

Shaw, S. R. (1968). Organization of the locust retina. *Symp. Zool. Soc. Lond.* **23**, 135–63.

Shaw, S. R. (1969). Interreceptor coupling in ommatidia of drone honeybee and locust compound eyes. *Vision Res.* **9**, 999–1029.

Shaw, S. R. (1972). Decremental conduction of the visual signal in barnacle lateral eye. *J. Physiol., Lond.* **220**, 145–75.

Shaw, S. R. (1975). Retinal resistance barriers and electrical lateral inhibition. *Nature, Lond.* **255**, 480–3.

Shaw, S. R. (1977). Restricted diffusion and extracellular space in the insect retina. *J. Comp. Physiol.* **113**, 257–82.

Shaw, S. R. (1978). The extracellular space and blood-eye barrier in an insect retina: An ultrastructural study. *Cell Tiss. Res.* **188**, 35–61.

Shaw, S. R. (1979). Signal transmission by graded slow potentials in the arthropod peripheral visual system. In *The Neurosciences: Fourth Study Program*, ed. F. O. Schmitt & F. G. Worden, pp. 275–95. Cambridge, Massachusetts: MIT Press.

Smith, T. G. & Baumann, F. (1969). The functional organisation within the ommatidia of the lateral eye of *Limulus*. *Prog. Brain Res.* **31**, 313–40.

Smith, T. G., Baumann, F. & Fuortes, M. G. F. (1965). Electrical connections between visual cells in the ommatidium of *Limulus*. *Science*, **147**, 1446–8.

Snyder, A. W., Laughlin, S. B. & Stavenga, D. G. (1977). Information capacity of eyes. *Vision Res.* **17**, 1163–75.

Strausfeld, N. J. (1976*a*). Mosaic organizations, layers, and visual pathways in the insect brain. In *Neural Principles in Vision*, ed. F. Zettler & R. Weiler, pp. 245–79. Berlin, Heidelberg, New York: Springer-Verlag.

Strausfeld, N. J. (1976*b*). *Atlas of an Insect Brain*. Berlin, Heidelberg, New York: Springer-Verlag.

Strausfeld, N. J. & Braitenberg, V. (1970). The compound eye of the fly (*Musca domestica*): connections between the cartridges of the lamina ganglionaris. *Z. Vergl. Physiol.* **70**, 95–104.

Strausfeld, N. J. & Campos-Ortega, J. A. (1972). Some interrelationships between the first and second synaptic region of the fly's (*Musca domestica* L.) visual system. In *Information Processing in the Visual System of Arthropods*, ed. R. Wehner, pp. 23–30. Berlin, Heidelberg, New York: Springer-Verlag.

Strausfeld, N. J. & Campos-Ortega, J. A. (1973). The L4 monopolar neurone: a substrate for lateral interaction in the visual system of the fly *Musca domestica* (L.). *Brain Res.* **59**, 97–117.

Strausfeld, N. J. & Campos-Ortega, J. A. (1977). Vision in insects: pathways possibly underlying neural adaptation and lateral inhibition. *Science*, **195**, 894–7.

Stuart, A. E. & Oertel, D. (1978). Neuronal properties underlying processing of visual information in the barnacle. *Nature, Lond.* **275**, 287–90.

Trujillo-Cenóz, O. (1965). Some aspects of the structural organisation of the intermediate retina of dipterans. *J. Ultrastr. Res.* **13**, 1–33.

Trujillo-Cenóz, O. & Melamed, J. (1966). Compound eye of dipterans: anatomical basis for integration – an electron microscope study. *J. Ultrastr. Res.* **16**, 395–8.

Trujillo-Cenóz, O. & Melamed, J. (1970). Light and electronmicroscope study of the one of the systems of centrifugal cells found in the lamina of muscoid flies. *Z. Zellforsch.* **110**, 336–49.

Trujillo-Cenóz, O. & Melamed, J. (1973). The development of the retina-lamina complex in muscoid flies. *J. Ultrastr. Res.* **42**, 554–81.

West, R. W. (1976). Light and electron microscopy of the ground squirrel retina: functional considerations. *J. Comp. Neurol.* **168**, 355–78.

Wilson, M. (1978*a*). The functional organisation of locust ocelli. *J. Comp. Physiol.* **124**, 297–316.

Wilson, M. (1978*b*). Generation of graded potential signals in the second order cells of locust ocellus. *J. Comp. Physiol.* **124**, 317–31.

Wilson, M. (1978*c*). The origin and properties of discrete hyperpolarising potentials in the second order cells of locust ocellus. *J. Comp. Physiol.* **128**, 347–58.

Wong, F. (1978). Nature of light-induced conductance changes in ventral photoreceptors of Limulus. *Nature, Lond.* **276**, 76–9.

Wood, M. R., Pfenninger, K. H. & Cohen, M. J. (1977). Two types of presynaptic configurations in insect central synapses: an ultrastructural analysis. *Brain Res.* **130**, 25–47.

Zeil, J. (1979). A new kind of neural superposition eye: the compound eye of male Biblionidae. *Nature, Lond.* **278**, 249–50.

Zettler, F. & Autrum, H. (1975). Chromatic properties of lateral inhibition in the eye of a fly. *J. Comp. Physiol.* **97**, 181–8.

Zettler, F. & Järvilehto, M. (1971). Decrement-free conduction of graded potentials along the axon of a monopolar neuron. *Z. Vergl. Physiol.* **75**, 402–21.

Zettler, F. & Järvilehto, M. (1972). Lateral inhibition in an insect eye. *Z. Vergl. Physiol.* **76**, 233–44.

Zettler, F. & Järvilehto, M. (1973): Active and passive axonal propagation of nonspike signals in the retina of *Calliphora. J. Comp. Physiol.* **85**, 89–104.

Zettler, F. & Weiler, R. (1976). Neuronal processing in the first optic neuropile of the compound eye of the fly. In *Neural Principles in Vision*, ed. F. Zettler & R. Weiler, pp. 227–37. Berlin, Heidelberg, New York: Springer-Verlag.

Zimmerman, R. P. (1978). Field potential analysis and the physiology of second-order neurones in the visual system of the fly. *J. Comp. Physiol.* **126**, 297–316.

IAN J. RUSSELL

The responses of vertebrate hair cells to mechanical stimulation

Introduction

For the purposes of this book the mechanoreceptive hair cells of the acoustico lateralis system have been given neuronal status. This is not altogether unjustified. They share a common embryological origin (in the auditory placode), with the second-order ganglion cells, which form afferent synapses with them, and they receive an efferent innervation which is integrated with the non-neural sensory input. Moreover, under certain physiological conditions action potentials have been elicited from them (Hudspeth & Corey, 1977; Fig. 7*d*). Thus they may be classified together with the neurones which form the subject of this book, in that their ability to produce impulses is facultative rather than obligative.

In some respects hair cells perform similar tasks to other 'neurones without impulses'. In particular they transduce and relay information to synapses in analogue form, without intermediate digital conversion. Regenerative action potentials are not an essential step in the conduction of the electrical signal over the surface of the hair cell, since the distances are short. Furthermore the isolation of the transducer membrane from rapid fluctuations in membrane conductance preserves the fidelity and frequency response of the cell and its moment-to-moment sensitivity.

This review is not an attempt to draw comparisons between the properties of hair cells and neurones without impulses, since any similarities which do exist are probably fortuitous and not part of a grand design. Instead the review will deal with the functional morphology of hair cells and their remarkable response properties, which would be largely degraded if they generated action potentials.

The basic morphology of hair cells

It is apparent, from a glance at Figs. 1, 2 and 3, that hair cells from different sensory epithelia in the acoustico-lateralis system vary in their form, the length and number of their ciliary processes and in their innervation. However they are all recognisable as hair cells in their conformation to a common design. They may be cylindrical or flask-shaped, and from their apical surfaces protrude two types of ciliary process, namely a group of stereocilia and a single kinocilium, lying to one side. The latter is a true cilium with an inner structure of nine double microtubules arranged around a core of two central tubules (Fig. 4). The outer tubule connects to triplet tubules comprising the wall of the basal body. The kinocilium may act as a focus for the development of the stereocilia which decrease stepwise in length with increasing distance from the kinocilium. The hair cells of the mammalian

Fig. 1. Schematic drawing showing the ultrastructure of the sensory epithelium of a lateral-line neuromast from the tail of a juvenile *Ambystoma mexicanum*. From Russell, 1976.

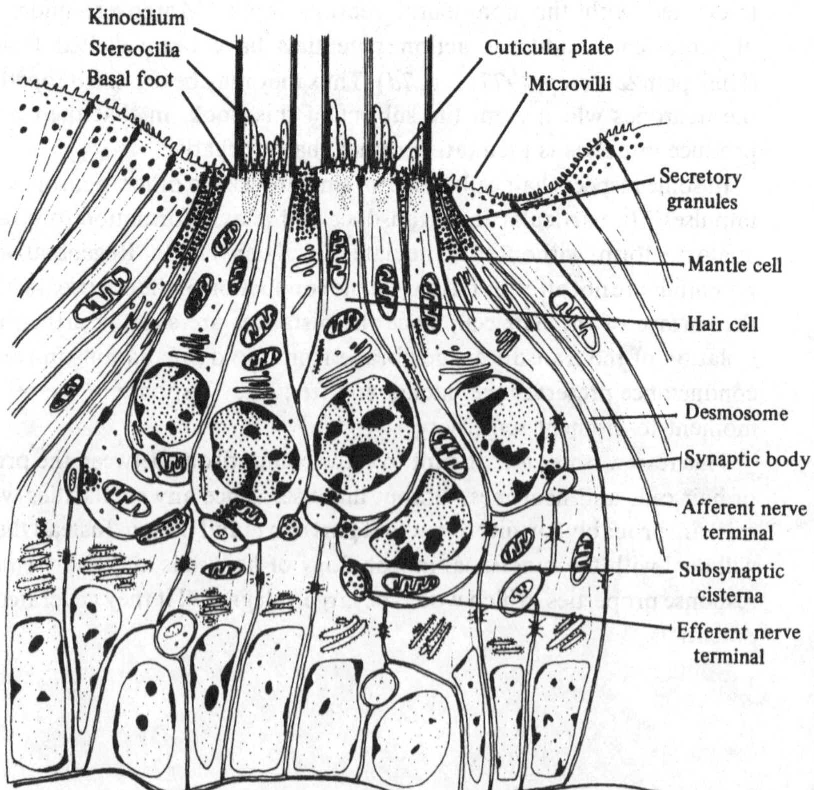

cochlea are slightly different. The kinocilia are present early during development (Smith, 1975) but disappear in the adult. The stereocilia, on the other hand, are not true cilia but microvilli with rootlets at their base which project into a tough, electron-dense structure, the cuticular plate. This structure occupies most of the apical region of the hair cell except for a region surrounding the basal body (Fig. 4).

Hair cells have been classified on the basis of their afferent innervation. Amphora-shaped cells in the vestibular systems of reptiles (Jørgensen, 1974), birds (Jørgensen & Anderson, 1972), and mammals (Wersäll, 1956) are enclosed up to their necks by a large nerve chalice which may envelope several

Fig. 2. Hair cells from sensory epithelium in the vestibular system of the pigeon. The striola are seen in cross-section. Four hair cells of type II with differing polarisation are seen and two hair cells of type I. Hc I, hair cell type I; Hc II, hair cell type II; NEn, afferent nerve ending; NEu, efferent nerve ending; Sc, supporting cell. From Jørgensen, 1970.

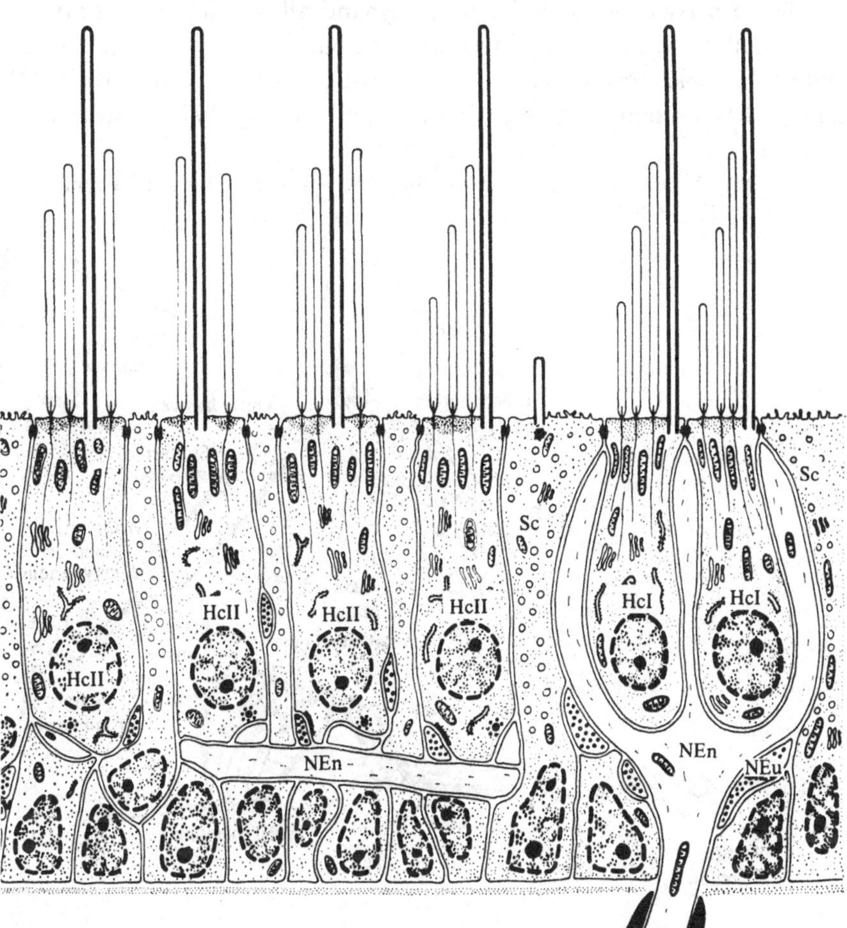

cells (Fig. 2). These have been classified as type I hair cells by Wersäll (1956), in contrast to the more common cylindrical or flask-shaped type II hair cells which are innervated by several small afferent endings.

The afferent nerve endings may be identified by a presynaptic body surrounded by vesicles resting on a thickened presynaptic plasma membrane, and by an absense of vesicles in the post-synaptic afferent terminal. The presynaptic body varies in form, being rod-like in vestibular hair cells of higher vertebrate, but spherical in the hair cells of the lateral line system. However, presynaptic bodies are absent from outer hair cells in the cochlea of the cat and monkey and occasional, very small presynaptic bodies are found in outer hair cells in the guinea-pig (Rodriquez-Echandia, 1967; Spoendlin, 1970; Bodian, 1978). Presynaptic bars are also associated with the nerve chalices of type II hair cells. Adjacent to these apparent chemical synapses are points of close apposition between pre- and post-synaptic membranes, which Spoendlin (1966) has suggested might be electrical synapses.

Efferent nerve endings are found (presynaptically) on hair cells, or (post-synaptically) on their afferent innervation throughout the acoustico-lateralis system. Notable exceptions are the amphibian papilla (Flock & Flock, 1966) and the basillar papilla in some lizard species (Mulroy, 1974; Baird, 1974).

Fig. 3. Drawing of the cochlear duct, second turn, guinea-pig. From Smith, 1975.

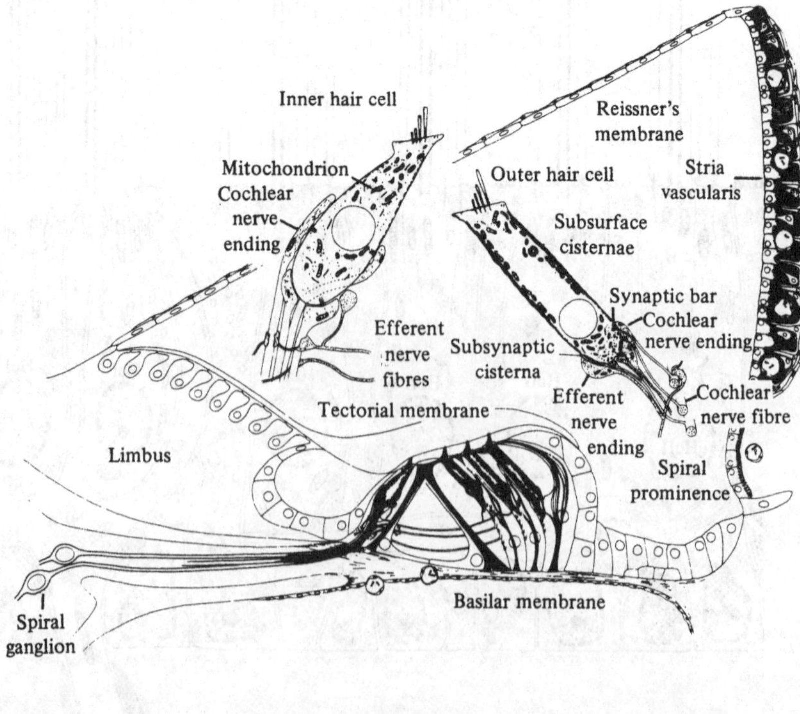

The efferent terminals are characterised by an abundance of vesicles, and by the presence of a large subsynaptic membraneous sac. This may extend to the cuticular plate in outer hair cells of the cochlea. The efferent innervation of these hair cells is so massive, relative to their afferent innervation, that the hair cells have been termed the 'effector organs of the olivo cochlear bundle' by Davis (1978).

The properties of the sensory hair bundles

The adequate stimulus for all hair cells of the acoustic-lateralis system is a shearing displacement of the sensory hair bundles (Flock, 1971, gives a review). So, as Pumphrey (1950) first pointed out, the sensory modalities of different acoustico-lateralis sense organs may be determined largely by the properties of accessory structures which are associated with the different sensory epithelia and the way in which these are coupled to the sensory hair bundles.

In acoustico-lateralis receptors, whose frequency response and modality appears to be dominated by the mechanical properties of the accessory structure, (e.g. lateral line, vestibular system, otolith receptors), the cupular

Fig. 4. Schematic drawing illustrating the ultrastructure of the stereocilium (left) with its rootlet and the kinocilium (right) with its basal body. The arrow indicates the direction of excitatory stimulation. From Flock, 1965.

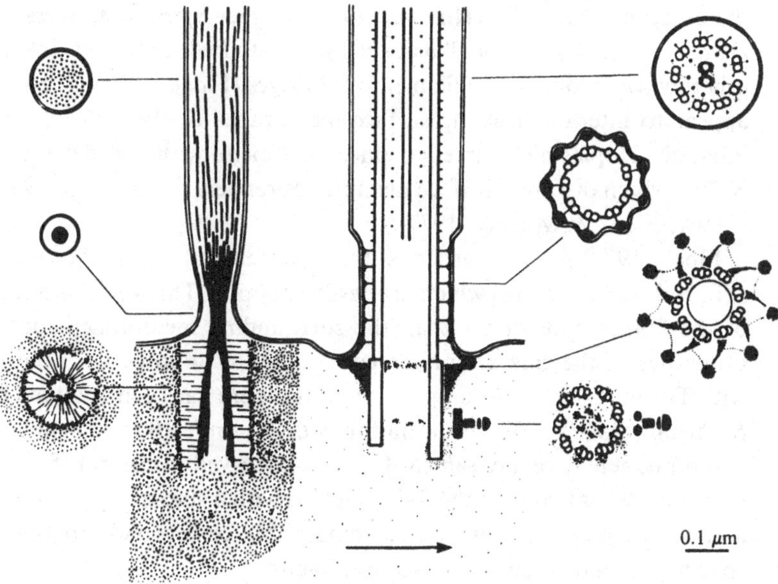

or otolithic membrane is tightly coupled either to the sensory hair bundle (Flock, 1965) or to the kinocilium which is usually well developed in these receptors (Hillman, 1972; Lewis & Li, 1975). In auditory receptors, like the lagena of reptiles and birds, or the mammalian cochlea, where the frequency responses of the hair cells are more sharply tuned than the mechanics of the basilar membrane, upon which they rest (Klinke, 1978; Weiss, Peake, Ling & Holton, 1978; Russell & Sellick, 1978; Sellick & Russell, 1978), the kinocilium has either regressed, or is absent (Mulroy, 1974; Baird, 1974; Smith, 1968), and the overlying tectorial membrane, if present, is supported by the tips of the stereocilia. In attempting to understand the significance of the considerable morphological variation which exists in the size, number and arrangement of sensory hair processes, recent work has concentrated on the mechanical properties of sensory hairs and their motion in response to mechanical disturbance.

Direct microscopic observation of hair cells (Hudspeth & Corey, 1977; Flock, 1977; Flock, Flock & Murray, 1977) reveals that the sensory hairs move as a bundle in response to mechanical motion, and that the movement of only a few cilia may be transmitted to the rest. For example, only the tallest of the stereocilia in cochlear hair cells and the single large kinocilium of hair cells in the crista ampularis (Hillman, 1972) are coupled to the overlying tectorial membrane and cupula respectively. Filamentous connections between sensory hairs transmit movements of the overlying structures to adjacent processes. These connections appear to be particularly well developed between the tops of the club-like kinocilia of some vestibular hair cells and the adjacent stereocilia (Hillman, 1972; Bagger-Sjöbäck & Wersäll, 1973). However, sensory cilia of all hair cells so far studied seem to be linked to each other by triplet or doublet filamentous bridges (Flock, 1977). These filaments appear to inter-connect the surface membranes of the cilia, and not their internal cytoplasmic filaments, since treatment with the detergent Triton X-100, which dissolves surface membranes renders the cilia readily separable (Flock, Flock & Murray, 1977).

Flock (1977) has also observed that the stereocilia of hair cells are rather stiff, rod-like structures which are easily snapped. This stiffness is a property of the fibrilla core rather than the surrounding membrane, since the stiff properties of the cilia are preserved after the membrane has been removed with Triton X-100. The composition of this core has been analysed by Flock & Cheung (1977). After denaturation with Triton X-100 and incubation with a subfragment-1 of myosin (S–11) they observed a chevron formation on filaments decorated with the S–11 fraction with a spatial separation of 35 nm and apices pointing toward the soma. This corresponds to the expected spacing of binding sites of myosin on actin.

There are further hints that active, mobile processes may be involved in sensory transduction. The ultrastructure of stereocilia in hair cells bears a close similarity to microvilli in cells of the intestinal mucosa (Moosekar & Tilney, 1975) which are mobile. Movements of the microvilli are caused by the sliding of myosin filaments attached to the terminal web (equivalent to cuticular plate). Moreover changes in the packing and attachment between the actin molecules of the microvilli induced by physiological reactions can alter their mechanical properties (Moosekar, 1975). Similar changes in the properties of stereocilia may account for some of the responses of the hair cells to alterations in their physiological environment.

The response properties of hair cells

The receptor potential is the causative link between the sensory input to the hair cell, a shear displacement of the ciliary bundle, and impulse initiation in the afferent fibres. It thus has the attribute of reflecting the remarkable response properties of hair cells.

Directional Sensitivity of Hair Cells

Extracellular receptor potentials have been recorded from the sensory epithelium of every type of acoustico-lateralis receptor. They resemble an electrical analogue of the mechanical stimulus delivered to the sensory epithelium over a vast frequency range from d.c. in otolith receptors to frequencies in excess of 150 KHz in the auditory systems of cetacians (Sales & Pye, 1974).

In the cochlea and cristae of the semicircular canals, only the fundamental frequency of the stimulus waveform is strongly represented in the extracellularly recorded receptor potential. In contrast, sinusoidal displacements of cupulae in lateral-line receptors and otolithic organs, for example, result in microphonic potentials which are twice the frequency of the applied stimulus (Fig. 5*a*). The important morphological differences between these receptors, and those which produce receptor potentials with a predominant single harmonic component is that the former have two populations of hair cells which are morphologically polarised in opposite directions, while the latter have a single population with a common polarity.

In an attempt to explain the appearance of the second harmonic component in extracellularly recorded receptor potentials from the lateral line organs of *Lota*, Flock & Wersäll (1962) proposed that the extracellular potential represents the summed activity of the oppositely polarised cells. They further predicted that displacement of the sensory hair bundles in one direction produced a depolarisation which was greater in magnitude than the hyperpolarisation produced by a corresponding but opposite displacement. The algebraic sum of extracellular potentials from oppositely polarised hair cells with these characteristics will be a potential with twice the frequency of the mechanical stimulus. The intracellular receptor potentials recorded from lateral-line hair cells are in fact asymmetrical, with the depolarising phase exceeding the hyperpolarising phase in amplitude (Fig. 5*b*).

Potentials were first recorded from hair cells in the superficial lateral-line receptors of the mudpuppy *Necturus* by Harris, Frischkopf & Flock (1970). The receptor potentials are small, maximally 0.8 mV, and in this respect they resemble receptor potentials which have subsequently been recorded from lateral line hair cells in other species (Flock, Jørgensen & Russell, 1972; Sand, Ozawa & Hagiwara, 1975). There is a very pleasing correlation between the morphological asymmetry of the hair cells and their directional response to

Fig. 5. *a*. Averaged extracellular receptor potential recorded from a lateral-line canal organ in *Acerina cernua*, the ruff, in response to sinusoidal oscillation of the cupula (lower trace). Downward deflections of the trace are depolarizations of the sensory epithelium. CM, extracellular microphonic potential; SP, extracellular summating potential. Frequency of vibration, 100 Hz; amplitude, 0.4 μm. Stimuli are presented at 600 ms intervals. Traces are averaged from 16 sweeps at 2056 counts per address. *b*. Averaged intracellularly recorded receptor potential from a hair cell in a lateral-line organ in *Lota lota*, the Burbot, in response to eight cycles of sinusoidal displacement of the cupula at 70 Hz. Traces are averaged from 10 sweeps and band pass filtered between 30 and 600 Hz. From Flock and Russell, 1976.

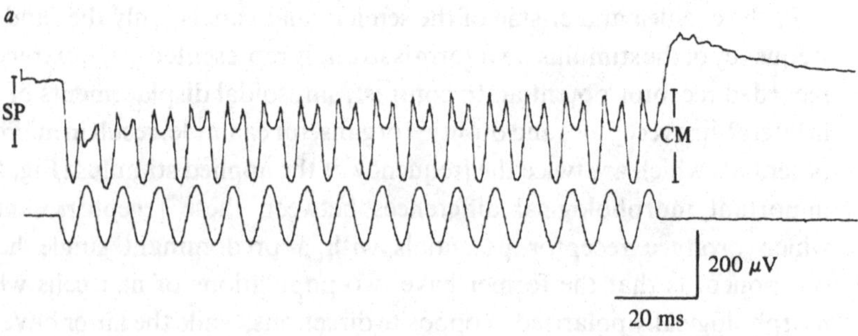

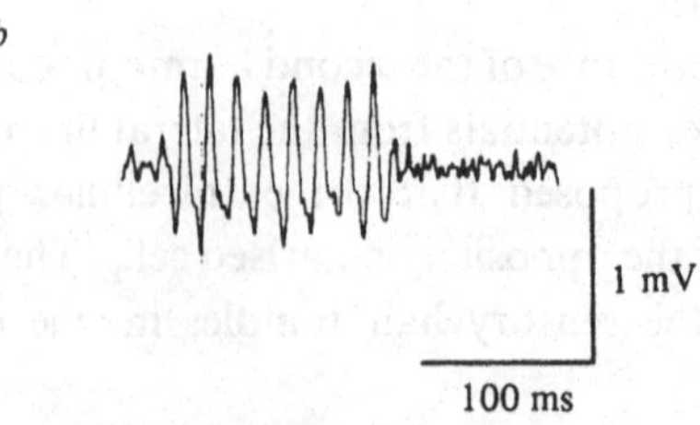

displacement. Within the same neuromast, cells fall into two groups with responses 180° out of phase. If one cell is, for instance, depolarised by a cranial displacement of the cupula, then a neighbouring cell is hyperpolarised by the same stimulus. Electron microscopy also reveals two groups of cells with kinocilia oriented in opposite directions.

Fig. 6. Schematic cutaway of the experimental preparation from bullfrog sacculus. Hair cells (HC) and supporting cells (SC) form an epithelial sheet supported by a layer of connective tissue. The otolithic membrane (OM), which normally couples stimuli to the hair bundles (HB) of the receptor cells, is shown partially dissected from the site of experimentation. While the intracellular potential is recorded through a glass microelectrode (ME), a hair cell is stimulated by a capillary stimulus probe (SP) slipped over the tip of its hair bundle and moved parallel to the epithelial surface (arrows). From Hudspeth & Corey, 1977.

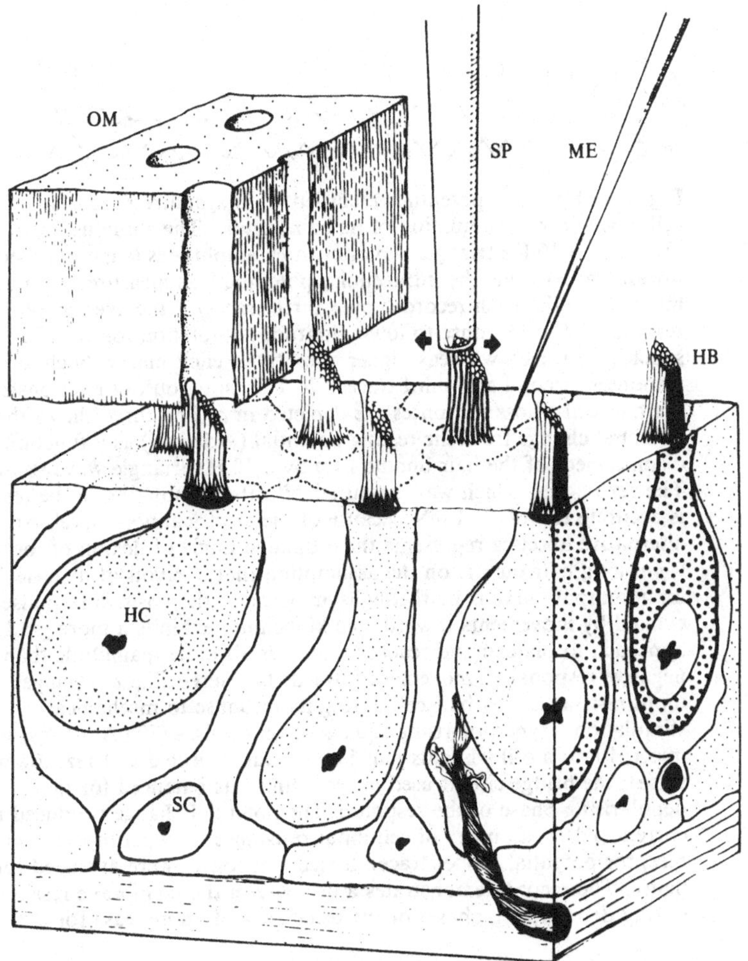

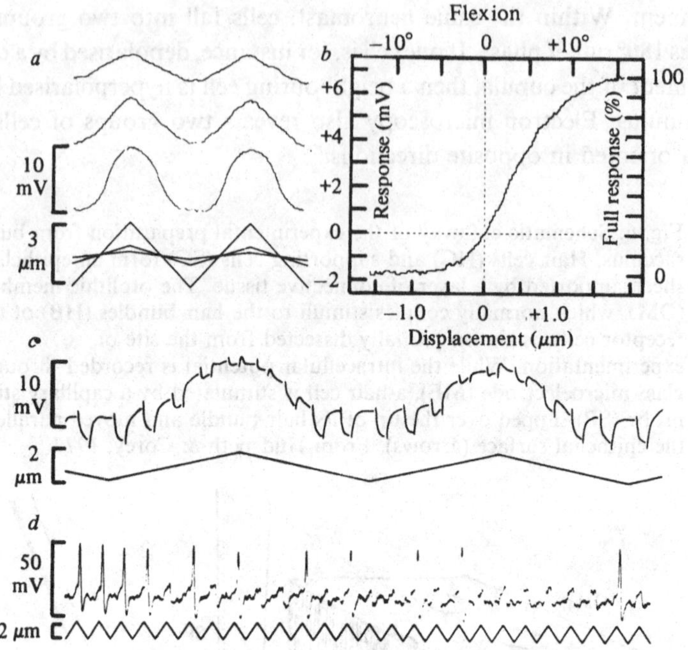

Fig. 7. *a.* Receptor potentials recorded from a bullfrog saccular hair cell during direct simulation of its hair bundle. The stimulus probe was driven with 10 Hz triangle waves of three amplitudes (superimposed lowest traces) while the intracellular potential was measured with a microelectrode; each record (upper three traces) is the average of 32 responses. The response to low-amplitude stimulation follows the driving signal (top trace), whereas higher amplitudes elicit markedly clipped responses (second and third traces). *b.* The input–output relationship for the hair cell whose responses are depicted in *a.* The curve shows the potential change from the resting potential (−58 mV) as a function of the displacement of the hair bundle's tip by a 10 Hz triangle wave stimulus. The zero point, which was measured with the stimulating probe removed, is accurate to only ±1 mV because of the electrical noise level. An alternative abscissa represents the estimated angle of flexion of the 9-μm-long hair bundle on the assumption that it pivots at its base. The input–output curve is markedly asymmetrical: the positive response evoked by movements towards the kinocilium exhibits a more gradual approach to plateau and reaches a greater absolute magnitude than the negative response of movement towards the stereocilia. *c.* Changes in the input resistance of a hair cell during its response to mechanical stimulation. Hyperpolarising square current pulses (70 pA, 45 ms) were injected into a cell while its hair bundle was deflected at 1 Hz (lower trace); the bridge circuit used in recording was balanced for the depolarising phase of the response. The potential change produced by the current pulses, an index of cell input resistance, is superimposed upon the receptor potential (upper trace). Sixteen responses were averaged for this figure. The record demonstrates a decrease in the membrane resistance during depolarising phases of the response and an increase for

The elegant experiments by Hudspeth & Corey (1977) on bullfrog saccular hair cells extended these earlier observations and lead to a direct correlation between the directional sensitivity of hair cells and their morphological polarisation. They watched the hair cells while applying controlled displacements to the sensory hair bundles and making intracellular recordings (Fig. 6). The receptor potentials were much larger than any previously reported (5–15 mV), and they were asymmetrical. A 0.5 μm displacement of the stereocilia towards the kinocilium yielded large depolarising potential changes, while the same amplitude displacement in the opposite direction caused a much smaller hyperpolarisation (Fig. 7). Displacement at right angles to the excitatory–inhibitory axis elicited no receptor potentials. From these data they were able to derive the transfer characteristics of the hair cell illustrated in Fig. 7*b*. Furthermore, deflection of the sensory hair bundle in the excitatory direction was associated with a conductance increase, while a decrease was associated with the hyperpolarisation (Fig. 7*c*). Thus, in common with other primary receptors from which receptor potentials have been measured, the potentials are associated with stimulus induced conductance changes in ion channels. Similar conductance changes associated with the receptor potential have also been recorded in hair cells of the mammalian cochlea (Russell & Sellick, 1978).

The site of transduction

According to a theory by Davis (1957) the voltage which provides the driving force for current flow across the apical membranes of hair cells in the mammalian cochlea is the different between a positive electrogenic potential of about +80 mV in the scala media outside the hair cell and the hair-cell resting potential of about −40 mV (Russell & Sellick, 1977, 1978). Current flow into the hair cells is modulated by changes in K+ conductance of the apical membrane to provide the receptor potential.

A similar battery exists across the apical membrane of hair cells in the superficial lateral line organs in the skin of aquatic amphibia. Measurement of potential, K^+, Cl^- and Ca^{2+} with ion-selective electrodes from the cupulae of lateral-line organs in *Xenopus laevis* (Russell & Sellick, 1976; McGlone, Russell & Sand, 1979) show that the cupulae maintain a K^+-rich micro-

---

hyperpolarising responses. *d*. Action potentials arising from the depolarising phases of receptor potentials in a saccular hair cell. The cell, which has a resting potential of −65 mV, was stimulated with a 10-Hz triangle wave motion (lower trace) applied to its hair bundle. Note that the action potential occasionally fails, leaving a pure receptor potential. From Hudspeth & Corey, 1977.

environment (about 100 mM KCl), and a positive endocupular potential (about 50 mV), adjacent to the outer surfaces of the hair cells. This potential, like the endocochlear potential recorded in the scala media of the mammalian cochlea (Johnstone & Sellick, 1972), is electrogenic in nature. It is not influenced by 100-fold changes in the K^+ level of the aquatic environment of the cupulae, and is drastically reduced by subcutaneous injections of the cardiac glycocide, ouabain. Since extracellular levels of K^+ at the apical surfaces of hair cells closely resemble their intracellular concentrations then, according to Davis' model, the reversal potential for the hair-cell receptor potential should be close to the endocupular potential. Intracellular recordings have been made from hair cells in the lateral organs in *Xenopus*, but they have been made from hair cells in the lateral-line organs of *Ambystoma mexicanum* where the endocupular potential reaches a maximum level of +20 mV (Russell & Sellick, unpublished results). The resting potentials of their hair cells are about −35 mV, and their receptor potentials reach a maximum of about 1 mV. These were reversed in nine cells when the potentials were depolarised to between +7 and +15 mV by intracellular current injection of about 0.1 nA through the single recording electrode. The close correspondence between the reversal potential for the hair-cell receptor potential, and the endocupular potential indicates that in these hair cells, at least, transduction is associated with an inflow of K^+ ions across the apical membrane and thus tends to substantiate Davis' theory.

For many years the kinocilium has been favoured as the site of the transduction process (Flock & Duvall, 1965; Flock, 1971; Hillman, 1972), principally because cilia appear to be the locus of transduction in many sensory cells including photoreceptors, olfactory receptors and invertebrate mechanoreceptors (see Thurm, 1974; Wiederhold, 1976 for reviews). The kinocilium is the focus of a hypothesis proposed by Hillman (1972) for transduction in hair cells. This was based on his observations with the electron microscope of hair cells in frog otolith organs, and the changes which were produced in the arrangement of sensory hairs when the otolithic membranes were mechanically displaced prior to fixation. The scheme is that the large cuticular plate is a relatively rigid structure supporting the stereocilia. Kinocilia are attached to adjacent stereocilia near their apical ends and are also linked to otolithic membranes or cupulae. When the stereocilia are displaced towards the kinocilium it is deformed and thrust into the cytoplasm. This results in a displacement of the cytoplasm adjacent to the kinocilium, and a bulging out of the plasma membrane which causes a change in its leakiness to the existing polarising current.

This theory of hair cell transduction is somewhat limited in that it relies upon morphological features of vestibular hair cells in the frog which are not found elsewhere in the acoustico-lateralis system. A wider variety of distortions of the apical processes of hair cells was observed by Flock *et al.* (1977) in living preparations and these are summarised in Fig. 8. However very recent evidence by Hudspeth & Jacobs (1979) indicates that in hair cells of the bullfrog sacculus, the kinocilium is not essential for transduction. Hudspeth & Jacobs made intracellular recordings from the hair cells, while mechanically stimulating portions of the sensory hair bundles (Fig. 9). Receptor potentials were not elicited when the kinocilium was isolated from the bundle of stereocilia and stimulated by itself, even when it was displaced by more than 3 μm, a distance considerably greater than the operating range of the cell. In contrast, deflection of the stereocilia bundle towards the position of the kinocilium produces depolarising receptor potentials. Maximum potentials of about 13 mV were evoked when the bundle was displaced about 0.5 μm, which corresponds to the maximum range of receptor potentials in intact cells (Fig. 7*a*). Smaller potentials were elicited from similar displacements of small bundles of stereocilia. Clusters of as few as ten stereocilia gave a response of about 2 mV.

Hudspeth & Jacobs (1979) have clearly indicated that an intact kinocilium is not necessary for transduction in vertebrate hair cells, and that this process is mediated by the stereocilia. They suggest that the role of the kinocilium in transduction may be limited to transmitting movements from cupulae or otolithic membranes to the stereocilia. The various forms that kinocilia assume (Lewis & Li, 1975) would then represent adaptations which are optimal for transmitting particular forms of stimulation to a relatively standardised transducer associated with the stereocilia. The loss of the kinocilium from cochlear hair cells, in mammals, and their regression in

Fig. 8. Theoretical modes of motion for sensory hairs. 0, resting position; 1, bending of flexible sensory hairs; 2, parallel shift when they are stiff and rigidly attached to their neighbours; 3, pivoting around their base when they are stiff; 4, separation of sensory hairs due to weak lateral binding. From Flock *et al.*, 1977.

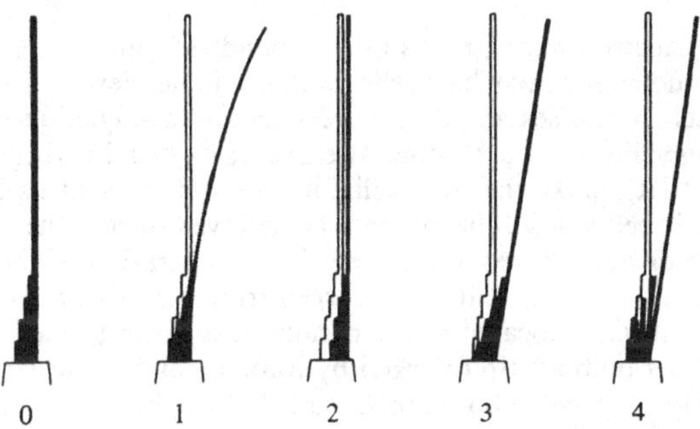

Fig. 9. Scanning electron micrographs of hair bundles from physiologically studied saccular hair cells. *a*. An oblique view of a normal hair bundle, which comprises about 50 stereocilia and a single kinocilium. Note that the kinocilium has a bulbous swelling at its distal end (marked by triangle) and that, unlike the stereocilia, it does not taper at its base. × 12000. *b*. A hair cell whose kinocilium was removed during the experiment; little or no stub remains at the site of separation of the cilium. × 12000. *c*. la normal hair bundle, seen from directly above. The kinocilium is eccentrically located in the bundle; it is distinguished from the stereocilia by its bulbous tip (marked by triangle) and its larger basal diameter. × 10000. *d*. A cell whose kinocilium (bulb at black triangle)

cochleae of birds, may serve to remove elastic linkage and permit improved responsiveness to high frequency stimuli.

Thus the type of transduction model for vertebrate hair cells proposed by Engström, Ades & Anderson (1966) and Hillman (1972), where the site of transduction is located at the base of the kinocilium, or the basal body (the proposed site of transduction in invertebrate mechanoreceptors (Thurm, 1974)), may have to be abandoned, unless the special conditions of the isolated saccular macula in Hudspeth & Jacob's experiments have somehow masked this site of transduction. It will be of interest to discover if transduction can be performed by only one stereocilium or if several are required.

The mechanosensitivity of hair cells

The sensitivity of hair cells is a difficult and variable quantity to measure. The most direct measurements are those made by Hudspeth & Corey (1977). They estimate, from the slope of input–output curves, that the hair cells of the isolated frog sacculus have a response of 20 mV per micron of displacement of the stereocilia (Fig. 7). They suggest further, that if threshold for neural excitation corresponds to a 10 μV change in receptor potential, as it does in photoreceptors (Fain, Granda & Maxwell, 1977), and electroreceptors (Bennett, 1978), then the threshold sensitivity of the saccular hair cells in 5Å.

This latter assumption is based on a variable quantity, namely the relationship between pre- and post-synaptic potentials at a synapse. For example, important differences exist in this relation between synapses of some electroreceptors and the classical squid giant synapse (Bennett, 1976, 1978). In the former there is a shift along the presynaptic voltage axis such that transmitter is being realeased in the absence of external stimuli, which considerably increases the sensitivity of the synapse (the slope of the input–output function). In the mammalian cochlea, neural threshold is associated with a 2 mV change in receptor potential (Russell & Sellick, 1978). If this is related to von Bekesy's (1960) measurement of basilar membrane vibration in the transverse plain at absolute threshold of hearing 10^{-12} m, then

was dissected free and held flat against the epithelial surface during electrophysiological recording. The cell gave normal responses to stimuli. × 10000. *e.* Hair bundle of a cell lacking a kinocilium; this cell also responded normally. × 10000. *f.* Apex of a cell with a bifid hair bundle. One component of the bundle lacks a kinocilium; the other portion has a normal kinocilium with a terminal bulb (marked by triangle). Mechanical stimulation of either component elicited responses. × 10000. From Hudspeth & Jacobs (1979).

hair cells in the mammalian cochlea have responses of about 9 mV per A° of displacement at centre frequency. However, there must be considerable doubt about the validity of these extrapolations from von Bekesy's and whether they bear any resemblance to the dimensions of shear displacement which actually occur at the apical surface of cochlear hair cells.

Quite clearly there is variation in the sensitivity of synapses in acoustico-lateralis hair cells since the threshold receptor potential for neural excitation in the mammalian cochlea (2 mV) exceeds the maximum receptor potentials recorded from hair cells in other receptors (Harris *et al.*, 1970; Flock *et al.*, 1973; Sand *et al.*, 1975). It may be that sensitivity in acoustico-lateralis systems may be achieved either by maximising the receptor potential in response to mechanical stimulation, or by raising the sensitivity of the synapse.

The responses of hair cells to high frequency sounds

The physiology of hair cells in vertebrate auditory receptors has attracted considerable interest, and the physiological basis for their frequency-

Fig. 10. Intracellular receptor potentials from a hair cell with a resting potential of -35 mV recorded in the 16 kHz region of the basilar membrane. *A–D*, responses to 300 Hz tone 90–50 dB in 10 dB steps; *E*, response to 3 kHz at 80 dB. Sound pressure levels in dB *re* 2×10^{-5} Nm^{-2}.

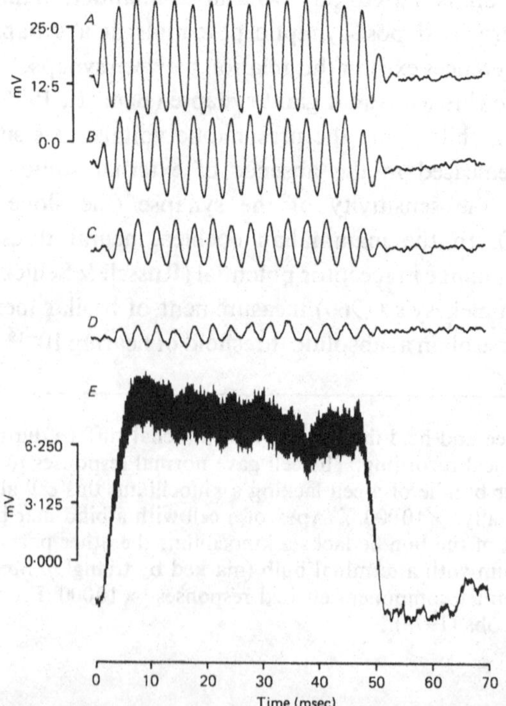

selective voltage responses has been a subject of recent investigation in mammals (Russell & Sellick, 1977, 1978; Sellick & Russell, 1978, 1979) and reptiles (Crawfurd & Fettiplace, 1978; Fettiplace & Crawfurd, 1978; Weiss *et al.*, 1978). However, of more relevance here is the fact that cochlear hair cells are capable of responding to high frequencies of acoustic stimulation. Notable examples are those in the cochleae of bats, Cetacea and other animals which respond to ultrasound. It is possible that the same transduction mechanism is capable of serving over the entire frequency range of acoustico-lateralis system from d.c. to ultrasonic frequencies; if so, what type of process are involved in hair cell transduction? These subjects have received an experimental approach in a recent study of response latency in receptor potentials recorded from an in vitro preparation of the sensory epithelium of the bullfrog by Corey & Hudspeth (1979).

They measured the response latency by applying a step mechanical

Fig. 11. The relation between the ratio a.c./d.c. components of the intracellular receptor potential for constant SPL of 70 dB versus frequency recorded from a cochlear hair cell. The curve is attenuated at 6 dB/Octave (thin line) above a cut-off frequency (arrow) between 200 and 300 Hz which corresponds to the electrical time constant of the hair cell.

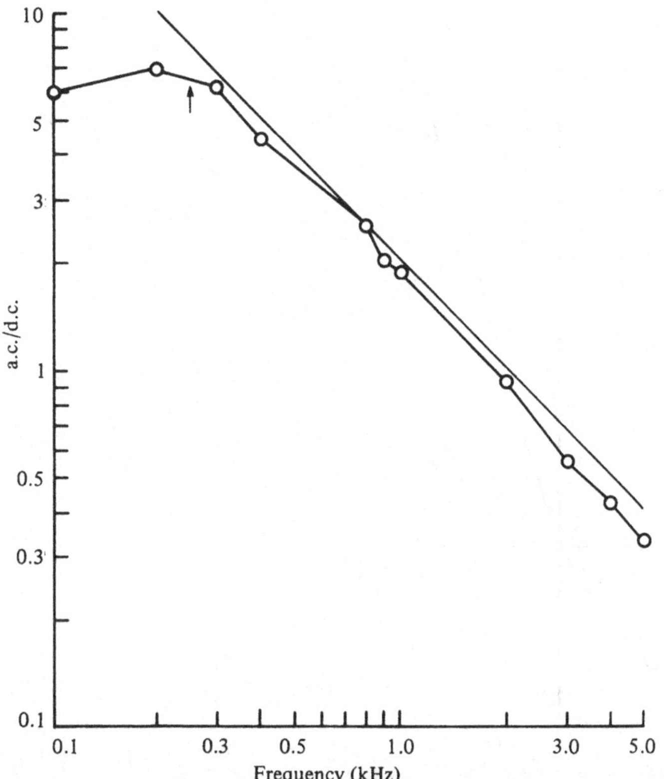

displacement of 0.5 μm amplitude and 150 μs duration to the otolithic membrane overlying the central group of hair cells in the sacculus, and accurately monitoring the displacement and microphonic potential. The time constant of the epithelium was taken into account, and with this correction, the latency in stimulus-response of the microphonic potential was found to be 40 ± 5 μs at 22 °C and 13 μs at 37 °C. This latency is quite adequate to deal with the responses of hair cells in the cochlea of Cetacia and bats, let alone the modest frequency response of the bullfrog sacculus. Therefore, it is feasable to consider a common transduction mechanism operating throughout the large frequency range over which different hair cells operate.

The latency is two orders of magnitude less than that measured for vertebrate photoreceptors (Brown & Murakami, 1964), and if a second

Fig. 12. Operating characteristic of a cochlear hair cell measured in response to a 200 Hz tone at 100 dB sound pressure level (SPL). The ordinate represents the amplitude of the receptor potential in millivolts, and the abscissa is the sinusoidal variation in pressure at the auditory meatus. The total excusion between $\sin\theta = \pm 1.0$ corresponds to basilar membrane displacement of about ± 1.0 μm at 200 Hz if it is assumed that this displacement is 10^{-12} m at 0 dB SPL (von Bekesy, 1960).

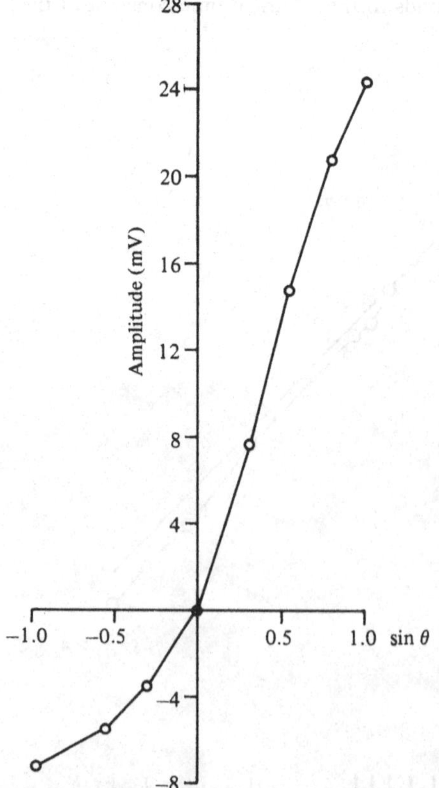

messenger is involved in hair-cell transduction (as it seems to be in vertebrate photoreception (Hagins, 1972)), the last mechanical stage must be close to the ionic conductance element to take into account the short diffusion paths permissible.

The Q_{10} of the latency over a range 38°–1° is 2.5, and this is larger than might be expected for a diffusion process ($Q_{10} = 1.2$). Therefore, diffusion probably accounts for only a small fraction of the latency. However, moderate temperature dependence and short delay argue against mechanisms of greater complexity, e.g. phosphorylation of membrane proteins (Greengard, 1976; Hartzell, Kuffler, Stickgold & Yoskikami, 1977).

Corey & Hudspeth suggest that the most likely mechanism for a rate-limiting step involves transition kinetics of the conduction element. The delay,

Fig. 13. *a*. Four successive traces of spontaneous EPSPs recorded intracellularly from a lateral-line nerve terminal. Vertical bar, 2 mV: time bar, 5 ms. *b*. The time delay between the externally recorded EPSPs in an afferent nerve terminal of a lateral-line organ which was mechanically stimulated at 100 Hz and at 40 dB above the threshold for excitation. Upper trace, externally recorded double microphonic potential; Lower trace, intracellularly recorded EPSPs. The traces are computor averages of 20 consecutive sweeps. The time delay between the peak of the excitatory phase of the microphonic potential and the peak of the EPSPs is indicated by d. Time bar, 5 ms. From Flock & Russell, 1976.

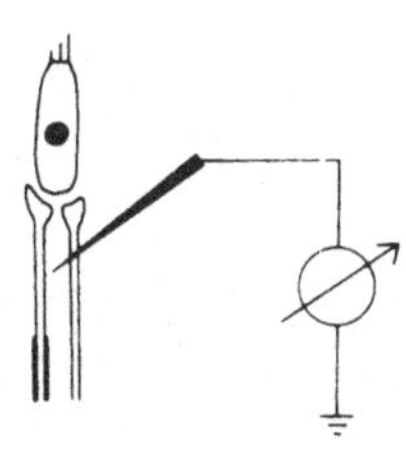

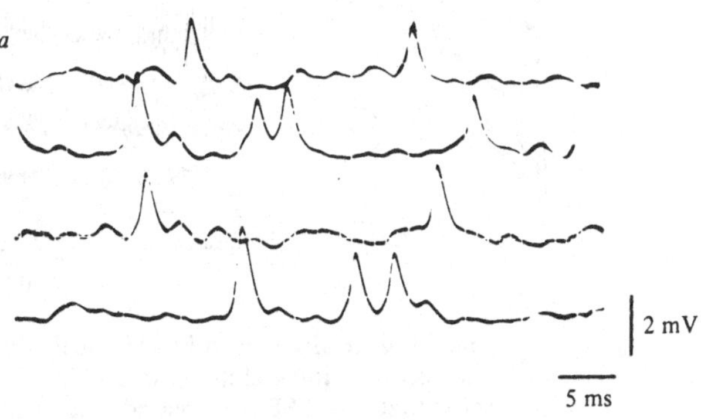

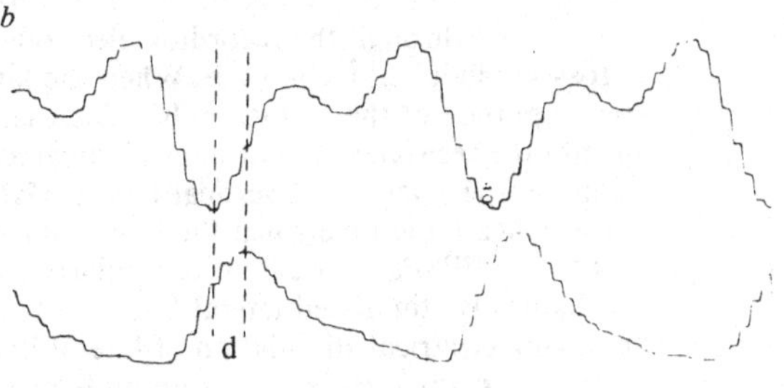

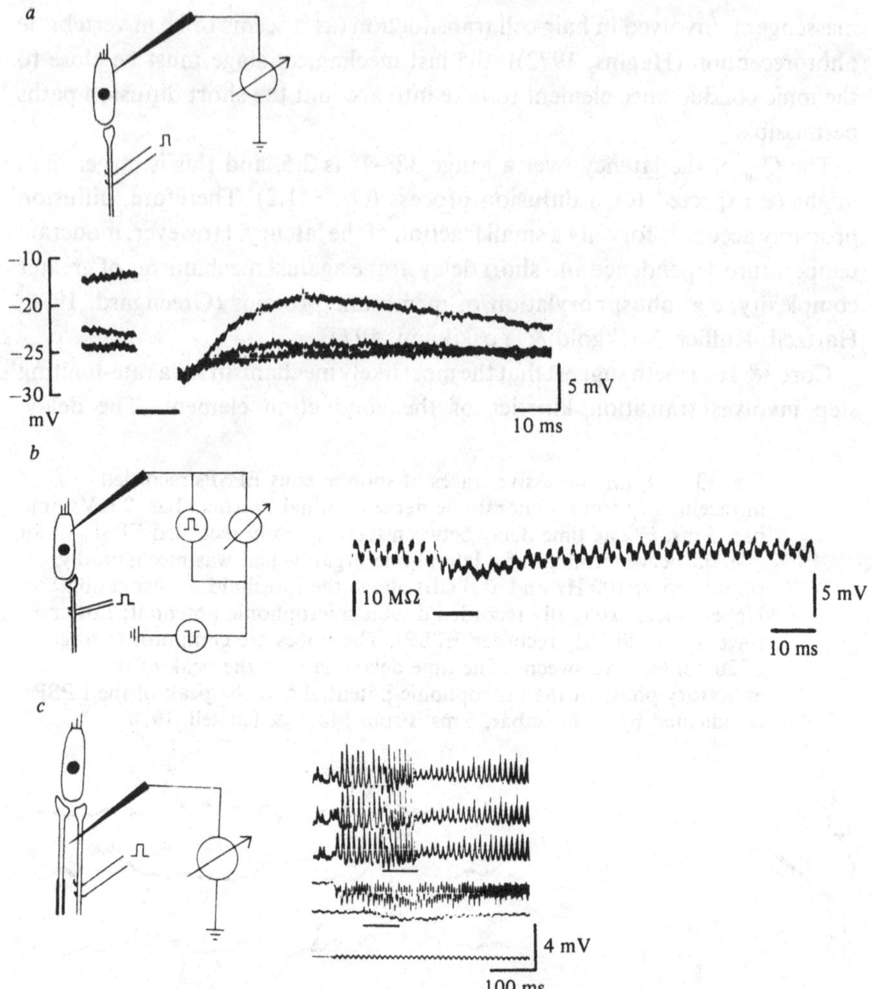

Fig. 14. *a*. IPSPs evoked by electrical stimulation of the lateral-line nerve and recorded intracellularly in a hair cell. The three successive traces show that the IPSP is increased when the membrane is depolarised. Horizontal and vertical bars, 10 ms and 5 mV respectively. *b*, Impedance change in a hair cell during the IPSP. Constant current pulses of 0.4 nA are injected through the recording electrode to produce voltage pulses across the hair-cell membrane. When the lateral-line nerve is stimulated the amplitude of the voltage pulses decreases indicating a decrease in membrane resistance. Right hand vertical and horizontal bar, 5 ms and 10 mV respectively. Left vertical bar 10 MΩ and horizontal bar beneath trace indicate the time when the lateral line nerve was stimulated at 200 Hz. *c*. EPSPs recorded intracellularly from afferent nerve terminals in mechanically stimulated lateral-line organs are reduced in amplitude following electrical stimulation of lateral line efferent fibres at 200 s^{-1} for 80 ms. First three traces, the amplitude of the EPSPs is reduced following

magnitude and temperature dependence are similar to the gating of Na^+ channels in nerve (Hodgkin & Huxley, 1952; Frankenhaeuser & Moore, 1963) or ACH receptor channels at the motor end-plate (Anderson & Stevens, 1973; Lester *et al.*, 1978). However, they point out that the receptor potential is too large to be the gating current, and it vanishes when permanent ions are removed.

Overwhelming evidence from behavioural and electrophysiological studies show that hair cells are capable of responding to high-frequency sounds; however, it is difficult to understand how hair cells signal this information to the afferent nerve fibres which form chemical synapses with them. Release of chemical transmitter is a voltage-dependent phenomenon and it seems reasonable to expect that the high-frequency potential changes associated with acoustic transduction in these hair cells would be filtered out by the capacitative impedance of the hair cell membrane. The electrical time constants of hair cells are 0.4–0.7 ms in the mammalian cochlea (Russell & Sellick, 1978) and about 10 ms in the frog sacculus (Hudspeth & Corey, personal communication).

The solution to this problem seems to be in the asymmetrical operating characteristic of hair cells. Intracellular receptor potentials which have been recorded from inner hair cells in the guinea-pig cochlea are large and may exceed 30 mV peak-to-peak (Russell & Sellick, 1978 and in preparation). At frequencies of auditory stimulation below the cut-off frequency of the hair cell (200–300 Hz) the waveform of the receptor potential resembles a rectified analogue of the stimulus (Fig. 10). When the frequency of stimulation exceeds the cut-off frequency the receptor potential falls-off at 6 dB/octave (Fig. 11) about a d.c. level which is determined by the asymmetrical operating characteristic and the stimulus intensity (Fig. 12). The d.c. level has been called the d.c. component of the receptor potential to distinguish it from the extracellularly recorded summating potential. The phasic component, which has been named the a.c. component of the receptor potential, is the intracellular equivalent of the cochlear microphonic.

At 1 KHz the ratio between the peak-to-peak amplitude of the receptor potential and the d.c. level is 2. This means that the hyperpolarising phase of the receptor potential is very close to the hair-cell resting potential and at frequencies above 1 KHz, the d.c. component will dominate the receptor

electrical stimulation of the efferent fibres during the horizontal bar. Fourth trace, extracellularly recorded microphonic potential is augmented during the inhibition. Fifth trace, the extracellular hyperpolarising potential in the absence of mechanical stimulation. Bottom trace, mechanical stimulus 70 Hz and 30 bB above threshold for excitation. Horizontal and vertical bar, 100 ms and 4 mV respectively.

potential. These findings correspond to measurements of phase locking of impulses in the auditory nerve to auditory stimulation, which declines at a rate of 6 dB/octave at frequencies above 1 KHz (Rose, Brugge, Anderson & Hind, 1968).

It is tempting to speculate that the asymmetrical operating characteristic, which seems to be a general characteristic of hair cells, is an adaptation which allows hair cells to respond to frequencies which would otherwise be attenuated by the low pass characteristics of their membranes.

Transmission at the afferent synapse

The site of impulse initiation in the sensory epithelia of acoustico-lateralis receptors is remote from the site of transduction. In lateral-line receptors it is probably located at the first node of Ranvier on the afferent fibres which innervate the hair cells (Pabst, 1977). Impulses have been recorded from the myelinated regions of these fibres in canal organs of *Lota* (Flock & Russell, 1973, 1976), but not in the unmyelinated nerve terminals. Similar, unpublished observations have also been made by Russell and Sellick in the mammalian cochlea where impulses have, on occasion, been recorded from the point where the afferent fibres become myelinated at the habenulae perforata, but impulses have not been recorded in nerve terminals beneath the inner hair cells. At the intervention between hair cells and afferent fibres are synapses which transmit through the release of a chemical transmitter of unknown composition. Failure to identify the transmitter by screening applied agents from the list of transmitters or blocking agents motivated Sewell, Norris, Tochibana & Guth (1978) to resort to Loewi's (1921) classical technique of using the animal as a detector. From their results Sewell *et al.* suggested that the excitatory substance was detectable in the perilymph of two widely different amphibian species, and that the substance might be common to both species. Further analysis should reveal whether or not this substance is the primary afferent transmitter of hair cells.

A causative relation between hair-cell receptor potentials and transmitter release has been elegantly demonstrated by Sand *et al.* (1975) who injected current intracellularly into hair cells of the superficial lateral line organs of *Necturus maculosus*, while recording impulses in the post–synaptic afferent fibres. They found that afferent impulse activity was modulated by this current injection, being increased when the hair cells were depolarised and decreased when they were hyperpolarised. Current injected into supporting cells or the afferent nerve terminals themselves did not modulate the afferent activity, even when the current was ten times more intense than that injected into the hair cells. Sand *et al.* concluded, from their observations, that mechanically evoked receptor potentials in hair cells have a direct effect on transmitter release at the afferent synapses.

Apart from its release from hair cells in response to mechanical stimulation the afferent transmitter is also released spontaneously and this continuous trickle of transmitter across the synapse is probably the basis for the spontaneous impulse activity recorded in afferent fibres of most acoustico-lateralis receptors. The quantal nature of this spontaneous and stimulus evoked transmitter release is evident in the excitatory post-synaptic potentials (EPSPs) which have been recorded intracellularly from afferent nerve terminals in the sacculus of the goldfish (Furukawa & Ishii, 1967), the lateral line organs of *Lota lota* (Flock & Russell, 1973, 1976) and *Necturus* (Sand *et al.*, 1975) (Fig. 13*a*). Measurement of these potentials indicates that the synapses introduce a delay of about 1.0 ms at 14 °C in the lateral line of *Lota* (Fig. 13*b*) and 0.64 ms at 24 °C in the sacculus of goldfish (Furukawa, Ishii & Matsuura, 1972).

The synapses are also the location of a type of sensory adaptation based on the depletion of the transmitter substance. Adapative run down in transmitter release has been observed in the afferent terminals of the goldfish sacculus and the lateral line canal organs of *Lota*, and the former preparation has been the subject of a recent analysis of this phenomenon by Furukawa & Matsuura (1978). They noticed that the rate of development of adaptation is about the same for different intensities of sound stimulation, and that a brisk discharge of new EPSPs is elicited in response to an increase in sound intensity even when the EPSPs are completely adapted to a continuous sound. From these findings they concluded that the size of the readily available store, and not the release fraction, is changed by a change in sound intensity. Thus the size of the readily available transmitter store in hair cells seems to depend on the intensity of the stimulus and the duration for which it has been applied. Furukawa *et al.* explained their results in terms of a model hair-cell synapse with multiple release sites. The more intense the stimulation of the hair cell, the greater the number of release sites involved in synaptic transmission. In the scheme for transmitter release proposed by Bennett, Florin & Pettigrew (1976) each release site corresponds to, at most, a single vesicle. The dense presynaptic body with its surrounding vesicles would, presumably, correspond to a multiple release site and the release of quanta would take place only at sites where synaptic vesicles are in very close contact with the presynaptic membrane (Heuser & Reese, 1973).

In the limited studies where it has been possible to compare pre- and post-synaptic events at hair-cell afferent synapses, it is apparent that synapses modify the information flowing across them by delaying, adapting and rectifying it. However, it is also conceivable that the afferent synapses may play a more profound role in determining the properties of hair cells, and in particular, those which are frequency selective. The frequency selectivity of electroreceptive cells in some species of South American electric fish has been tentatively attributed to voltage-selective Ca^{2+} channels in the presynaptic membrane of afferent synapses in these cells (Hopkins, 1976; Viancour, 1978). Voltage-dependent channels are also believed to form the basis for frequency tuning in cochlear hair cells of the terrapin (Crawfurd & Fettiplace, 1978; Fettiplace & Crawfurd, 1978). However, the location and identity of these channels remain to be discovered.

The post-synaptic action of efferent fibres on hair cells

The post-synaptic action of efferent fibres has been recorded intracellularly from hair cells in the lateral line system (Flock & Russell, 1973, 1976), where it is inhibitory. The inhibitory post-synaptic potentials (IPSPs, Fig. 14*a*) are long-lasting (50–250 ms) and associated with a conductance increase in the hair cell membrane (Fig. 14*b*). The IPSPs hyperpolarise the hair cells and this leads to an increase in the driving force for ion movement across the apical surface of the hair cell which is manifested as an augmentation of the voltage response to mechanical stimulation.

Associated with the hyperpolarisation of the hair cell is a decrease in transmitter release and a consequent reduction in the size of excitatory post-synaptic potentials recorded in the afferent terminals of the lateral line nerves (Fig. 14*c*). These intracellular events are reflected in extracellular recordings from sensory epithelia in the acoustico-lateralis system, where stimulation of the efferent system causes an increase in the microphonic and summating potentials and inhibition of impulse activity in the afferent nerve fibres (see Klinke & Galley (1974) for a review).

Efferent inhibition of hair-cell activity is an enigmatic phenomenon, since not all hair cells receive efferent synapses (Flock & Flock, 1966) and in most cases the normal activity of efferent neurones is unknown. In the lateral line system efferent activity precedes and accompanies active body movements associated with the contraction of fast muscles (Russell, 1971; Roberts & Russell, 1972; Russell, 1976), and it has been proposed that efferent fibres have a protective role against unwanted self-stimulation and that they shift the operating range of the hair cells. Other roles have been suggested, for the efferent system, which could be performed centrally (Klinke & Galley, 1974). In fact, closer scrutiny of the post-synaptic action of the efferent fibres on hair cells reveals that their activity is not simply integrated with the mechanically evoked voltage responses, but transforms them. This is illustrated in the mammalian cochlea where stimulation of the crossed olivo-cochlear bundle, whose fibres terminate as efferent endings located predominantly on outer hair cells, inhibits stimulus evoked activity, but not spontaneous activity, in the afferent fibres of the auditory nerves, the vast majority of which innervate inner hair cells (Spoendlin, 1978). The efferent terminals are believed to increase the conductance of the post-synaptic membrances of the outer hair cells which then short-circuits the driving voltage across the apical surfaces of the inner hair cells (Fex, 1967). However, this cannot be their only role, since the efferents influence the frequency selectivity of the auditory nerve fibres in a manner not predicted by a simple decrease in driving voltage across the inner hair cells. They broaden the neural frequency tuning curves, increase their thresholds and shift their centre frequencies to lower values (Wiederhold & Kiang, 1970; Robertson & Johnstone, 1978). The neural frequency tuning curves reflect the frequency selectivity of the inner hair cells (Russell & Sellick, 1978), and the receptor potentials of these cells are believed to reflect their mechanical input (e.g. Fig. 10). Can it be that the post-synaptic activity of efferent fibres on outer hair cells has somehow transformed the mechanical input to the inner hair cells?

Figs. 10, 11 and 12 contain data from unpublished experiments done in collaboration with Dr P. M. Sellick. I am grateful to Dr T. Collett for his valued criticism of the manuscript. This work is supported by a grant from the M.R.C.

References

Anderson, C. R. & Stevens, C. F. (1973). Voltage clamp analysis of acetylcholine produced end-plate current fluctuations at frog neuromuscular junction. *J. Physiol., Lond.* **235**, 655–91.

Bagger-Sjöbäck, D. & Wersäll, J. (1973). The sensory hairs and tectorial membrane of the basilar papilla in the lizard *Calotes versicolor. J. Neurocytol.* **2**, 329–50.

Baird, I. L. (1974). Anatomical features of the inner ear in submammalian vertebrates. In *Handbook of Sensory Physiology. V/I Auditory System: Anatomy, Physiology (Ear)*, ed. W. D. Keidel & W. D. Neff, pp. 159–212. Berlin, Heidelberg, New York: Springer-Verlag.

Bekesy, G. von (1960). *Experiments in Hearing*, ed. E. G. Wever. New York: McGraw-Hill.

Bennett, M. R., Florin, T. & Pettigrew, A. G. (1976). The effects of calcium ions on the binomial statistic parameters that control acetylcholine release at preganglionic nerve terminals. *J. Physiol., Lond.* **257**, 597–620.

Bennett, M. V. L. (1976). Transmission at receptor synapses. In *Mechanisms in Transmission of Signals for Conscious Behaviour*, ed. T. Desiraju, pp. 345–66. Amsterdam: Elsevier.

Bennett, M. V. L. (1978). Mechanism of afferent discharge from electroreceptors: implications for acoustic reception. In *Evoked Electrical Activity in the Auditory Nervous System*, ed. R. Naunton & C. Fernandez, pp. 83–9. New York: Academic Press.

Bodian, D. (1978). Synapses involving auditory nerve fibres in primate cochlea. *Proc. Natl. Acad. Sci.*, **75**, 4582–6.
Brown, K. T. & Murakami, M. (1964). A new receptor potential of the monkey retina with no detectable latency. *Nature, Lond.* **201**, 626–8.
Corey, D. P. & Hudspeth, A. J. (1979). Response latency of vertebrate hair cells. *Biophys. J.* **26**, 499–506.
Crawfurd, A. C. & Fettiplace, R. (1978). Ringing responses in cochlear hair cells of the turtle. *J. Physiol., Lond.* **284**, 120–22*P*.
Davis, R. (1957). Biophysics and physiology of the inner ear. *Physiol. Rev.* **37**, 1–49.
Davis, H. (1978). Symposium summary. In *Evoked Electrical Activity in the Auditory Nervous System*, ed. R. F. Naunton & C. Fernandez, pp. 575–82. New York: Academic Press.
Engström, H., Ades, H. W. & Anderson, A. (1966). *Structural pattern of the organ of Corti.* Stockholm: Almgrist & Wiksell.
Engström, H. & Ades, H. (1973). The ultrastructure of the organ of Corti. In *The Ultrastructure of Sensory Organs*, ed. J. Friedmann. Amsterdam: North Holland.
Fain, G. L., Granda, A. M. & Maxwell, J. H. (1977). Voltage signal of photoreceptors at visual threshold. *Nature, Lond.* **265**, 181–3.
Fettiplace, R. & Crawfurd, A. A. (1978). The coding of sound pressure and frequency in cochlear hair cells of the terrapin. *Proc. R. Soc. Lond. Series B*, **203**, 209–18.
Fex, J. (1967). Efferent inhibition in the cochlea realted to AC DC activity. *J. Acoust. Soc. Amer.* **41**, 666–7.
Flock, Å. (1965). Electronmicroscopical and electrophysiological studies on the lateral line canal organ. *Acta oto-laryngol.* (*Stockh.*) *Suppl.* **199**, 1–90.
Flock, Å. (1971). Sensory transduction in hair cells. In *Handbook of Sensory Physiology*, Vol. I, ed. W. Loewenstein, pp. 396–441. Berlin: Springer.
Flock, Å. (1977). Physiological properties of sensory hair cells in the ear. In *Psychophysics and Physiology of Hearing*, ed. E. F. Evans & J. P. Wilson, pp. 15–25. London: Academic Press.
Flock, Å. & Cheung, H. C. (1977). Actin filaments in sensory hairs of inner ear receptor cells. *J. Cell Biology*, **75**, 339–43.
Flock, Å. & Duvall, A. J. (1965). The ultrastructure of the kinocilium of the sensory cells in the inner ear and lateral line organs. *J. Cell Biol.* **25**, 1–8.
Flock, Å. & Flock, B. (1966). Ultrastructure of the amphibian papilla in the bullfrog. *J. Acoust. Soc. Amer.* **40**, 1262.
Flock, Å., Flock, B. & Murray, E. (1977). Studies on the sensory hairs of receptor cells in the inner ear. *Acta Oto-Laryngol.* **83**, 85–91.
Flock, Å., Jørgensen, M. & Russell, I. J. (1973). The physiology of individual hair cells and their synapses. In *Basic Mechanisms in Hearing*, ed. A. R. Moller, pp. 273–306. New York: Academic Press.
Flock, Å. & Russell, I. J. (1973). The post-synaptic action of efferent fibres in the lateral line organ of the burbot *Lota lota*. *J. Physiol., Lond.* **235**, 591–605.
Flock, Å. & Russell, I. J. (1976). Inhibition by efferent nerve fibres: action on hair cells and afferent synaptic transmission in the lateral line canal organ of the Burbot, *Lota lota*. *J. Physiol., Lond.* **257**, 45–62.
Flock, Å. & Wersäll, J. (1962). A study of the orientation of the sensory

hairs of the receptor cells in the lateral line organ of a fish with special reference to the function of the receptors. *J. Cell Biol.* **15**, 19–27.

Frankenhaeuser, B. & Moore, L. E. (1963). The effects of temperature on the sodium and potassium permeability changes in myelinated nerve fibres of *Xenopus laevis*. *J. Physiol., Lond.* **169**, 431–7.

Furakawa, T. & Ishii, Y. (1967). Neurophysiological studies on hearing in goldfish. *J. Neurophys.* **30**, 1377–403.

Furukawa, T., Ishii, Y. & Matsuura, S. (1972). Synaptic delay and time course of post-synaptic potentials at the junction between hair cells and eighth nerve fibres in the goldfish. *Jap. J. Physiol.* **22**, 617–35.

Furukawa, T. & Matsuura, S. (1978). Adaptive run down of excitatory post-synaptic potentials of synapses between hair cells and eighth nerve fibres in the goldfish. *J. Physiol., Lond.* **276**, 193–210.

Greengard, P. (1976). Possible role for cyclic nucleotides and phosphorylated membrane proteins in postsynaptic actions of neurotransmitters. *Nature, Lond.* **260**, 101–8.

Hagins, W. A. (1972). The visual process: excitatory mechanisms in the primary receptor cells. *Ann. Rev. Biophys. Bioengr.* **1**, 131–58.

Harris, G. G., Frischkopf, L. S. & Flock, Å. (1970). Receptor potentials from hair cells of the lateral line. *Science*, **167**, 76–9.

Hartzell, H. C., Kuffler, S. W., Stickgold, R. & Yoskikami, D. (1977). Synaptic excitation and inhibition resulting from direct action of acetylcholine on two types of chemoreceptors on individual amphibian parasympathetic neurones. *J. Physiol., Lond.* **271**, 817–46.

Heuser, J. E. & Reese, T. S. (1973). Evidence for recycling of synaptic vescicle membrane during transmitter release at the frog neuromuscular junction. *J. Cell Biol.* **57**, 315–44.

Hillman, D. E. (1972). Observations on morphological features and mechanical properties of the peripheral vestibular receptor system in the frog. In *Basic Aspects of Central Vestibular Mechanisms*, ed. A. Brodal & O. Pompeiano, pp. 69–73. Amsterdam: Elsevier Publishing Company.

Hodgkin, A. L. & Huxley, A. F. (1952). A quantitative description of membrane current and its application to conductance and excitation in nerve. *J. Physiol., Lond.* **117**, 500–44.

Hopkins, C. D. (1976). Stimulus filtering and electroreception: tuberous electro-receptors in three species of Gymnotid fish. *J. Comp. Physiol.* **111**, 171–207.

Hudspeth, A. J. & Corey, D. P. (1977). Sensitivity, polarity and conductance change in the response of vertebrate hair cells to controlled mechanical stimuli. *Proc. Natl. Acad. Sci.* **74**, 2407–11.

Hudspeth, A. J. & Jacobs, R. (1979). Stereocilia mediated transduction in vertebrate hair cells. *Proc. Nat. Acad. Sci.* **76**, 1506–9.

Johnstone, B. M. & Sellick, P. M. (1972). The peripheral auditory apparatus. *Quarterly Rev. Biophys.* **5**, 1–57.

Jørgensen, J. M. (1970). On the structure of the macula lagena in birds with some notes on the avian maculae utriculi and sacculi. *Vidensk. Meddel. f. Dansk NaturHist. For.* **133**, 121–47.

Jørgensen, J. M. (1974). The sensory epithelia of the inner ear of two turtles *Testudo graeca* L. and *Pseudomys scripta* (Schoepff). *Acta Zool.* **55**, 289–98.

Jørgensen, J. M. & Anderson, T. (1972). On the structure of the avian maculae. *Acta Zool.* **54**, 121–30.

Klinke, R. (1978). Frequency analysis in the inner ear of mammals in comparison to other vertebrates. *Verh. Dtch. Zool. Ges.* 1–15.

Klinke, R. & Galley, N. (1974). Efferent innervation of vestibular and auditory receptors. *Physiol. Rev.* **54**, 316–57.

Lester, H. A., Nass, M. M., Krouse, M. E., Wasserman, N. H. & Erlanger, B. F. (1978). ACh receptor-channels begin to open 30 μsec after agonist is applied. *Neurosci. Abstr.* **4**, 370.

Lewis, E. R. & Li, C. W. (1975). Hair cell types and distributions in the otolithic and auditory organs of the bullfrog. *Brain Res.* **83**, 35–50.

Loewi, O. (1921). Über humerale Ubertragbarkeit der Herznerven wirkung. *Pflügers Arch.* **189**, 239–42.

McGlone, F. P., Russell, I. J. & Sand, O. (1979). Measurement of calcium ion concentration in the lateral line of *Xenopus laevis*. *J. Exp. Biol.* **83**, 123–30.

Mooseker, M. S. (1975). Brush border motility. *J. Cell Biol.* **71**, 417–33.

Mooseker, M. S. & Tilney, L. G. (1975). Organisation of an actin filament-membrane complex. *J. Cell Biol.* **67**, 725–43.

Mulroy, M. J. (1974). Cochlear anatomy of the alligator lizard. *Brain Behav. Evol.* **10**, 69–87.

Pabst, A. (1977). Number and location of the sites of impulse generation in the lateral line afferents of *Xenopus laevis*. *J. Comp. Physiol.* **114A**, 51–67.

Pumphrey, R. J. (1950). Hearing. In *Physiological Mechanisms in animal behaviour* (4th Symposium of the Society for Experimental Biology), pp. 3–18. Cambridge University Press.

Roberts, B. L. & Russell, I. J. (1972). The activity of lateral line efferent neurones in stationary and swimming dogfish. *J. Exp. Biol.* **57**, 435–48.

Robertson, D. & Johnstone, B. M. (1978). Efferent transmitter substance in the mammalian cochlea: single neuron support for acetylcholine. *Hearing Research*, **1**, 31–4.

Rodriquez-Echandia, E. L. (1967). An electron microscopic study on the cochlear innervation. I. The recepto neural junctions at the outer hair cells. *Z. Zellforschung.* **78**, 30–46.

Rose, J. E., Brugge, J. E., Anderson, D. J. & Hind, J. E. (1968). Patterns of activity in single auditory nerve fibres of the squirrel monkey. In *Hearing Mechanisms in Vertebrates*, ed. A. V. S. de Reuck & J. Knight, pp. 144–56. London: Churchill.

Russell, I. J. (1971). The role of the lateral line efferent system in *Xenopus laevis*. *J. Exp. Biol.* **54**, 621–41.

Russell, I. J. (1976). Amphibian lateral line receptors. In *Frog Neurobiology*, ed. R. Llinás & W. Precht. Berlin, Heidelberg: Springer-Verlag.

Russell, I. J. & Sellick, P. M. (1976). Measurement of potassium and chloride ion concentrations in the cupulae of the lateral lines of *Xenopus laevis*. *J. Physiol., Lond.* **257**, 245–55.

Russell, I. J. & Sellick, P. M. (1977). Tuning properties of cochlear hair cells. *Nature, Lond.* **267**, 858–60.

Russell, I. J. & Sellick, P. M. (1978). Intracellular studies of hair cells in the mammalian cochlea. *J. Physiol., Lond.* **284**, 261–90.

Sales, G. & Pye, D. (1974). Ultrasonic Communiation by Animals. London: Chapman & Hall.

Sand, O., Ozawa, S. & Hagiwara, S. (1975). Electrical and mechanical stimulation of hair cells in the mudpuppy. *J. Comp. Physiol. A.* **102**, 13–26.

Sellick, P. M. & Russell, I. J. (1978). Intracellular studies of cochlear hair cells: filling the gap between basilar membrane mechanics and neural excitation. In *Evoked Electrical Activity in the Auditory Nervous System*, ed. F. Naunton & C. Fernandez, pp. 113–40. New York: Academic Press.

Sellick, P. M. & Russell, I. J. (1979). Two-tone suppression in cochlear hair cells. *Hearing Res.* **1**, 227–36.

Sewell, W. F., Norris, C. H., Tachibana, M. & Guth, P. S. (1978). Detection of an auditory nerve-activating substance. *Science*, **202**, 910–12.

Smith, C. A. (1968). Electron microscopy of the inner ear. *Ann. Otol. Rhinol. Laryngol.* **77**, 269–84.

Smith, C. A. (1975). The inner ear: its embryological development and microstructure. In *The Nervous System, Vol. 3: Human Communication and its Disorders*, ed. D. B. Tower, pp. 1–18. New York: Raven Press.

Spoendlin, H. H. (1966). The Organ of the Cochlear Receptor. Basel: Karger.

Spoendlin, H. (1970). Auditory, vestibular, olfactory and gustatory organs. In *Ultrastructure of the Peripheral Nervous System and Sense Organs*, ed. Bischoff, pp. 173–263. Stuttgart: Georg. Thieme.

Spoendlin, H. (1978). The afferent innervation of the cochlea. In *Evoked electrical activity in the auditory nervous system*, ed. F. Naunton & C. Fernandez, pp. 21–42. New York: Academic Press.

Thurm, U. (1974). Basics of the generation of receptor potentials in epidermal mechanoreceptors of insects. In *Mechanoreception*, ed. J. Schwartzkopt. *Abh. Rhein-Westf. Acad. Wiss.* **53**, 355–85.

Viancour, T. A. (1978). Electroreception of weakly electric fish. Ph.D. Thesis, University of California, San Diego.

Weiss, T., Peake, W. T., Ling, A. & Holton, T. (1978). Which structures determine frequency selectivity and tonotopic organization of vertebrate nerve fibres? Evidence from the aligator lizard. In *Evoked Electrical Activity in the Auditory Nervous System*, ed. R. F. Naunton & C. Fernandez, pp. 91–112. New York: Academic Press.

Wersäll, J. (1956). Studies on the structure and innervation of the sensory epithelium of the cristae ampullaris in the guinea pig. *Acta Otolaryngol. Suppl.* **126**, 1–85.

Wersäll, J., Björkroth, B., Flock, Å. & Lundquist, P. G. (1973). Experiments on ototoxic effects of antibiotics. *Adv. Otorhinolaryngol.* **20**, 14.

Wiederhold, M. L. (1976). Mechanosensory transduction in 'sensory' and 'motile' cilia. *Ann. Rev. Biophys. Bioeng.* **5**, 39–62.

Wiederhold, M. L. & Kiang, N. Y. S. (1970). Effects of electric stimulation of the crossed olivo cochlear bundle on single auditory nerve fibres in the cat. *J. Acoust. Soc. Amer.* **48**, 950–65.

BRIAN M. H. BUSH

Non-impulsive stretch receptors in crustaceans

A unique type of muscle receptor organ is found at the base of each leg in crustaceans (Fig. 1*a*; Alexandrowicz & Whitear, 1957; Alexandrowicz, 1958, 1967; Whitear, 1965). Like the single sensory neurone of the crayfish abdominal muscle receptor, and in contrast to the vertebrate muscle spindle afferents, the two afferent nerve fibres of each thoracic–coxal muscle receptor organ (TCMRO) are large enough to impale with microelectrodes (Fig. 1*b*). Unlike both the former muscle receptors, however, the sensory signal evoked by stretch of the TCMRO is conducted to the central nervous system without the intervention of all-or-none impulses (Ripley, Bush & Roberts, 1968; Bush, 1976, 1977). Instead the depolarising receptor potentials are conducted electrotonically over the whole distance from receptor to central nervous system, that is in 'analogue' form rather than by frequency coded pulse trains (Fig. 1*d*). Nevertheless this 'amplitude-modulated' signal is able to elicit impulse discharges in motoneurones supplying the 'extrafusal' promotor muscle in parallel with the TCMRO (Bush & Roberts, 1968). We thus have a stretch reflex which, in the general features of its motor output and overt muscular response, is indistinguishable from that of either the abdominal muscle receptor organ or the mammalian muscle spindle.

How is it that two such similar, and apparently simple, motor response patterns can be mediated by such evidently different afferent mechanisms? What are the pertinent morphological, ultrastructural and biophysical distinguishing features which might underlie such different forms of neuronal signalling? What, if any, are the functional advantages making for selection of one or the other mode, and what are the *dis*advantages? These are some of the principal questions to which this chapter is addressed, taking the crab's coxal stretch receptor as a simple model system.

Fig. 1. Morphology of thoracic–coxal muscle receptor organ (TCMRO) of the crab, *Carcinus*. *a*. Anatomy of TCMRO of right posterior leg *in situ* (dorsal view). *b*. Base of TCMRO and its innervation. *c*. Central projections and cell bodies of sensory and promotor neurones (upper) and T fibre (lower drawing; broken lines indicate mid-line of the ganglion). *d*. Promotor reflex (top trace) and intracellularly recorded T fibre response (middle) to stretching the receptor muscle (bottom trace). *e*. Endings of sensory dendrite (*d*): *df*, dendrite finger; *m*, mitochondria; *sc*, string cell; *vs*, 'vacuolated string'. *f*. Fibres of sensory nerve in cross-section. *a*, *b*, *d*, from Bush & Roberts, 1968, 1971; *c*, *f*, from Bush, 1976; *e*, from Whitear, 1965.

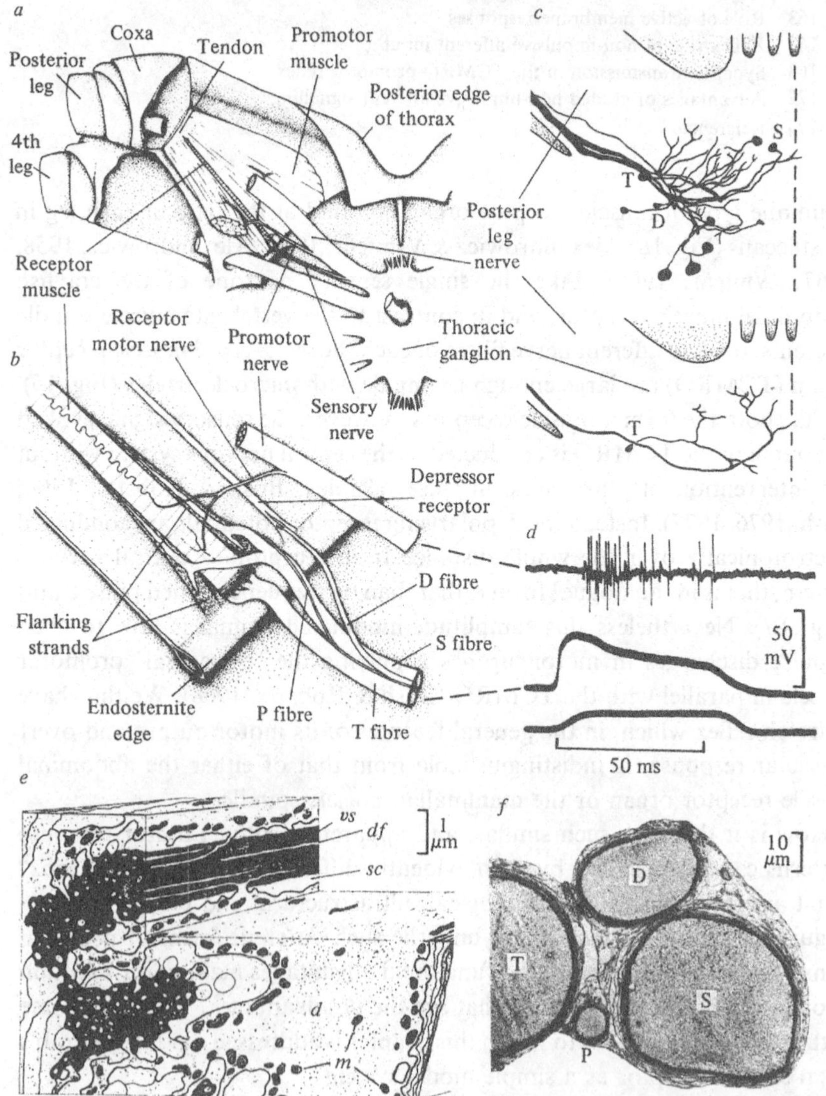

Morphology of the coxal receptor neurones

In contrast to almost all other known types of primary sensory neurones in Crustacea or other arthropods, the cell bodies of the coxal afferent neurones lie within the central nervous system (Fig. 1*c*). Each soma is connected via a thin neurite to a thicker synaptic zone (of 40–60 μm diameter in the Atlantic blue crab, *Callinectes sapidus*: Fig. 12*d*; Blight & Llinás, 1980). From here several branches may arise (at least in the European shore crab, *Carcinus maenas*) before the fibre enlarges steadily until, some 1–3 mm outside the ganglion, it reaches its maximum diameter (40–70 μm in *Carcinus*, or 60–100 μm in *Callinectes* and the Pacific mud crab, *Scylla serrata*: Mirolli, 1979*a*). Thence it continues as a uniform cylinder for several millimetres (over a centimetre in the posterior leg of large crabs) before branching to innervate the base (T fibre) or flanking strands (S fibre in crabs) of the receptor muscle (Fig. 1*b*).

The nerve containing the large S and T fibres in crabs (Fig. 1*f*) also contains a much finer 'P fibre' which terminates peripherally alongside the S fibre, and a fourth non-impulsive 'D fibre' of intermediate diameter (*c*. 30 μm in *Carcinus*) innervating the non-muscular elastic depressor receptor strand (Alexandrowicz & Whitear, 1957; Bush, 1976). Further afferents of varying diameter supply the levator receptor and, in decapod crustaceans other than Brachyura (crabs), one or two 'additional innervated strands' in the coxal region. All of these, presumably non-impulsive, sensory neurones have relatively short nerve fibres and centrally located cell bodies.

Four somewhat similar non-impulsive sensory neurones innervate a single elastic stretch receptor organ at the base of the uropods in the sand crab, *Emerita analoga* (Paul, 1972). Their 3–4 mm long 'giant' (40–50 μm diameter) afferent fibres branch extensively within the last abdominal ganglion where the large somata lie. Graded, decrementally conducted signals in these afferents evoke reciprocal reflexes in antagonistic muscles, which appear to contribute phasically towards one of two alternative locomotor patterns (Paul, 1976). Thus non-impulsive mechanoreceptive afferent inputs may have widespread reflex effects, and are not necessarily limited to mediating simple stretch reflexes.

On the basis of their unusual morphology and particularly the central location of their cell bodies, it might be argued that the afferent fibres of these various stretch receptors in Crustacea are in fact dendrites rather than axons. The lack of impulses in these neurones might then seem somewhat less surprising. However, action potentials occur in processes generally accepted as dendrites in a variety of neurone types in both vertebrates and invertebrates, including the distal processes of the ubiquitous bipolar sensory neurones of arthropods (e.g. Mellon & Kennedy, 1964; Mendelson, 1966). Thus even if the coxal receptor afferents are considered to be dendrites – a question which is surely debatable (cf. vertebrate somatosensory afferents and their dorsal root ganglia) – the lack of impulses cannot be explained in terms of a simple differentiation between dendrites and axons.

Nor is the fact that many of these coxal receptor afferents are quite short and/or of relatively large diameter an adequate basis for their non-impulsive behaviour, since other, shorter or larger diameter nerve fibres conduct impulses in the normal way, even within the same animal. Thus the thoracic–coxal (TC) chordotonal organ in crayfish and lobsters, for example, conveys similar sensory information to the adjoining TCMRO over the same short distance (a few millimetres), but by means of conventional impulse trains in the afferent fibres of some fifty or more bipolar sensory neurones (Alexandrowicz & Whitear, 1957; Bush, 1976: Fig. 2*c*). The presence of this duplicate, 'impulsive' TC organ in the Astacura is the more surprising in

Fig. 2. Membrane potential of afferent fibres. *a*. Electrogenic sodium pump contribution in S fibre of *Cancer*: cooling or 10^{-3} M ouabain depolarise the fibre from its initial transmembrane potential of −63 mV. *b*. T and S fibre potentials and receptor muscle tension variation with 0.5 mm increments in length in *Carcinus*. *c*. Chordotonal organ (top trace) responds only to shortening, while T and S fibres depolarise on stretching, in *Homarus*. *a*, from M. Mirolli, personal communication, published in Bush, 1976; *b*, B. M. H. Bush, and *c*, B. M. H. Bush & A. J. Cannone, unpublished.

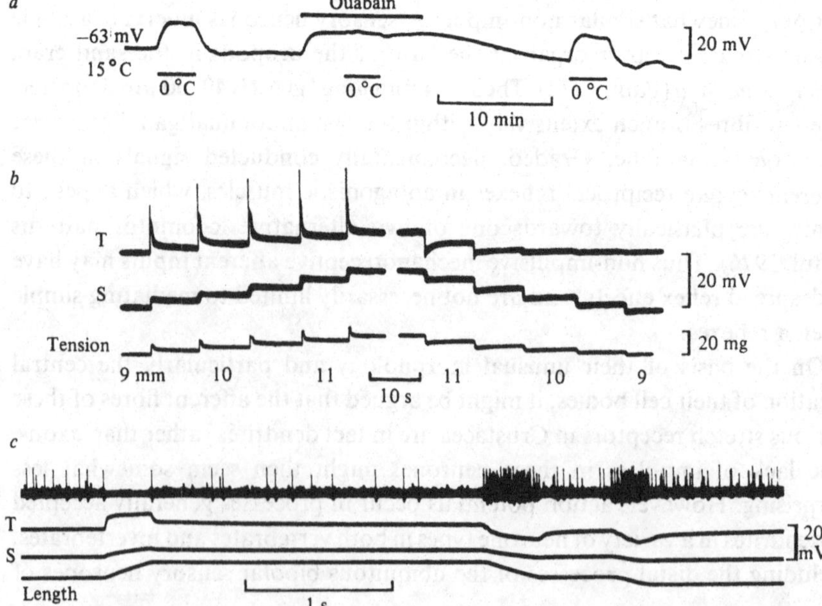

view of its total absence in other decapod Crustacea. Apparently its role has been assumed in the latter groups by the two or three non-impulsive afferents of the single thoracic–coxal muscle receptor, even where the afferent nerve is many times longer as in the back legs of crabs, especially in large species like *Callinectes*, *Cancer* and *Scylla*.

'Resting' potentials and receptor strand length

A microelectrode impaling any of the three large diameter afferent nerve fibres (T, S or D fibres) records a transmembrane potential of −40 to −80 mV, though the true resting potential when the receptor muscle or strand is completely slack generally lies between −60 mV and −70 mV. Although potassium dependent, the low slope of about 30 mV per tenfold change in external potassium ion concentration (Roberts & Bush, 1971) implies that other factors are involved. A substantial fraction of the membrane potential evidently stems from an electrogenic sodium–potassium pump, since the fibres depolarise in the presence of ouabain, low external potassium ion concentration, when extracellular sodium ion is replaced by lithium ion, or when the temperature is lowered (Fig. 2*a*; Mirolli, 1979*b*).

The transmembrane potential of the S and T fibres decreases with increasing length or tension of the muscle receptor strand (Fig. 2*b*). These changes in membrane potential depend upon extracellular sodium, since they are reduced by its replacement with relatively impermeant choline or tris ions. Such replacement also results in an increased input resistance of the cell, and often also an initial hyperpolarisation, indicating a significant resting permeability to sodium ions, even in the relaxed receptors. That is, there appears to be a permanent but variable 'short circuit' condition in these neurones, probably in the input or transducer region of the dendrites (see below). The absence of mitochondria from these 'dendritic fingers' is consistent with the idea of two spatially separated current sources, an inward 'leak' of sodium ions peripherally, and an outward active sodium ion transport throughout the rest of the afferent fibre, where mitochondria abound beneath the plasma membrane (Krauhs & Mirolli, 1975).

It may be concluded, then, that the 'resting' membrane potential of these sensory neurones is a compound function of the extracellular potassium and sodium ion concentrations, in conjunction with a significant contribution from an electrogenic sodium pump. In addition, these extracellular cation concentrations also affect the cable properties of the afferent fibres, since replacement of either with impermeant cations results in an increase in input resistance, and also a decrease in the decrement along the fibre of the response to receptor stretch or current injection (see below: Mirolli, 1979*b*).

Receptor potentials evoked by stretch

The steady-state membrane potential of the S fibre varies approximately linearly with receptor strand length, while that of the T fibre depends primarily upon the tension within the receptor muscle with which its sensory terminals lie in series (Bush & Roberts, 1971; Bush & Godden, 1974; see also Fig. 2*b*). Sudden stretching of the receptor muscle, either directly or by remotion at the thoracic–coxal joint, evokes characteristic depolarising

Fig. 3. Receptor potentials recorded intracellularly from T and S fibres in *Carcinus* in response to ramp-function and sinusoidal stretching of the receptor muscle. *a*. Characteristic waveforms and response components: *d* dynamic, *s* static; *a* initial (or 'acceleration'), *v* velocity, *l* length – components. *b*. Two frequencies of sinusoidal length changes. *c*. Three velocities of ramp stretch: note separation of α and β components at low velocity. *d*. Dynamic responses and tension changes (*t*) to identical stretches at three initial lengths; length range *in situ*: 8–12 mm; t_0, zero tension level. *Calibrations* (*in this and subsequent figures, except where otherwise stated*): 20 mV (T and S fibres), 20 mg (tension traces); 1 s. *a*, from Bush & Roberts, 1971; *b–d*, from Bush, 1977 and *c*, Cannone & Bush, 1980*a*.

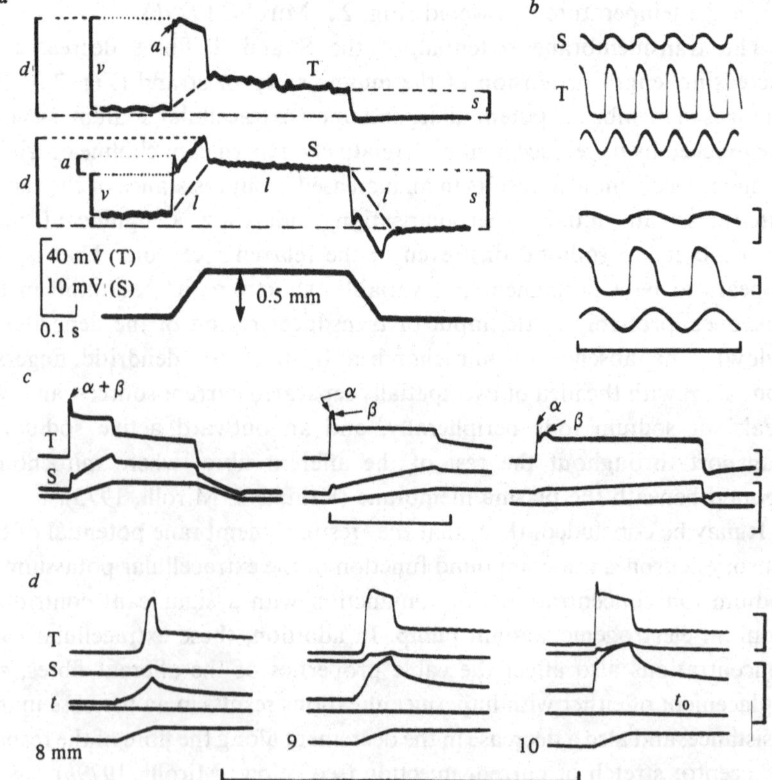

receptor potentials in both S and T fibres (Fig. 3). The waveform and amplitude of these responses vary systematically with the time course and amplitude of the stimulus. In response to constant-velocity stretch and hold stimuli, for example, each afferent fibre shows a characteristic 'dynamic component' related in amplitude and form to the velocity of stretch, followed by a decline towards a less depolarised level related to the new length. The T fibre shows a much greater velocity sensitivity than the S fibre and, being in series with the receptor muscle, is also strongly influenced by its initial tension (and hence also length). Thus, particularly at short initial lengths, the efferent supply to the receptor muscle plays an important role in maintaining the 'gain' or sensitivity of the T fibre's dynamic response, both during stretching and also during release (see Figs. 3*d*, 4*d*, 10*a*, *b*).

These receptor potential waveforms in response to ramp function stretches have been simulated on a digital computer with a simple non-linear mechanical model (Berger & Bush, 1979). This model assumes that the sensory response

Fig. 4. T and S fibre and receptor muscle responses to receptor efferent stimulation and/or stretch. *a–d*. Repetitive motor stimulation (at 100 Hz during bars in *a* and *d*) evokes facilitating and summating excitatory junction potentials in a receptor muscle fibre (*RM* in *b*, *c*), and tetanic (isometric) contraction (*t*), which depolarises the T fibre (*a*), and modifies its response to stretch and especially release (*d*), cf. *e*, without motor stimulation; graded intensity stimulation recruits two receptor motor axons (*c*). *f*. Small twitch-like contractions elicit brief T fibre responses. *g*. 'Tension clamped' stretches produce S fibre responses resembling the length change (*l* in *g*). Thus the T fibre responds primarily to tension and the S fibre to length. From Bush & Godden, 1974; Bush, 1976 and (*f*, *g*) unpublished.

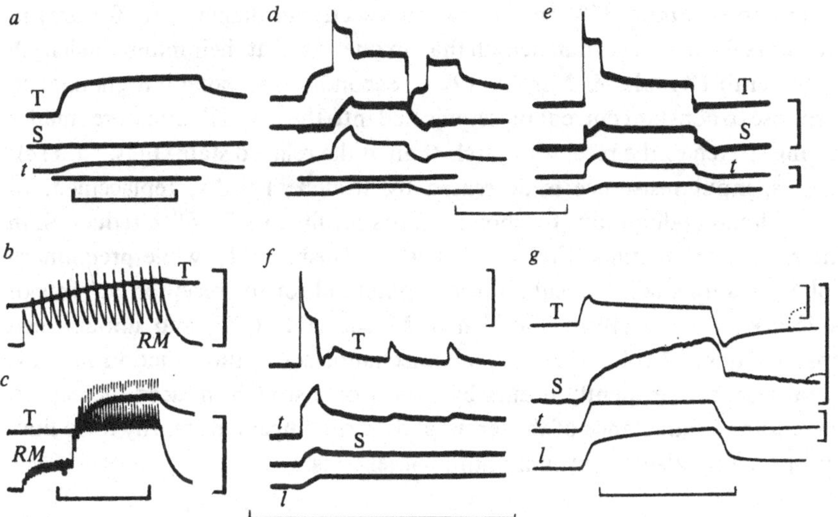

is proportional to the deformation of the respective dendritic terminals, or more simply the change in length of the immediately surrounding connective tissue. With appropriate parameter values, the resulting model responses closely reflect the experimentally recorded receptor potentials, and their variation with the parameters of the stimulus. It may be inferred, then, that the responses of the S and T fibres can be explained largely in terms of the mechano-electric transduction at the sensory terminals, without any significant modification by the electrical properties of the afferent fibres.

This conclusion is consistent with several other observations on these receptors, including responses to sinusoidal (Fig. 3*b*) and pseudo-random white noise inputs (Bush, DiCaprio & French, 1978, and unpublished observations), and recordings of tension changes in the receptor muscle during various forms of imposed length change and evoked contraction (Fig. 4; Bush & Godden, 1974; Bush, 1976, 1977). Isometric contraction of the receptor muscle, for instance, results in graded depolarisation of the in-series T fibre, with little or no effect on the S fibre membrane potential (Fig. 4*a–c*). Occasionally small twitch-like contractions appear in the recorded tensions, probably in response to paired or grouped discharges in one or both receptor motoneurones; such twitches are correlated with small transient depolarisations of the T fibre, again with little sign of any concomitant S fibre response (Fig. 4*f*).

How are these receptor potentials generated? Several lines of evidence suggest that, as in other mechanoreceptors, sensory transduction occurs by means of an increase in ionic conductance of the receptive membrane of the stressed dendritic terminals. First, the supposed transducer region of the dendrites, i.e. the 'dendrite fingers' (Fig. 1*e*), were found in the S fibre of *Cancer* to be about 25% smaller in cross-sectional diameter (*c.* 0.1 μm) in receptors fixed at maximum length than in ones fixed at their minimum length in the crab (Krauhs & Mirolli, 1975). Secondly, the potential changes in response to constant current pulses injected into the T and S fibre were smaller during stretch of the receptor muscle than in the relaxed state (Figs. 5*a*, 11*a*); that is, input resistance is decreased by stretch. Thirdly, replacement of extracellular sodium ions by choline or tris results in a 70–80% reduction in the receptor potentials (Fig. 5*b*: Roberts & Bush, 1971), while preliminary voltage-clamp studies reveal a reversal potential for the receptor currents of +25 mV or more (Bush, Godden & Macdonald, 1975, and unpublished observations; Mirolli, 1979*a*). Thus mechanoelectric transduction in these coxal receptors evidently occurs by means of a stretch-induced increase in membrane conductance of the terminal dendrite fingers, primarily to sodium and probably also to calcium and/or other ions.

Decremental conduction within the afferent fibres

If two microelectrodes are inserted into either S or T fibre at different distances from the receptor muscle, both record similar receptor potentials to a given stretch stimulus (Figs. 5*a*, 6*a*). However the response recorded at the proximal electrode is somewhat smaller and slower than that from the more distal site. Assuming infinite cable properties, the decrement of the stretch-evoked response in either S or T fibre of *Carcinus* corresponds to a length constant of the order of 1–2 cm (Bush, 1976; Cannone & Bush, 1980*b*). On this basis the response recorded at the point where the afferent fibre enters

Fig. 5. Transduction and conduction in the afferent fibres. *a*. T fibre responses in *Scylla* recorded with two electrodes at 1 mm (upper trace) and 6.3 mm (lower trace) from its distal branch point. Two long stretches of the receptor muscle, each applied in two steps, evoked depolarising receptor potentials, and constant hyperpolarising pulses were injected through a third microelectrode 0.2 mm from the proximal recording electrode. Note large decrease in input resistance during stretch, and greater decrement in distally directed current response than proximally conducted receptor potential. Starting membrane potentials are shown; slower time-base applies to repeating current pulses. *b*. Effect on S and T receptor potentials in *Carcinus* of replacing external sodium with choline. Calibrations 10 mV and 20 ms per division. *a*, from Mirolli, 1979*a*; *b*, from Roberts & Bush, 1971.

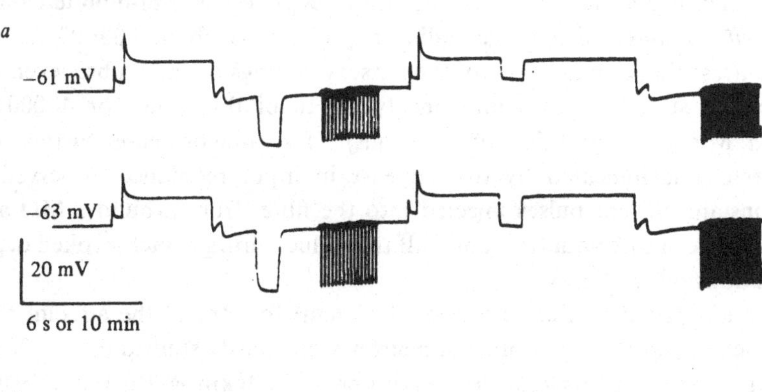

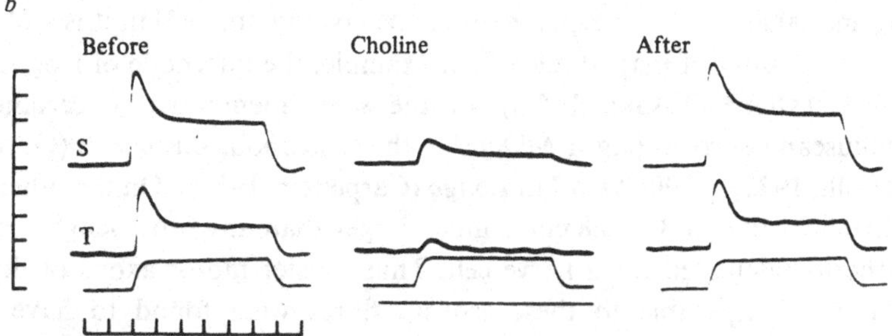

the thoracic ganglion would be about 80–90% of its amplitude at the receptor muscle, some 5 mm distally (in the posterior leg segment of a 6 cm wide crab).

Using more rigorous quantitative methods on large mud crabs (*Scylla*) of about 16 cm width, Mirolli (1979*a*) determined the cable properties of S fibres averaging about 9 mm long by 90 μm diameter. The mean decrement of the steady-state component of the response to stretch or injected current flowing proximally over about 6 mm was only some 9%. Since this decrement was similar for (small) current steps of either polarity, conduction along this main (middle) section of the S fibre was passive. However, a significantly greater decrement (20%) over this section was observed with current flowing in a distal direction, from a proximally inserted stimulating electrode. This is consistent with the notion that the fibre terminates distally in a 'leaky' ending, even in the unstretched condition.

This last conclusion is supported by Mirolli's theoretical analysis of the spatial decrement of the steady-state response to stretch, as determined by two electrodes impaling the S fibre at different distances from its distal bifurcation. The best fit to his experimental data on this decrement, after correction for the error introduced by the microelectrode shunt, was given by a semi-infinite cable model with length constant, $\lambda = 5.8$ cm. From this, and assuming a specific axoplasmic resistance, R_i, of 60 Ωcm (similar to that found in other crustacean nerve fibres: Hodgkin & Rushton, 1946), the specific membrane resistance, R_m ($= 2\lambda^2 R_i/r$, for a semi-infinite cable), of *Scylla* S fibres of average radius $r = 45$ μm, is about 900000 Ωcm². By contrast the estimated R_m of the sensory endings of the S fibre, even in the relaxed state, is only about one twentieth of this value, or 40000 Ωcm² (M. Mirolli, unpublished observations). This value decreases further during stretch, as indicated by the decrease in input resistance, observed with constant current pulses injected into the fibre, from around 2 MΩ at rest to between three-quarters and half this value during stretch-evoked depolarisations of 15–20 mV.

Although the value of around 1 MΩcm² for Rm of the afferent fibre is much higher than that of most neurones previously studied (e.g. 5000 Ωcm² for 30 μm diameter crab leg nerve fibres (Hodgkin & Rushton, 1946); or around 9000 Ωcm² for *Sepia* giant axons (Weidman, 1951)), it is still within the same order of magnitude as, for example, the internode of frog A fibres (100000 Ωcm² (Tasaki, 1955)), or the *soma* membrane of certain large molluscan neurones (e.g. 1 MΩcm² in the gastropod, *Anisodoris* (Gorman & Mirolli, 1972); 27000 Ωcm² in *Loligo* (Carpenter, 1973)). On the other hand, Mirolli's value of $\lambda = 5.8$ cm is much larger than any other length constant hitherto calculated for a nerve cell. Thus lobster motor axons of diameter (75 μm) comparable to these sensory fibres were found to have length constants averaging 1.6 mm (Hodgkin & Rushton, 1946), while even the giant axons of squids (with diameters around 200 μm in *Sepia* and 500 μm in *Loligo*) and of *Anisodoris* (*c*. 150 μm), have length constants only up to 1.0 cm and 2.0 cm, respectively (Weidman, 1951; Gorman & Mirolli, 1972).

These sensory fibres are thus well adapted to conducting slow graded membrane potential changes over distances of a centimetre or more, with minimal decrement. This is readily appreciated when considering their

Fig. 6. Decremental conduction and active membrane responses in the afferent fibres. *a*. Receptor potential recorded at two points 2.4 mm apart in a 3.5 mm long S fibre in *Carcinus*. *b*. Spiky dynamic component in a *Carcinus* T fibre. *c*. Gradation of spiky transient in a T fibre of *Potamon* with stretch velocity. Calibrations, 20 mV; 100 ms. *d*. Large decrement and slowing of initial fast transient response to a step stretch: ratio of proximal/distal transient amplitudes = 0.44, cf. 0.86 for the ratio of the respective static responses. *e*. Distal localisation of 'partial action potential' evoked by a strong depolarising current step injected 0.3 mm from the proximal electrode: note the larger size and faster onset and rise-time of the spike at the distal electrode, despite the larger steady-state response at the proximal electrode. *d* and *e* are from *Scylla* S fibres; electrode distances from the distal bifurcation are indicated on each trace. *a*, from Bush & Roberts, unpublished; *b*, from Roberts & Bush, 1971; *c*, from Bush, 1976; *d*, *e*, from Mirolli, 1979*a*.

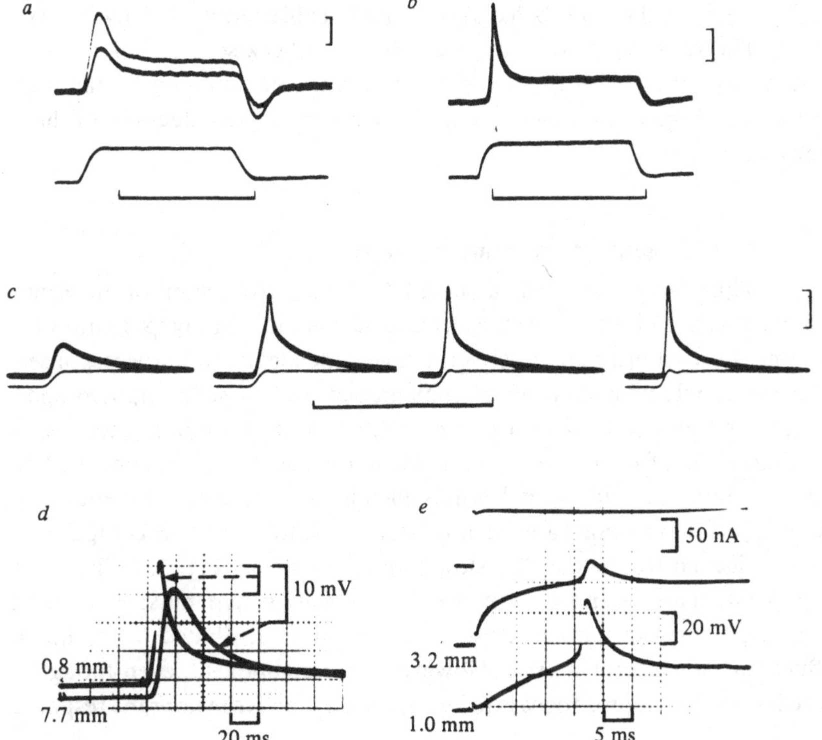

electrotonic lengths (L = fibre length/λ (Rall, 1969)). Values down to L = 0.4 and 0.3 for S and T fibres, respectively, were found in *Carcinus* (Cannone & Bush, 1980*b*), while Mirolli's (1979*a*) figure for λ = 5.8 cm in S fibres in *Scylla* averaging 0.87 cm long gives an electrotonic length of only 0.15. That is, the output (synaptic) zone of the fibre within the thoracic ganglion may be as little as a fifth of a length constant away from the base of the receptor muscle. A further factor contributing to the efficiency of this analogue signalling is the presence of an electrogenic sodium ion pump which, by maintaining a relatively high membrane potential in the face of a continual inward leak of sodium ions peripherally, thereby also maintains a large 'driving force' for the receptor potentials.

A potentially serious drawback of the high membrane resistance of these afferent fibres is the consequent large membrane time constant ($\tau_m = R_m C_m$). Mirolli (1979*a*) measured the times for the voltage responses to injected current steps to reach 84% of their final steady level, obtaining values around 160 ms for eight S fibres and 120 ms for one T fibre in *Scylla*. These are smaller than would be expected for an infinite cable (for which $\tau_m = t_{84\%}$: Hodgkin & Rushton, 1946), and once again the experimental data were most closely modelled by a semi-infinite cable terminated by a short circuit. This is consistent with Mirolli's analysis of the spatial decrement indicating that the distal portion of the fibre behaves as a short cable terminated by a leaky ending. The resulting reduction in the fibre's otherwise even longer time constant no doubt helps to overcome the limitations imposed by the high specific membrane resistance upon the electrotonic conduction of high frequency signals.

Graded active membrane responses

Thus far it has been assumed that the coxal receptor neurones respond in an entirely passive, electrotonic manner. Many preparations, however, show definite signs of active, voltage-dependent membrane responses. These commonly take the form of small graded 'spikes' superimposed upon the basic receptor potential, at or near the onset of the response to stretch or riding on top of small oscillations in membrane potential (Fig. 6*b*, *c*). Even quite slow ramp-function stretches may elicit in the T fibre a prominent initial response which, in some cases, can be resolved into a spiky 'α component' closely followed by a somewhat slower 'β component' (Fig. 3*c*). The latter component, like the initial response in the mammalian muscle spindle's receptor potential (Hunt & Ottoson, 1976), can be attributed to the initial stiffness in the receptor muscle. However the α component, even though it precedes the β response, appears to be a *secondary*, active membrane response of the afferent fibre itself (Bush, 1976; Cannone & Bush, 1980*a*). Under certain conditions, as when the receptor muscle is slack at the start of the stretch, the β component may be absent altogether, while in some afferent fibres an α component is totally lacking (Fig. 6*a*).

Small graded spikes are sometimes also seen in response to depolarising current steps (Fig. 7). At low current strengths the depolarising response resembles a mirror image of the smooth, exponential response to an equal hyperpolarising current step. Above a certain threshold level, however, the rate of depolarisation exceeds that of the corresponding hyperpolarising response, so that the membrane potential reaches a plateau sooner than it would otherwise do. With increasing current strengths small 'bumps' appear near the beginning of the plateau, and these sharpen progressively into spiky transients with further increase in amplitude and rate of depolarisation.

Fig. 7. Graded spike responses to depolarising currents injected into *Carcinus* S and T fibres. *a*. Responses of an S fibre to three graded depolarising pulses and a large hyperpolarising pulse. *b–d*. Depolarising and (*b*) hyperpolarising responses of another S fibre, from a resting membrane potential of -55 mV (*b*, *c*), and (*d*) following a 20 s period of imposed hyperpolarisation to -67 mV. *e–g*. Depolarising responses of a T fibre from a previously depolarised membrane potential (*e*), and at a resting potential of -56 mV before (*f*) and after (*g*) addition of 3 μM tetrodotoxin. Calibrations: 50 mV, 100 nA; 100 ms. *a*, from Roberts & Bush, 1971; *b–g*, from Bush, DiCaprio & Taylor, 1980 and unpublished.

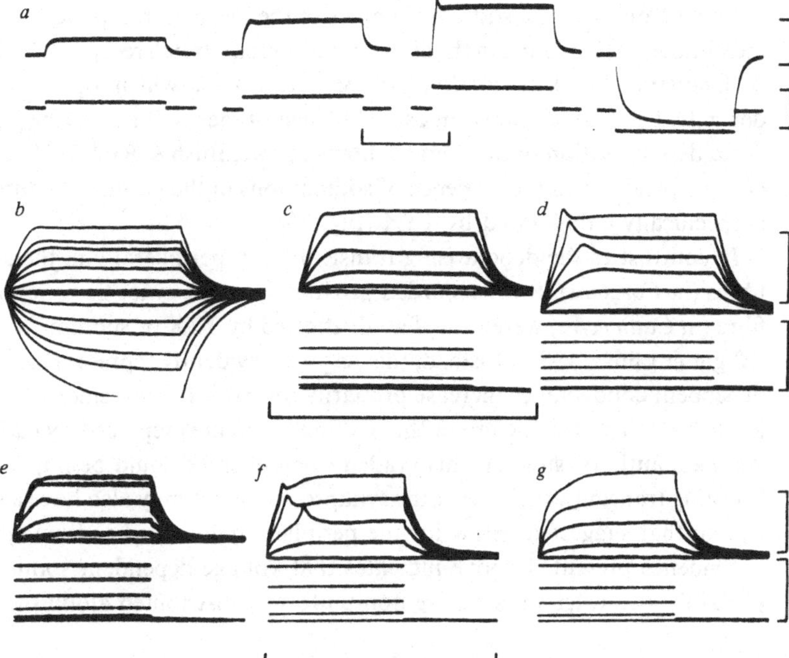

Experiments with two or three microelectrodes impaling the fibres at different points suggest that these apparently active membrane responses may not be evenly distributed along their length. Rather, at least in the crabs *Carcinus* (Cannone & Bush, 1980*a*) and *Scylla* (Mirolli, 1979*a*), they tend to be most prominent in the distal regions of the fibre. Thus the decrement in the initial spiky (α) component of a stretch-evoked response is greater than that of the remainder of the receptor potential (Fig. 6*d*), while any transient in the response to a depolarising current pulse is larger and occurs earlier at the distal electrode, even when the current is injected proximally (Fig. 6*e*). On the other hand, recordings from the synaptic zone of the T fibre in the crab *Callinectes* indicate that similar, graded, depolarisation transients may also arise in this intraganglionic region of the fibre, even when apparently absent or significantly smaller in the fibre immediately outside the ganglion (Blight & Llinás, 1980).

What is the ionic basis of these graded spikes? Two lines of evidence indicate that those occurring peripherally in the afferent fibres depend upon fast, voltage-sensitive sodium channels. First, the graded spiky components of the responses to receptor stretch or depolarising current pulses are blocked by 10^{-7} M tetrodotoxin (Figs. 8*a*, 7*g*: Roberts & Bush, 1971; Bush, 1977; Bush, DiCaprio & Taylor, 1980 and unpublished). Whether or not graded spikes are present, however, tetrodotoxin (TTX) has no effect on the primary, stretch-evoked receptor potentials, even though these are themselves sodium dependent. Thus the fast, TTX-sensitive sodium channels are evidently different from the slow sodium channels of the mechanoreceptive transducer membrane, as also found in the crayfish abdominal stretch receptor (Nakajima & Onodera, 1969). Secondly, 10^{-5} M veratridine, which opens voltage-dependent sodium channels in excitable membranes (Ulbricht, 1969), causes rapid depolarisation of the S and T fibres (Lowe, Bush & Ripley, 1978). This effect depends upon the presence of sodium ions in the bathing medium, and is specifically antagonised by TTX (Fig. 8*b*).

In contrast to the depolarising transients seen peripherally in the afferent fibres (in *Carcinus*), those recorded proximally in the synaptic zone of the T fibre (in *Callinectes*) were little if at all affected by TTX or sodium-free saline (Blight & Llinás, 1980). Instead, this response evidently represents a voltage-dependent conductance increase primarily for calcium ions, since it is largely abolished when the calcium in the bathing medium is replaced by cadmium. As these authors' suggest, this graded calcium spike could be important in boosting transmitter release at the synapse, though it may also have a similar role to that suggested below for the peripherally localised sodium spikes.

Evidence presented above indicates that voltage-dependent sodium (and probably also calcium) sites are non-uniformly distributed along the length of the fibres. Moreover the observed variability in 'spikeness' within and particularly between species indicates considerable variability in the density as well as the distribution of the active sites. Whether the fast sodium channels overlap or are spatially separated from the 'slow' mechanosensitive sodium channels of the transduction membrane, however, remains uncertain. If, as in other, spiking neurones, the fast channels are contiguous with membrane exhibiting sodium pump activity, then a more-or-less complete separation of fast and slow channels would seem likely, since there is some evidence that the transducer and electrogenic regions of the S fibre at least are spatially separated (Krauhs & Mirolli, 1975; Mirolli, 1979*b*).

Given that fast sodium channels of some kind are present in these

Fig. 8. Tetrodotoxin and veratridine effects on receptor and membrane potentials of two T fibres in *Carcinus*. *a*. Graded spikes are blocked by 0.3 μM tetrodotoxin without affecting the underlying receptor potentials; calibrations: 20 mV and 20 ms per division. *b*. Veratridine (50 μM) depolarises the membrane potential, eventually reversing the receptor potentials, and both effects are abolished by 1 μM tetrodotoxin. Representative receptor potentials (*c*, *d*) and the stretch stimulus are shown on a faster time-base; those before veratridine were similar to (*d*). *a*, from Roberts & Bush, 1971; *b–d*, from Lowe, Bush & Ripley, 1978.

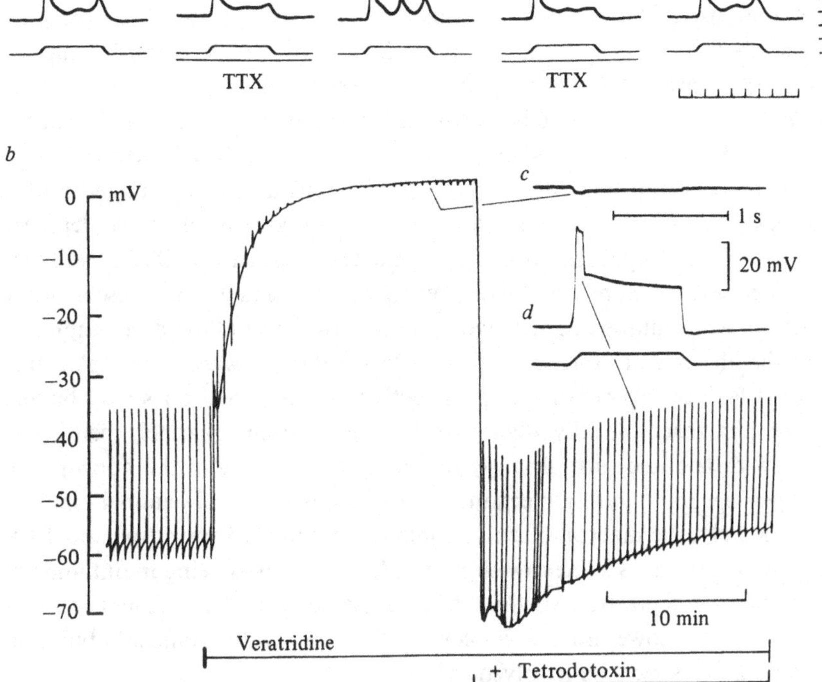

non-impulsive neurones, what distinguishes their membranes from those of 'normal' neurones which support all-or-none sodium action potentials? One possibility is that the fast sodium channels are permanently inactivated, partially in preparations showing small graded spikes or completely in other preparations. Evidence in support of this view is provided by the appearance of graded spikes on the rising phase of T or S fibre responses to depolarising current steps applied after a period of imposed hyperpolarisation (Fig. 7*d*), or their disappearance after preceding depolarisation of the fibre (Fig. 7*e*: Bush *et al.* (1980) and unpublished observations). However, even after several minutes of hyperpolarisation to levels as great as 100 mV or more, the S and T fibres still fail to produce all-or-none overshooting action potentials, though in some preparations there is an increase in spontaneous irregular fluctuations of membrane potential during the hyperpolarising currents. Thus, although there may indeed be some degree of sodium inactivation at normal membrane potentials in these neurones, this can only partially explain their non-regenerative nature.

The further possibility that persistent potassium activation, due to the often relatively low prevailing membrane potentials in these neurones, could underlie the lack of regenerative impulses, is also unlikely to be the whole explanation in view of the evidence from prolonged hyperpolarising currents. The known potassium channel blocking agents, tetrethylammonium or 3- or 4-aminopyridine, have no obvious effect on the stretch-evoked receptor potentials in *Carcinus* (B. M. H. Bush, D. A. Lowe & A. Roberts, unpublished observations) or on the T fibre responses to depolarising current pulses in *Callinectes* (Blight & Llinás, 1980) when applied extracellularly. The latter authors did observe a small, graded, presumably calcium-mediated, regenerative response in the fibre's synaptic region after prolonged iontophoretic injection of TEA *into* the T fibre, but only with very large depolarisations. Thus it appears improbable that there are any sigificant voltage-dependent (or calcium activated) potassium conductances in these sensory neurones which alone might prevent the generation of all-or-none impulses.

In the absence of more definitive evidence, then, it would seem plausible that the lack of regenerative impulse activity in these neurones may be due to the pharmacologically demonstrated fast sodium channels being too sparsely distributed, insufficiently differentiated, lacking in the appropriate voltage-dependent gating mechanisms, or having modified channel kinetics. Alternatively, the macromolecular complex containing veratridine and TTX sensitive receptors is somehow suppressed, the channels being maintained in the closed configuration so that the membrane is in a permanent state of refractoriness (Lowe, Bush & Ripley, 1978). Conceivably some combination of these factors may be involved.

Role of active membrane responses

Given that voltage-dependent membrane responses appear to contribute to the afferent signals, albeit to a limited extent, do they play any significant role in the normal physiological functioning of these non-impulsive sensory neurones? Since they appear to occur sporadically even within a particular species, and when present are generally quite small, it could be argued that these active responses are unlikely to have any important function, but simply represent an evolutionary remnant of once 'normal' nerve membrane.

Support for such a thesis might be claimed on the basis of two experimental observations. First, the evident localisation of any graded sodium spikes towards the periphery of the afferent fibres, and their relatively larger attenuation along the fibre than the primary receptor potential, noted above, suggest that they may not in fact reach the central nervous system. Secondly, the promotor reflex response to either stretch or current evoked depolarisation of the T fibre shows little or no apparent effect of the early spike (see below). With slow ramp-function stretches in particular, any high frequency burst at the onset of the reflex discharge appears to be a direct response to the β component of the receptor potential, rather than the preceding α component (see Figs. 9*b*, 10*a*, *b*).

There remains the possibility that any voltage-dependent responses in the afferent fibres could be of physiological value in 'sharpening up' the rising phase of rapid depolarising signals. Such an effect has already been noted when comparing the response to depolarising and hyperpolarising current steps, even when subthreshold for graded spikes (Fig. 7*b*). This property may well play a significant role in compensating for the filtering effect of the large membrane resistance, and hence time constant, upon the transmission of rapid transients or high frequency signals. That the graded calcium spikes observed in the synaptic region of the T fibre in *Callinectes* may be of functional importance in this way, is indicated by their effect in speeding up the post-synaptic response to strong presynaptic depolarisation, and hence reducing the promotor reflex latency (Fig. 12*c*; Blight & Llinás, 1980.

Reflex role of non-impulsive afferent input

Can the non-impulsive activity of the S and T fibres observed in isolated preparations be considered representative of their normal physiological behaviour? While no direct test of this question has yet been devised in an intact animal, it is possible to record reflex activity simultaneously with the afferent response *in situ* in a perfused dissected preparation. Imposed movement of the most proximal thoracic–coxal joint of any leg results in modulation of the motor output to the basal muscles of that leg. Passive remotion most commonly results in reflex excitation of the motoneurones innervating the leg promotor muscle, and simultaneous inhibition of the remotor motoneurones (Bush & Roberts, 1968; Bush, 1977). This is the efferent basis of the 'resistance reflex', analogous to the Sherringtonian myotatic reflex. In certain conditions, however, particularly in lively preparations, imposed joint movements may evoke 'assistance reflexes', in phase with and thus tending to reinforce rather than resist the imposed movements (DiCaprio & Clarac, 1980). The mechanism of this switching between resistance and assistance modes remains to be determined. Its occurrence may reflect endogenous changes between postural and locomotory 'states', respectively, of the central nervous system.

Fig. 9. Promotor reflex and simultaneous afferent responses to receptor muscle stretch at different velocities in *Carcinus*. *a*, *b*. Instantaneous frequency displays (top) for the small promotor unit, Pm1, together with T and S receptor potentials and recordings from the whole promotor nerve (bottom traces), showing impulses in Pm1 and two large units, Pm2 and 3. *c*. T fibre and reflex responses, showing parallel adaptation; only Pm1 and 2 continue to discharge after the dynamic component. *d*. Dynamic reflex and afferent responses to a fast stretch. Note that Pm2 and 3 impulses are often indistinguishable (*a*, *b*, *d*), or Pm3 impulses may be slightly smaller that those of Pm2 (*c*). Calibrations: 100 Hz (*a*, *b*); 20 mV (*a–d*), 1 s (*a–c*) or 100 ms (*d*). From Bush & Cannone, 1973 and unpublished; Bush, 1977; Cannone & Bush, 1980*a*.

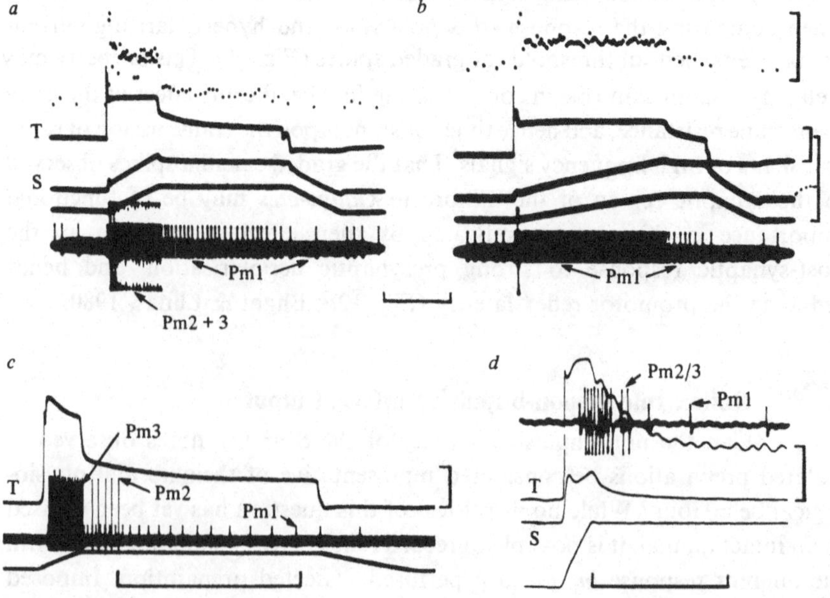

The sensory input for the promotor resistance reflex (and also the reciprocal assistance reflexes) is mediated by the single TCMRO of that limb. Thus stretching the receptor muscle directly in a posterior leg preparation of the shore crab (*Carcinus maenas*) may excite up to 8 or 9 promotor motoneurones (Bush, 1976; Cannone & Bush, 1980*a*). Cutting the TCMRO sensory nerve evokes strong and prolonged discharge of several or all of the promotor motoneurones, and abolishes all reflex response to subsequent passive movement of the thoracic–coxal joint or receptor muscle stretch. A similar result follows cutting the T afferent fibre alone. By contrast, cutting only the S fibre elicits a relatively weak promotor discharge, and appears to have little or no effect on the reflex response to subsequent joint movement or receptor stretch. These observations suggests that the promotor stretch reflex is mediated largely by the T fibre, the S fibre playing at most a minor, perhaps supplementary role in this reflex.

Fig. 10. Correlated influences of the receptor muscle on T fibre and promotor reflex responses in *Carcinus*. *a*. Dynamic responses to identical stretches of a slack receptor muscle following different intervals (as indicated). *b*. Effect of brief 100 Hz stimulation (bars) of receptor motor nerve upon subsequent dynamic response to stretch at 5 s intervals. Note separate development of α and β components. *c*. Adaptation in promotor (Pm1–3) reflex responses to a sustained T fibre depolarisation evoked by 100 Hz stimulation of the motor nerve, which elicits summating excitatory junction potentials in the (isometric) receptor muscle (RM); dots indicate occurrence of Pml impulses. Calibrations: 20 mV, 1 s. From Bush, 1977; Cannone & Bush, 1980*a*, *c*.

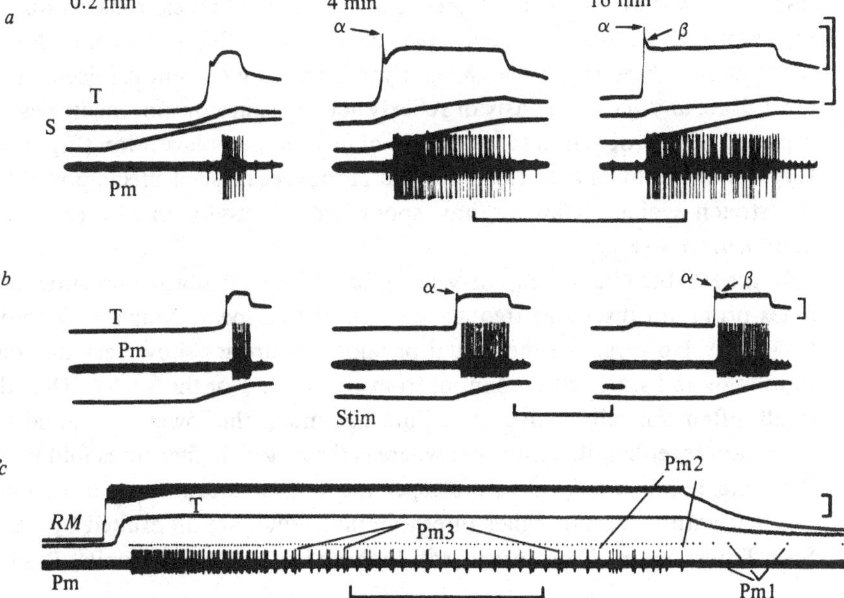

This conclusion is reinforced by experiments involving intracellular recording from and/or current injection into the T or S fibre, while simultaneously recording (extracellularly) from the promotor motor nerve. The promotor response to constant velocity ramp–hold–release stimuli, for example, shows a clear correspondence with the waveform of the concurrent T fibre receptor potential, and accordingly differs substantially from the S fibre's response waveforms (Bush & Cannone, 1973; Cannone & Bush, 1980*a*: Fig. 9). For a given state of the preparation, the number of active promotor units and their individual discharge frequencies vary directly with the instantaneous level of depolarisation of the T fibre, which in turn varies with the parameters of the stimulus and with other time-, length- and tension-dependent factors (Figs. 9, 10*a*, *b*). Further, endogenous or evoked receptor motor activity leading to an overt increase in receptor muscle tension and consequent depolarisation of the T fibre, or simply to an enhancement in the dynamic response to subsequent passive stretch, results in corresponding increments in the reflex response (Fig. 10*b*, *c*; Cannone & Bush, 1980*c*).

When either the T or S fibre is depolarised directly by current injection, either via an intracellular microelectrode (Bush & Roberts, 1968; Bush & Cannone, 1973; Bush, 1977), or through the cut central end of the isolated fibre across a sucrose gap (Blight & Llinás, 1980), one or more promotor motoneurones are excited (Fig. 11). Both the number of motoneurones involved and their individual discharge frequencies increase with the level of imposed depolarisation, up to eight being activated by strong depolarisation of the T fibre in *Carcinus* (Cannone & Bush, 1980*b*). This current-induced response can sum with a simultaneous stretch-evoked reflex, thus producing high motor discharge frequencies, particularly on T fibre stimulation during the dynamic phase of stretch. When hyperpolarising current is injected into the T fibre, any spontaneously or reflexly active promotor motoneurones are suppressed, in proportion to the degree of membrane polarisation (Fig. 11*d*). Hyperpolarisation of the S fibre, however, produces no such effect upon either the stretch-evoked reflex or any spontaneous activity in the promotor motoneurones (Fig. 11*e*).

In general the relationship between afferent fibre membrane potential and reflex promotor discharge frequency in *Carcinus* appears roughly sigmoidal (Fig. 11*b*). However the individual promotor neurones show very different thresholds and sensitivities to input from either the T or the S fibre. Thus the small, often tonically active unit, Pml, has much the lowest threshold for excitation by either afferent fibre, whereas the larger, higher threshold units, Pm2 and Pm3, usually have a steeper maximum slope. These three units together with a fourth, higher threshold unit, Pm4, are all excited by either S or T fibre input. Four larger, still higher threshold phasic units, Pm5–8,

Fig. 11. Reflex promotor and receptor motor responses to current injected into the T or S fibre in *Carcinus*. *a*. Promotor responses to identical constant current pulses (not shown) in the T fibre at different receptor muscle lengths (as indicated). Note progressive decrease in T fibre potential change with increase in receptor length, and recruitment of Pm1–3 with increasing *levels* of depolarisation. *b*. Input–output relationships for T fibre-induced firing of motoneurones Pm1–3. The arrow indicates the resting membrane potential (at a receptor length of 9 mm), and the maximum slopes or 'gains' are also shown. *c*. A large T fibre depolarisation, produced here by strong current and concurrent receptor stretch, may excite both receptor motoneurones, Rm1 and Rm2, and one or more high threshold promotor motoneurones (Pmx). *d*. Graded hyperpolarisation of the T fibre progressively suppresses tonic promotor activity and their reflex response to stretch. *e*. Similar hyperpolarisation of the S fibre (bottom record) does not do so, even though strong S fibre depolarisation can excite Pm1–3 (top). Calibrations: 20 mV, 1 s. From Bush, 1977; Cannone & Bush, 1980*b*.

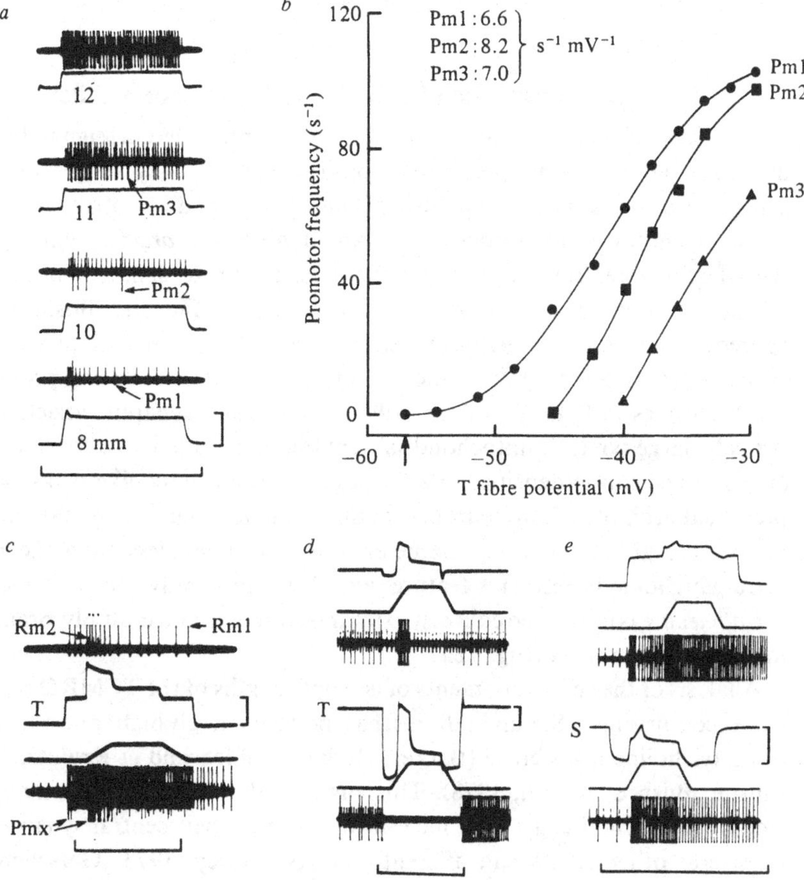

are recruited only by the T fibre, while a further very small unit, Pm9, is activated by strong S fibre depolarisation but not by the T fibre.

In addition to providing a continuously graded input to up to eight motoneurones innervating the promotor muscle, the T fibre (but not the S fibre) also exerts reflex control over the two excitatory motoneurones of its own receptor muscle (Bush & Cannone, 1974; Bush, 1976, 1977). Thus strong depolarisation of the T fibre, and in a lively preparation stretching the receptor muscle, gradedly excites first the larger receptor motor unit and, with very large depolarisations, also the smaller unit (Fig. 11*c*). This 'autogenic reflex', like the main promotor stretch reflex, is abolished by hyperpolarisation (or section) of the T fibre, but not of the S fibre. As to its functional significance, a plausible hypothesis in view of its slow mechanical time course is that it may provide reinforcement or 'gain control' of the main stretch reflex, for instance by helping to offset any unloading of the receptor as a consequence of reflex promotor shortening.

Synaptic transmission in the TCMRO-promotor reflex

The close reflex following of the T fibre afferent signal by the discharge pattern of the promotor motoneurones strongly suggests monosynaptic transmission in this pathway. This is supported by cobalt 'backfills' of the promotor motoneurones and S and T fibres in *Carcinus*, which show the afferent fibres running through the main region of dendritic branching of these motoneurones in the thoracic ganglion (Fig. 1*c*; Bush, 1976). Moreover in *Callinectes*, light- and electron-microscope sections reveal close apposition between both the T and S fibres and dendrites of several promotor motoneurones (Blight & Llinás, 1980). Numerous synaptic vesicles and typically large synaptic mitochondria were found in both fibres, concentrated around presynaptic densities in areas adjoining fine spine-like processes of presumed promotor motoneurones. In this putative synaptic zone the afferent fibres were about 50 μm in diameter, only 20–30 μm less than the main extra-ganglionic fibres. This feature, and their apparently sparse branching in *Callinectes*, is in marked contrast to the more tapered and multiply-branched motoneurone profiles (Fig. 12*d*).

Analysis of the cellular contents of isolated lengths of the TCMRO sensory nerve (comprising T, S, P and D fibres) has shown relatively high concentrations of acetylcholine transferase (Barker, Herbert, Hildebrand & Kravitz, 1976; Emson, Bush & Joseph, 1978). This suggests that these sensory neurones employ acetylcholine as a chemical transmitter at their central synapses, as at several other crustacean afferent endings (Florey, 1973; Gerschenfeld, 1973). In *Callinectes*, however, no effect was observed upon the T fibre-promotor transmission characteristics of conventional blockers of cholinergic synapses (Blight & Llinas, 1980), so the question remains open.

In the context of this book, one of the most interesting and significant aspects of the crab coxal receptor system is the continuously graded, *tonic* nature of the synaptic transmission in the stretch reflex pathway. One or more promotor motoneurones may be tonically active due to steady presynaptic input, and small (less than 10 mV) increments in T fibre membrane potential can evoke powerful reflex effects (Fig. 11*b*; Cannone & Bush, 1980*b*, *c*). How is this tonic control of efferent output, over a wide frequency range in up to eight promotor motoneurones, achieved by the membrane potential changes of a single sensory neurone (the T fibre)?

In their recent detailed study of the reflex pathway in *Callinectes*, Blight & Llinas (1980) recorded intracellularly from the synaptic zone of the T fibre and simultaneously from one of the three largest of the seven promotor motoneurones. Depolarising current pulses (50–100 ms long) injected across a sucrose gap into the cut central end of the isolated T fibre evoked graded, smoothly rising, excitatory post-synaptic potentials (EPSPs) in the motoneurones (Fig. 12*a*, inset). Both the rate of rise and peak amplitude of the post-synaptic response, recorded in the motoneurone soma, increased sigmoidally with presynaptic depolarisation, reaching a maximum around zero potential (Fig. 12*a*). The responses showed an apparent threshold presynaptic level, 10–15 mV depolarised from the imposed transmembrane potential of −80 to −90 mV. The minimum latency between the presynaptic depolarisation reaching threshold and the onset of the post-synaptic response (recorded in a motoneurone neurite, closer to the synapse than the soma) was about 1.7 ms. With very large presynaptic depolarisations (over 60 mV) the post-synaptic responses in some cases became progressively 'suppressed', declining during a 100–200 ms pulse but giving a transient 'off response' at the end, indicating that this suppression is not due to transmitter depletion.

These transfer characteristics of the crab T fibre-promotor neurone synapse closely resemble those found previously for the chemically mediated giant synapse of the squid. For example, the transmission delay calculated for the squid synapse from records comparable to those of Blight & Llinás is about 1.5 ms (Katz & Miledi, 1970). Even the values of the presynaptic potential producing threshold and maximal post-synaptic responses, the 'suppression' potential, and the slope of the logarithmic relationship between pre-potential and post-response (*c*. 12.5 mV presynaptic depolarisation per tenfold increase in post-synaptic response) are all similar to those found for the squid synapse (e.g. Katz & Miledi, 1967; Llinás, 1977). However, in contrast to the squid synapse, which is adapted for the transmission of large *phasic* signals of the kind involved in rapid escape responses, the crab's T fibre motoneurone

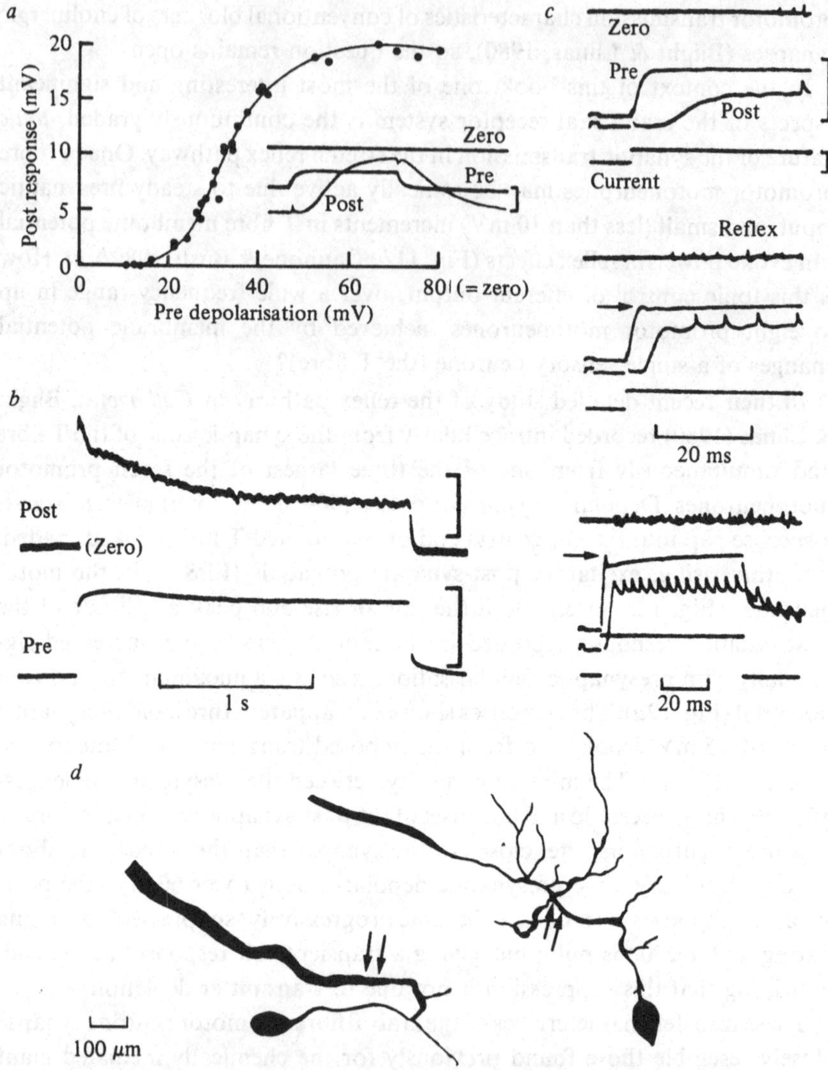

Fig. 12. Transfer characteristics of T fibre–promotor motoneurone synapse in *Callinectes*, determined by current injection into the T fibre across a sucrose gap. *a*. Relationship between peak amplitude of postsynaptic potential in a type I promotor neurone, and depolarisation of the T fibre from an imposed membrane potential of −80 mV; inset shows sample response to a 100 ms constant current pulse. *b*. Type II motoneurone and T fibre responses to a long 1.3 μA pulse. *c*. Promotor nerve reflex (top traces) and concurrent pre- and post-synaptic responses to three different current pulses (bottom traces); note reduction of reflex latency following presynaptic 'depolarisation transient' (arrows). *zero*, reference level for the presynaptic potential (*pre*) recorded in the synaptic zone of the T fibre (arrows in *d*); *post*, post-synaptic potential recorded in

synapse can continue to transmit over many minutes (Bush & Roberts, 1968; Cannone & Bush, 1980*b*; Blight & Llinas, 1980). With large sustained presynaptic depolarisations, however, some depression of the post-synaptic response may occur over the first second or so (Fig. 12*b*). This is attributed by Blight & Llinás to depletion of 'reserve' stores of transmitter, over and above the basic turnover capacity, which is sufficient to maintain transmission indefinitely at the normally lower levels. Recovery from this depression occurs in about 1.5 s, compared to 4–5 s in the squid synapse or rat neuromuscular junction. Evidently the vesicular turnover rate is high in the crab compared to the squid synapse, which shows depletion within a few milliseconds (Katz & Miledi, 1967; Llinás, 1977). Moreover the crab T fibre synapse shows no evidence of facilitation or habituation, as found in many phasic synapses (e.g. that mediating the gill-withdrawal reflex of *Aplysia*; Kandel, 1976). These properties of the crab's T fibre motoneurone synapse are all appropriate to its role in mediating the promotor stretch reflex, the tonic component of which is undoubtedly of prime importance in postural control.

As noted previously, the promotor reflex response to a maintained stretch of the TCMRO shows a decrease in frequency which parallels the decline in T fibre membrane potential (Fig. 9*c*). In many cases, however, the decline in impulse frequency manifestly exceeds that of the receptor potential, both during the initial dynamic response and during the 'static' component (Figs. 9*a*, 10*c*). This may be at least partly explained by the depression in the post-synaptic response to large and prolonged depolarisations of the T fibre, noted above (Fig. 12*b*). Thus the early fast component of this synaptic depression has a similar time course (100–200 ms) to the decay of the initial high frequency burst in the impulse discharge, while the ensuing slower component is reflected in the frequency decline over the first second or two of the maintained stretch. In addition, however, there is a much slower decline in reflex discharge frequency, possibly resulting from a process resembling 'accommodation' in the spike-generating membrane (Cannone & Bush, 1980*b*).

A further question which the experiments of Blight & Llinás (1980) help to resolve is the origin of the differences in thresholds for activation of the various promotor units. These authors found no differences in the presynaptic (T fibre) threshold for EPSP generation in the three largest promotor

---

the motoneurone soma (*a*, *b*) or a neurite close to the synaptic zone (*c*). Calibrations: 50 mV presynaptically, 5 mV post-synaptically; and (*c*) 1 μA current. *d*. Camera lucida drawings of a cobalt stained T-fibre cell (left) and a type I promotor motoneurone (right): arrows indicate the region of closest apposition as determined by histological sections. From Blight & Llinás, 1980.

motoneurones (their types I–III). Yet the threshold for *impulse* initiation in the largest, phasically discharging type I motoneurone was some 20 mV more depolarised than the presynaptic threshold for its own EPSP, at which point the EPSP was already about 60% of its maximal amplitude. By contrast, the recorded presynaptic thresholds for both the EPSP and the resulting impulse discharge in the more tonic type III motoneurone were close together, and almost identical to that of the type I EPSP threshold. Thus at least among these three or four promotor units in *Callinectes*, the threshold differences between the phasic and relatively tonic motoneurones seem to reside mainly in the spike generating mechanisms, rather than in their respective EPSPs. Although most of the promotor responses studied previously in *Carcinus* (Bush, 1976, 1977; Cannone & Bush, 1980*a*, *b*, *c*) were in units other than the two or three largest, most phasic motoneurones, a similar basis for the observed differences in reflex response thresholds may well apply here too.

Advantages of graded non-impulsive afferent signalling

Simple intuitive considerations suggest that non-impulsive transmission of signals within and between neurones might have several functional advantages over the more conventional impulse coded mode of nerve signalling. In particular, electrotonic conduction does not involve the encoding of the input signal into a frequency modulated spike train and subsequent decoding to produce a graded post-synaptic response, in both of which there is scope for loss of information. However, the important disadvantage, as indicated previously, is the effect of such signalling upon the high frequency response of the conducting fibre. Thus the high membrane resistance required for a large length constant implies also a large time constant, and this has a limiting effect upon the range of frequencies which can be reliably transmitted. This limitation can to some extent be overcome by mechanisms of the kind described in the present case, namely reduction of the effective membrane time constant by reducing the input resistance of the cell, and by the provision of some degree of voltage-dependent conductance. The former device is presumably more effective in neurones of relatively short length, since the effect upon the membrane time constant decreases with increasing distance from the low resistance membrane. Furthermore, the resultant reduction in membrane potential and hence signal amplitude needs to be offset, as in this case, by a significant electrogenic component to the membrane potential.

Given the possibility for the major drawbacks of non-impulsive signalling, namely the inter-related problems of signal attentuation and loss of high frequency response, to be overcome by appropriate devices such as those illustrated here, it remains important to demonstrate that the supposed positive advantages of non-spike signalling are in fact valid and realistic. One approach to this problem is to utilise modelling techniques (Pearson, 1976; R. A. DiCaprio & K. G. Pearson, personal communication). Using such an approach, it can be shown that single spiking inputs lead to discontinuities in the output signal. This is easy enough to visualise, since at low input discharge frequencies the post-synaptic potentials would appear as discrete events unless the time constant of the post-synaptic membrane was unduly long, in which case high frequency information transfer would again be lost. Only by a multiplicity of spiking inputs is it possible to provide a smoothly graded output signal over a wide dynamic range. This is an important function of the large numbers of afferent neurones commonly present in sensory systems, including for example, the chordotonal proprioceptor organs at most crustacean leg joints (Mill, 1976).

One exception to the rule that spiking afferent inputs generally occur in numbers is the stretch reflex of the crayfish abdominal muscle receptor organ. Whereas the stretch reflex of vertebrate skeletal muscles in most cases involves large populations of muscle spindle afferents feeding monosynaptically into a common pool of motoneurones, the crayfish abdominal stretch reflex involves only a single afferent channel (the slowly adapting receptor, MRO1) feeding into a single (slow extensor) motoneurone, probably monosynaptically. In contrast to the monosynaptic reflex output of the several promotor neurones in the TCMRO system, however, the single extensor motoneurone excited by the abdominal MRO does not show close, high-sensitivity following of the afferent input. Rather it appears to be a somewhat irregular reflex output at a low frequency, the input–output relation varying with a ratio of 5:1 to 10:1 up to a maximum output frequency of around 15 impulses per second (see Fields, 1976). No doubt this low gain reflex following in the abdominal MRO case does not present any serious functional disadvantages, since the reflex appears to be part of a general abdominal postural control system, which in itself is a relatively simple behavioural pattern. In the case of the walking legs, however, it would seem important that any reflex feedback control should be precise and rapid in its effect, particularly since the coordination of the multi-articulated legs must of necessity involve a much more complex control system.

The computer model referred to above demonstrates clearly that the output from a single spiking input channel is necessarily discontinuous, at least at low frequencies. This, together with the various experimental observations referred to above, seems to favour non-spiking transmission if only a single or limited number of input channels is available. As shown in the TCMRO-promotor reflex, precise and close following by the motoneurones of the non-impulsive input from the T fibre is clearly evident. This particular example, therefore, appears to provide a good illustration of the advantages of non-impulsive transmission within and between neurones.

References

Alexandrowicz, J. S. (1958). Further observations on proprioceptors in Crustacea and a hypothesis about their function. *J. Mar. Biol. Assocn.* **37**, 379–96.

Alexandrowicz, J. S. (1967). Receptor organs in the coxal region of *Palinurus vulgaris. J. Mar. Biol. Assocn*, **47**, 415–32.

Alexandrowicz, J. S. & Whitear, W. (1957). Receptor elements in the coxal region of decapod Crustacea. *J. Mar. Biol. Assocn*, **36**, 603–28.

Barker, D. L., Herbert, E., Hildebrand, J. G. & Kravitz, E. A. (1972). Acetylcholine and lobster sensory neurones. *J. Physiol., Lond.* **226**, 205–29.

Berger, C. S. & Bush, B. M. H. (1979). A non-linear mechanical model of a non-spiking muscle receptor. *J. Exp. Biol.* **83**, 339–43.

Blight, A. R. & Llinás, R. (1980). The non-impulsive stretch-receptor complex of the crab: a study of depolarisation-release coupling at a tonic sensorimotor synapse. *Phil. Trans. R. Soc. Series B*, **290**, 219–76.

Bush, B. M. H. (1976). Non-impulsive thoracic-coxal receptors in crustaceans. In *Structure and Function of Proprioceptors in the Invertebrates*, ed. P. J. Mill, pp. 115–51. London: Chapman & Hall.

Bush, B. M. H. (1977). Nonimpulsive afferent coding and stretch reflexes in crabs. In *Identified Neurons and Behavior of Arthropods*, ed. G. Hoyle, pp. 439–60. New York: Plenum Press.

Bush, B. M. H. & Cannone, A. J. (1973). A stretch reflex in crabs evoked by muscle receptor potentials in nonimpulsive afferents. *J. Physiol., Lond.* **232**, 95–7*P*.

Bush, B. M. H. & Cannone, A. J. (1974). A positive feed-back reflex to a crustacean muscle receptor. *J. Physiol., Lond.* **236**, 37–9*P*.

Bush, B. M. H., DiCaprio, R. A. & French, A. S. (1978). White noise analysis of a non-spiking stretch receptor. *J. Physiol., Lond.* **277**, 15*P*.

Bush, B. M. H., DiCaprio, R. A. & Taylor, P. S. (1980). A preparation to investigate voltage-dependent changes in a non-spiking nerve membrane. *J. Physiol.* **303**, 20–1*P*.

Bush, B. M. H. & Godden, D. H. (1974). Tension changes underlying receptor potentials in nonimpulsive crab muscle receptors. *J. Physiol., Lond.* **242**, 80–2*P*.

Bush, B. M. H., Godden, D. H. & Macdonald, G. A. (1975). Voltage clamping of non-impulsive afferents of the crab thoracic-coxal muscle receptor. *J. Physiol., Lond.* **245**, 3–5*P*.

Bush, B. M. H. & Roberts, A. (1968). Resistance reflexes from a crab muscle receptor without impulses. *Nature, Lond.* **218**, 1171–3.

Cannone, A. J. & Bush, B. M. H. (1980*a*). Reflexes mediated by non-impulsive afferent neurones of thoracic-coxal muscle receptor organs in the crab, *Carcinus maenas*. 1. Receptor potentials and promotor motoneurone responses. *J. Exp. Biol.* **86**, 275–303.

Cannone, A. J. & Bush, B. M. H. (1980*b*). Reflexes mediated by non-impulsive afferent neurones of thoracic-coxal muscle receptor organs in the crab, *Carcinus maenas*. 2. Reflex discharge evoked by current injection. *J. Exp. Biol.* **86**, 305–31.

Cannone, A. J. & Bush, B. M. H. (1980*c*). Reflexes mediated by non-impulsive afferent neurones of thoracic-coxal muscle receptor organs in the crab, *Carcinus maenas*. 4. Motor activation of the receptor muscle. *J. Comp. Physiol.* (in press).

Carpenter, D. O. (1973). Electrogenic sodium pump and high specific resistance in nerve cell bodies of the squid. *Science*, **179**, 1336–8.

DiCaprio, R. A. & Clarac, F. (1980). Reversal of a walking leg reflex elicited by a muscle receptor. *J. Exp. Biol.* **88**.

Emson, P. C., Bush, B. M. H. & Joseph, M. H. (1978). Transmitter metabolising enzymes and free amino acid levels in sensory and motor nerves and ganglia of the shore crab (*Carcinus maenas*). *J. Neurochem.* **26**, 779–83.

Fields, H. L. (1976). Crustacean abdominal and thoracic muscle receptor organs. In *Structure and Function of Proprioceptors in the Invertebrates*, ed. P. J. Mill, pp. 65–114. London: Chapman & Hall.

Florey, E. (1973). Acetylcholine as sensory transmitter in Crustacea. *J. Comp. Physiol.* **83**, 1–16.

Gerschenfeld, H. M. (1973). Chemical transmission in invertebrate central nervous systems and neuromuscular junctions. *Physiol. Rev.* **53**, 1–119.

Gorman, A. L. F. & Mirolli, M. (1972). The passive electrical properties of the membrane of a molluscan neurone. *J. Physiol., Lond.* **227**, 35–49.

Hodgkin, A. L. & Rushton, W. A. H. (1946). The electrical constants of a crustacean nerve fibre. *Proc. R. Soc. Lond. Series B*, **133**, 444–79.

Hunt, C. C. & Ottoson, D. (1976). Initial burst of primary endings of isolated mammalian muscle spindles. *J. Neurophysiol.* **39**, 324–30.

Kandel, E. R. (1976). *Cellular Basis of Behavior*. San Francisco: W. H. Freeman & Company.

Katz, B. & Miledi, R. (1967). A study of synaptic transmission in the absence of nerve impulses. *J. Physiol., Lond.* **192**, 407–36.

Katz, B. & Miledi, R. (1970) A further study of the role of calcium in synaptic transmission. *J. Physiol., Lond.* **207**, 789–801.

Krauhs, J. M. & Mirolli, M. (1975). Morphological changes associated with stretch in a mechano-receptor. *J. Neurocytol.* **4**, 231–46.

Llinás, R. R. (1977). Calcium and transmitter release in squid synapse. In *Society for Neuroscience Symposia*, vol. II, ed. W. M. Cowan & J. A. Ferendelli, pp. 139–60. Bethesda: Society for Neuroscience.

Lowe, D. A., Bush, B. M. H. & Ripley, S. H. (1978). Pharmacological evidence for 'fast' sodium channels in nonspiking neurones. *Nature, Lond.*, **274**, 289–90.

Mellon, DeF. & Kennedy, D. (1964). Impulse origin and propagation in a bipolar sensory neuron. *J. Gen. Physiol.* **47**, 487–99.

Mendelson, M. (1966). The site of impulse initiation in bipolar receptor neurones of *Callinectes sapidus* L. *J. Exp. Biol.* **45**, 411–20.

Mill, P. J. (1976). Chordotonal organs of crustacean appendages. In *Structure and Function of Proprioceptors in the Invertebrates*, ed. P. J. Mill, pp. 243–97. London: Chapman & Hall.

Mirolli, M. (1979*a*). The electrical properties of a crustacean sensory dendrite. *J. Exp. Biol.* **78**, 1–27.

Mirolli, M. (1979*b*). Electrogenic Na^+ transport in a crustacean coxal receptor. *J. Exp. Biol.* **78**, 29–45.

Nakajima, S. & Onodera, K. (1969). Membrane properties of the stretch receptor neurones of crayfish with particular reference to mechanisms of sensory adaptation. *J. Physiol.* **200**, 161–85.

Paul, D. H. (1972). Decremental conduction over 'giant' afferent processes in an arthropod. *Science*, **176**, 680–2.

Paul, D. H. (1976). Role of proprioceptive feedback from nonspiking mechanosensory cells in the sand crab, *Emerita analoga. J. Exp. Biol.* **65**, 243–58.

Pearson, K. G. (1976). Nerve cells without action potentials. In *Simpler Networks and Behavior*, ed. J. C. Fentress, pp. 99–110. Sunderland, Massachusetts: Sinauer Associates, Inc.

Rall, W. (1969). Time constants and electrotonic length of membrane cylinders and neurons. *Biophys. J.* **9**, 1483–508.

Ripley, S. H., Bush, B. M. H. & Roberts, A. (1968). Crab muscle receptor which responds without impulses. *Nature, Lond.*, **218**, 1170–1.

Roberts, A. & Bush, B. M. H. (1971). Coxal muscle receptors in the crab: the receptor current and some properties of the receptor nerve fibres. *J. Exp. Biol.* **54**, 515–24.

Tasaki, I. (1955). New measurements of the capacity and the resistance of the myelin sheath and the nodal membrane of the isolated frog nerve fibre. *Am. J. Physiol.* **181**, 639–50.

Ulbricht, W. (1969). The effect of veratridine on excitable membranes of nerve and muscle. *Reviews of Physiology, Biochemistry and Pharmacology*, **61**, 18–71.

Weidman, S. (1951). Electrical characteristics of *Sepia* axons. *J. Physiol., Lond.* **114**, 372–81.

Whitear, M. (1965). The fine structure of crustacean proprioceptors. II. The thoracico-coxal organs in *Carcinus, Pagurus and Astacus. Phil. Trans. R. Soc. Series B*, **248**, 437–56.

A. JOHN SIMMERS

Non-spiking interactions in crustacean rhythmic motor systems

Introduction

Neurones without action potentials are becoming recognised as intrinsic components not only of many sensory systems, but also of the central pathways in several different motor systems (see Pearson, 1976, for a review). Examples of non-spiking neurones controlling motor activity are so far confined to rhythmic behaviours of arthropods; viz. ventilatory, swimmeret and food grinding movements in crustaceans, (Mendelson, 1971; Maynard & Walton, 1975; Graubard, 1978; Heitler & Pearson, 1980; Simmers & Bush, 1980), and walking in insects (Pearson & Fourtner, 1975; Burrows & Seigler, 1978; see also Burrows, this volume). The purpose of this chapter is to review briefly the evidence for non-spiking central neurones in crustacean motor systems and to discuss their respective roles in the production and control of rhythmic behaviour.

The three systems examined here represent well established examples of centrally determined neural rhythmicity which have been used frequently as cellular models for studying the genesis of patterned motor activity, in both vertebrates and invertebrates. The ventilatory system of hermit crabs and lobsters was the first in which central non-spiking oscillator neurones were reported to control cyclic motor output (Mendelson, 1971). It remained the only example of this type until Pearson & Fourtner (1975) described cells with similar properties in the walking system of the cockroach. Although the neuronal organisation of the crayfish swimmeret system has been explored extensively over the past 19 years (see Davis, 1973, for a review), only recently have non-spiking neurones been shown to play an important role in generating the swimmeret motor programme (Heitler & Pearson, 1980). Perhaps the best-described rhythmic motor systems, in terms of neuronal connectivity and synaptic interaction, are the pyloric and gastric pattern generator networks of the lobster stomatogastric ganglion (see review,

Selverston, Russell, Miller & King, 1976). Although non-spiking neurones have been demonstrated in this system (Maynard, 1972; Maynard & Walton, 1975; Graubard, 1978), the role of these cells in the operation of the stomatogastric ganglion remains unknown. Of greater functional significance, however, has been the finding that impulse-producing motoneurones which comprise the pyloric pattern generator, use graded synaptic transmission as part of their normal function (Maynard & Walton, 1975; Raper, 1979; Graubard, Raper & Hartline, 1980). The discovery that spiking cells can also participate in local circuit interactions without the need for action potentials presents a further interesting complexity in our attempts to characterise neuronal behaviour. If different regions of the same cell are involved in different integrative functions, then these regions must be considered separately, no longer regarding the cell as a single functional unit (Shepherd, 1972). It is on this basis, therefore, that the 'non-spiking' properties of 'spiking' neurones within the stomatogastric system warrant consideration in the context of this volume.

Gill ventilation system

The gill chambers of decapod crustacea are irrigated by the rhythmic sculling movements of the bilateral pair of scaphognathites or gill bailers. Each gill bailer is controlled mainly by five levator and five depressor muscles innervated by two motor branches of a single ventilatory nerve root arising from the suboesophageal (macrura) or thoracic (brachyura) ganglion (Pasztor, 1968; Young, 1975). During normal ventilatory behaviour, the muscles of a bailer are activated by alternating bursts of impulses in the two populations of antagonistic motoneurones, so as to draw water forward across the gills and out through the exhalent channel of the branchial chambers (Pasztor, 1968; Pikington & Simmers, 1973; Young, 1975). As in other arthropod systems, the ventilatory motor programme can be generated spontaneously by the central nervous system in the complete absence of feedback from peripheral sense organs (Mendelson, 1971).

Detailed descriptions of the ventilatory motor output pattern have yielded some insight into the neuronal organisation underlying its production (Pasztor, 1968; Pilkington & Simmers, 1973; Young, 1975). But the most persuasive evidence for cellular mechanisms of control has come from direct intracellular recordings from ventilatory neurones within the central pattern generating circuitry itself (Mendelson, 1971; Simmers, 1979; Simmers & Bush, 1980). The remainder of this section will be devoted to the results of the latter studies and in particular the evidence for non-spiking neurones controlling ventilatory behaviour.

Intracellular recordings have been made from the central neuropilar processes of ventilatory motoneurones within the semi-isolated thoracic ganglion of the shore crab, *Carcinus maenas* (Simmers & Bush, 1980). During periods of spontaneous cyclic motor output, these cells display large oscillations in membrane potential phase-locked to the peripherally recorded rhythm, with bursts of action potentials superimposed upon the depolarised peaks (Fig. 1). Identification of motoneurones was dependent upon a 1 : 1 correspondence

Fig. 1. Intracellular recordings from the neuropilar segment of a crab bailer levator motoneurone during rhythmic ventilatory motor output. *a*. The motoneurone fires bursts of action potentials riding on the depolarised peaks of smooth membrane potential oscillations (third trace). The first and second traces are simultaneous extracellular recordings from the levator (LN) and depressor (DN) branches of the ventilatory nerve root. Each intracellular spike is followed at a constant latency by an impulse in LN. Injected hyperpolarising current (2 nA, fourth trace) decreases the amplitude of membrane oscillation and prevents spike discharge in the cell. *b*. With stronger hyperpolarising current (5.5 nA), the membrane wave becomes phase-inverted, indicating that the cell is being driven by a periodic inhibitory synaptic input. A. J. Simmers & B. M. H. Bush, unpublished data.

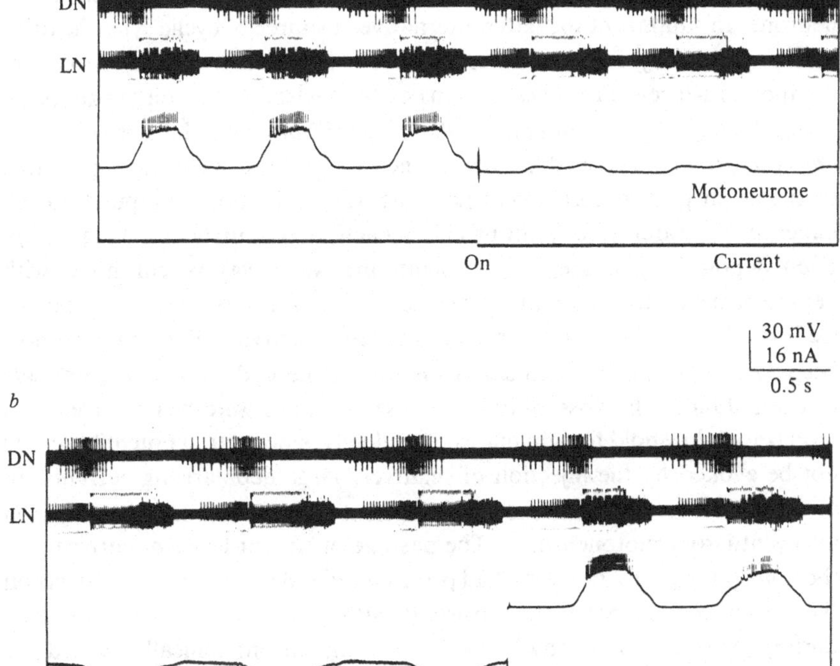

between the intracellularly recorded spikes and impulses in an axon in the peripheral nerve discharge.

Inspection of all intracellular records so far obtained from ventilatory motoneurones has failed to reveal any discrete, unitary synaptic potentials, either immediately before or after each burst, which might be contributing to the production of the underlying slow waves. There is direct evidence, however, that the slow oscillations result from synaptic input to the motoneurones. Small hyperpolarising currents injected into the levator motoneurone of Fig. 1 decreased the amplitude of membrane oscillation and selectively blocked spike discharge in this cell. At sufficiently high levels of hyperpolarising current, however, the membrane wave became phase-inverted (Fig. 1*b*), indicating that the cell membrane was being held beyond the reversal potential of an inhibitory synaptic drive. The complete reversal of the membrane wave suggests that for this levator motoneurone, oscillation is produced largely by synaptic inhibition between spike bursts, rather than by periodic excitation during each burst.

The smooth repolarisation following these levator bursts could result from a summation of many small IPSPs which remain undetected. Alternatively, the synaptic input could be mediated by a graded chemical interaction in which periodic depolarisation of a non-spiking presynaptic element causes a corresponding hyperpolarisation (and inhibition) of the post-synaptic motoneurone. In support of the latter alternative, a source of cyclic synaptic drive to the ventilatory motoneurones has been identified in several species of decapod crustacea. The investigations of Mendelson (1971) on the suboesophageal ganglion of hermit crabs (*Pagurus pollicarus*) and lobster (*Homarus americanus*) first revealed a type of neurone whose membrane potential oscillated in precise correspondence with rhythmic motor output to a gill bailer, but apparently without producing action potentials itself (Fig. 2*a*(*i*)). Each depolarising phase of the membrane wave was synchronised with depressor nerve activity, and the repolarising phase with the levator bursts. Cells displaying similar non-spiking oscillatory activity (Fig. 2*b*) have now been located within the thoracic ganglion of the crab, *Carcinus* (Simmers & Bush, 1980). The possibility that these neurones were being driven at a level below threshold for spiking seems unlikely, since action potentials could not be evoked by the injection of relatively large depolarising currents. In contrast, low levels of current were sufficient to elicit spiking when injected into ventilatory motoneurones. The passage of similar levels of current into the non-spiking cells, however, had pronounced polarity-dependent effects on the motor output patterns, consistent with the phase-relations observed during the free-running rhythm. Depolarising current tonically excited the population of depressor motoneurones while inhibiting the antagonistic levator motoneurones (Fig. 2*a*(*ii*), *b*(*i*)). Conversely, imposed hyperpolarisation caused continuous levator discharge and a relative decrease in depressor activity (Fig. 2*a*(*iii*), *b*(*ii*)). During these periods of imposed polarisation there was no sign of membrane potential oscillation at the free-running rhythm (e.g. Fig. 2*a*(*ii*)). Furthermore, current pulses of either polarity had a clear resetting effect on the timing of the motor pattern (e.g. Fig. 2*b*(*i*, *ii*)).

These results indicate that each of the non-spiking cells is itself an intrinsic part of the central nervous oscillator controlling ventilation, and is not merely an element interpolated between the rhythm generator and the motoneurones. The effects of injecting current into this type of neurone could be demonstrated repeatedly and consistently throughout the course of impalement. They are not due to extracellular spread of current since withdrawal of the microelectrode from the cell immediately abolished any response to applied current. Furthermore, these effects have never been observed when injecting current into other types of ventilatory interneurone or motoneurones so far encountered.

It remains conceivable that the non-spiking neurones described here do in fact produce action potentials, but that their ability to spike was abolished by the microelectrode penetrating the cell membrane, or that spikes generated elsewhere in the cell were undetectable at the recording site. Neither possibility seems likely, however, for the following reasons. Although spiking cells typically elicit an initial high frequency discharge following electrode impalement, no such injury response has been observed in non-spiking neurones (Simmers & Bush, 1980). Penetrating these cells, nevertheless, has a marked effect upon the ongoing ventilatory motor pattern. In most preparations, rhythmic activity ceased altogether, while in others it persisted erratically after penetration, returning gradually to normal over several minutes. Throughout this recovery period, the timing of motor bursts remained closely correlated with the slow changes in membrane potential of the impaled cell. In addition to the effects produced by subsequent current injection, the above observations suggest that even if a spiking mechanism is present normally in these cells, it is not required for the modulation of motoneuronal activity.

It also seems improbable that the microelectrode had penetrated a part of the neurone which showed no evidence of action potentials being generated at a distant locus. This situation would require membrane properties which allow the unattenuated passage of small imposed potential changes away from the site of stimulation, while completely blocking the electrotonic spread of impulses back to the electrode. If such a mechanism does exist, then it is highly probable that on some occasions at least, the probing electrode would have penetrated a region of the cell which does support spike activity. However,

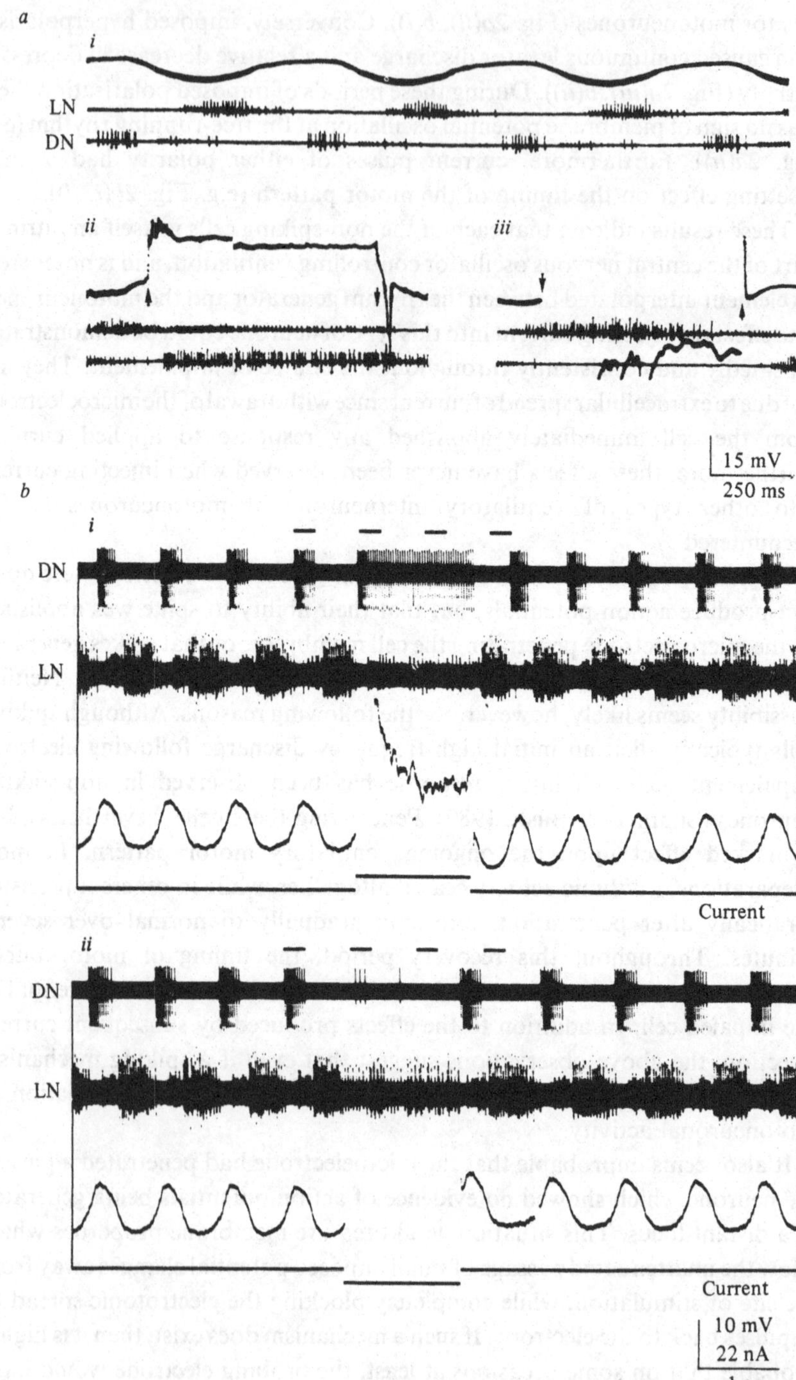
a
i
LN
DN
ii
iii
15 mV
250 ms
b
i
DN
LN
Current
ii
DN
LN
Current
10 mV
22 nA
1 s

no recordings have been obtained to date from spiking neurones with physiological properties similar to those described for non-spiking cells. Furthermore, the lack of discrete post-synaptic activity in the ventilatory motoneurones argues that, even if spikes continue to be generated in an isolated region of an apparently non-spiking driver neurone, they play no functional role in the synaptic interaction between the two.

Mendelson (1971) suggested that a single non-spiking oscillator neurone within each hemi-ganglion is entirely responsible for generating the ipsilateral ventilatory motor programme. This hypothesis was based on the observation that depolarising an oscillator cell excited the depressor motoneurones and inhibited the antagonistic levators, while hyperpolarisation had the opposite effect. Reciprocal post-synaptic activation could be achieved by a direct excitatory coupling between the oscillator and depressor motoneurones, with the levator population negatively coupled either directly, or via a separate inhibitory interneurone. Thus the depressor cells would be excited during the depolarising phase of the oscillator cycle, while the levator motoneurones are inhibited and remain silent. The latter units would in turn become activated during each repolarising phase by post-inhibitory rebound and/or an independent source of tonic excitation. Consistent with this scheme, but without excluding other possibilities, is the finding that the levator motoneurones are indeed subject to an inhibitory synaptic drive during rhythmic activity (Fig. 1).

A single oscillator cell could in principle control an entire motor rhythm. Alternatively, as in the cockroach walking system (Pearson & Fourtner, 1975), the ventilatory pattern generator could comprise a number of non-

---

Fig. 2. Ventilatory rhythm generation by non-spiking neurones in the lobster and shore crab. *a, i.* Oscillations in membrane potential of a non-spiking neurone (top trace) in the lobster suboesophageal ganglion during spontaneous ventilatory motor output. The second and third traces are extracellular recordings from the levator (LN) and depressor nerves (DN) respectively. *a, ii.* Imposed depolarisation of the cell excites the depressor motoneurones while inhibiting levator motoneurones. *a. iii.* Imposed hyperpolarisation causes sustained levator nerve discharge and inhibits the depressor motoneurones. *b.* A similar non-spiking neurone in the thoracic ganglion of the shore-crab. The membrane potential of this neurone (third trace) also oscillates at the same frequency as the ventilatory motor pattern. *b, i.* Depolarising current (6 nA, fourth trace) injected into the cell (bridge circuit unbalanced) continuously excites a depressor motoneurone and inhibits the levator motoneurones. *b, ii.* Hyperpolarising current (6 nA) has the opposite effect. Pulses of either current polarity reset the timing of the ventilatory rhythm. The bars above each record in *b* indicate the expected time of occurrence of depressor bursts in the absence of injected current. *a*, from Mendelson, 1971; *b*, from Simmers & Bush, 1980.

spiking neurones. The discovery of more than one type of oscillator neurone controlling crab ventilation has provided evidence to support the latter possibility. One of these appeared to have properties functionally antagonistic to the cell type described above. During rhythmic motor output, the depolarising phase of its membrane oscillation coincided with levator bursts and repolarisation with depressor bursts. Accordingly, depolarising the cell with injected current caused tonic excitation of the levator motoneurones and inhibition of the depressor motoneurones, while hyperpolarising current had the opposite effect.

Inspection of the recordings from the non-spiking cell of Fig. 2*b*, provides additional evidence to indicate the existence of an oscillator network rather than a single cell. First, weak phasic activity invariably persisted briefly in the depressor motoneurone after the start of a hyperpolarising current pulse. If periodic excitation of the depressor motoneurones is derived solely from rhythmic depolarisation in a single presynaptic neurone, then hyperpolarising it (at least with strong current) would be expected effectively to block this input, and immediately terminate depressor motor activity. Second, although hyperpolarising current evoked sustained discharge in the levator motoneurones, the firing rate during negative current pulses never reached that of levator bursts during the free-running rhythm.

A further type of neurone encountered within the crab thoracic ganglion also showed no evidence of either spontaneous or evoked action potentials (Simmers & Bush, 1980). Again, its membrane potential oscillated, albeit with a relatively small (< 2 mV) amplitude, in phase with bursts in the ventilatory motoneurones (Fig. 3*a–c*). In this cell type, however, imposed changes in membrane polarisation were found to exert a graded control on the overall frequency of the ventilatory rhythm. Depolarising current immediately evoked a faster ongoing cycle rate which was sustained only for the duration of the injected current (Fig. 3*a*), whereas hyperpolarisation has the opposite effect (Fig. 3*b*). At higher levels of negative current, however, rhythmic activity ceased altogether, but resumed immediately, at the same phase in the cycle and at the original frequency, when current injection ceased (Fig. 3*c*). The graded nature of this frequency modulation is represented graphically in Fig. 3*d*. The maximum relative change in cycle period induced by injecting current into this cell covers most of the range of output period observed in the normal ventilating animal.

A similar type of neurone has been found in the suboesophageal ganglion of the hermit crab and in the lobster (Mendelson, 1971). Depolarising this element initiated rhythmic ventilatory motor input, the rate of bursting being directly proportional to the level of membrane depolarisation. Non-spiking interneurones with homologous physiological properties have also been described in the walking system of the cockroach (Fourtner, 1976).

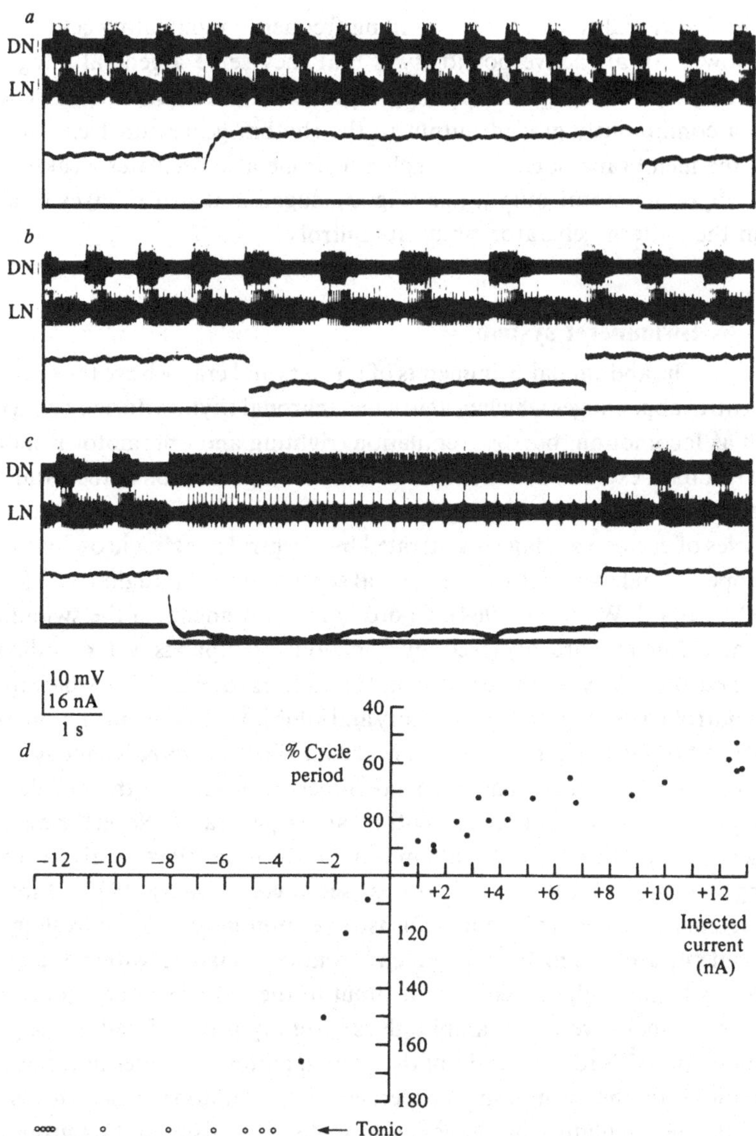

Fig. 3. Graded control of the crab ventilatory pattern generator by a non-spiking neurone. *a.* Depolarising current (fourth trace) injected into the cell (third trace, bridge circuit unbalanced) accelerates reciprocal burst activity in the ventilatory motoneurones (DN, depressor nerve; LN, levator nerve). *b*, Hyperpolarising current has the opposite effect. *c.* Beyond a critical level of hyperpolarising current, rhythmic discharge ceases altogether. *d.* Relationship between amount of injected current of either polarity and cycle period (expressed as a percentage of the unmodulated period). Each point is an average determined from at least four consecutive cycles. Open circles denote levels of hyperpolarising current at which rhythmic activity was suppressed. From Simmers & Bush, 1980.

The functional role of these switching/frequency modulating cells remains unknown. An attractive possibility is that they serve a general integrating function, converting various spiking inputs (whether command or sensory) into a continuously graded output to the rhythm generator itself. The fact that the membrane potential of such a neurone also oscillates weakly along with rhythmic motor output (e.g. Fig. 3), suggests that it receives feedback from the pattern generator circuit it controls.

Swimmeret system

The abdominal swimmerets of lobsters and crayfish are the four pairs of ventral appendages which beat in a metachronal rhythm during behaviours such as locomotion, burrow ventilation, righting and optomotor responses. The results of extensive studies on the neuronal organisation of the swimmeret system have shown that the rhythmic alternating motor output to the eleven muscles of each appendage is generated by a central ganglionic oscillator that can operate independently of peripheral sensory inputs (Hughes & Wiersma, 1960; Ikeda & Wiersma, 1964). Coordinated movements of the swimmerets in each segment are achieved by functional couplings via coordinating interneurones between these segmental oscillators (Stein, 1971), which in turn are controlled by descending excitatory and inhibitory command interneurones in the ventral nerve cord (Wiersma & Ikeda, 1964; Davis & Kennedy, 1972).

Until recently, there has been no direct evidence for the cellular and synaptic basis of the swimmeret central pattern generator. Nevertheless, some general properties have been implicated from detailed analyses of the temporal motor patterns it produces (see review, Davis, 1973). First, the antagonistic powerstroke and returnstroke motoneurones controlling each swimmeret are driven by a single endogenous oscillator within the corresponding hemiganglion. Second, the input to the motoneurone pool is in the form of a sine-wave whose amplitude is inversely proportional to the period of oscillation. Third, the details of the efferent pattern are determined by small variations in the membrane properties of the motoneurones themselves. Fourth, the oscillator simultaneously excites synergistic motoneurones and inhibits antagonists.

These inferred properties now require interpretation and extension in the light of more recent information on the swimmeret motor system. Intracellular studies have shown that the swimmeret motoneurones are themselves an integral part of the central pattern generator, and are not simply passive output elements of a promotor oscillator (Heitler, 1978). Further, there is now direct evidence that, as in the other Arthropod motor systems already mentioned, non-spiking interneurones play an important role in the swimmeret pattern-generating system (Heitler & Pearson, 1980).

Intracellular recordings from the neuropile of isolated crayfish abdominal ganglia have revealed a type of neurone that lacked any sign of impulses, but the membrane potential of which oscillated in phase with the swimmeret motor output pattern (Fig. 4). Each depolarising phase coincided with periodic bursts in the powerstroke motoneurones, and the repolarising phase with antagonistic returnstroke bursts. Small depolarising currents injected into one of these cells excited the powerstroke motoneurones and inhibited the returnstroke motoneurones (Fig. 4*a*). Conversely, hyperpolarising current inhibited powerstroke activity and caused continuous discharge in returnstroke motoneurones (Fig. 4*b*). During these periods of imposed polarisation, there was a concomitant reduction in the amplitude of the membrane

Fig. 4. A non-spiking interneurone which is part of the swimmeret central pattern generator in the 4th abdominal ganglion of the crayfish. *a*. Depolarising current (2 nA, solid bar) injected into the nonspiking cell (Int) causes subthreshold depolarisation of a 4th ganglion powerstroke motoneurone (Mn), excites a number of powerstroke motoneurones recorded extracellularly from the posterior branch of the 1st root (third trace), and inhibits returnstroke motoneurones in the anterior branch (bottom trace). *b*. Hyperpolarising current (2 nA) has the opposite effect. Short pulses of depolarising (*c*) or hyperpolarising current (*d*) injected into the same cell (top traces) reset the timing of rhythmic output from the 5th ganglion (second traces, whole 1st root recordings), 4th ganglion (powerstroke, third traces; returnstroke, fourth traces), and 3rd ganglion (bottom traces, whole 1st root recording). The bars below *c* and *d* indicate the expected time of occurrence of 4th ganglion powerstroke bursts in the absence of injected current. From Heitler & Pearson, 1980.

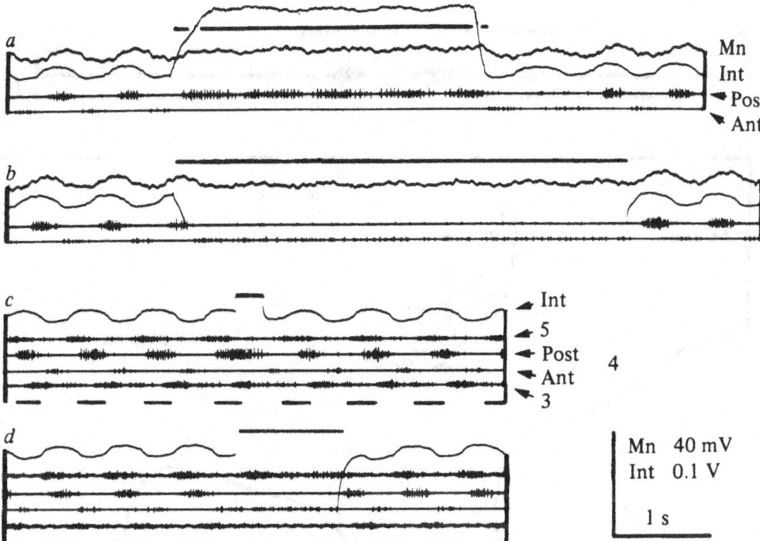

potential oscillations recorded simultaneously in a powerstroke motoneurone (Fig. 4*a*, *b*). Depolarising current caused an overall depolarisation of the motoneurone (Fig. 4*a*), while hyperpolarising current had the opposite effect (Fig. 4*b*).

The importance of this non-spiking interneurone in the rhythm-generating circuit was further evident from the effects of brief current pulses on the timing of the motor programme. Injection of either depolarising or hyperpolarising current pulses could reset the period of rhythmic output from the appropriate ganglion, as well as that from other ganglia (Fig. 4*c*, *d*). This, together with the foregoing data, indicates that the cell is part of the segmental oscillator, and furthermore, it must have access to those neural elements which coordinate the oscillators of different segments.

The conclusion that neurones with these functional properties are in fact 'non-spiking interneurones', was based on their apparent inability to generate action potentials, either spontaneously, or in response to: 1, membrane disturbance when first penetrated; 2, injection of relatively high

Fig. 5. A non-spiking interneurone which initiates the swimmeret motor programme in the crayfish. *a*. 6 nA depolarising current (solid bar) injected into the cell (first trace, bridge circuit unbalanced) produces reciprocal burst activity in the powerstroke (second trace) and returnstroke motoneurones (third trace). *b*. Outline of the left 4th hemiganglion showing the structure of the neurone determined by intracellular staining with Lucifer Yellow. A single process from the cell body (diameter 10 μm) gives rise to a dendritic arborisation which remains wholly within the ipsilateral neuropile. From Heitler & Pearson, 1980.

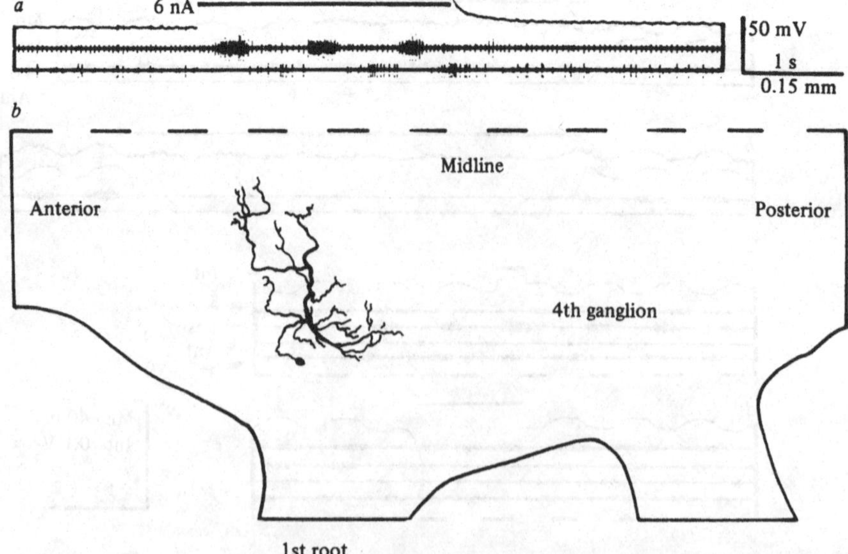

levels (15 nA) of depolarising current, whereas small currents (5 nA) of either polarity strongly influenced the activity of motoneurones; 3, release from imposed hyperpolarisation; and 4, antidromic stimulation of the peripheral motor roots.

Other cell types have been encountered within the swimmeret system which satisfy these physiological criteria for non-spiking interneurones. Two of these have been identified morphologically by intracellular staining. One example is shown in Fig. 5. Tonic depolarisation of this neurone initiated rhythmic burst activity in the swimmeret motoneurones of an otherwise quiescent preparation (Fig. 5*a*). Iontophoretic injection of the fluorescent dye, Lucifer Yellow (Stewart, 1978), revealed a cell structure with no axon and a dendritic arborisation confined to a localised region of the ipsilateral hemiganglion (Fig. 5*b*). Consistent with its apparent non-spiking behaviour, therefore, the cell is an intraganglionic interneurone in which neural information would have to be transmitted over relatively short distances only. It is of further interest that this interneurone appears to have functional properties similar to the 'integrating neurone' reported in the cockroach walking system (Fourtner, 1976) and the ventilatory systems of crabs and lobsters (Mendelson, 1971; Simmers & Bush, 1980).

Stomatogastric system

Graded non-impulse mediated transmission is also a property of many synaptic connections in the lobster stomatogastric ganglion. Maynard (1972), and Maynard & Walton (1975) described both non-spiking *and* spiking presynaptic neurones within this system whose postsynaptic effects rely either totally or in part on the release of transmitter as a continuously graded function of presynaptic membrane polarisation. Recent studies have examined more closely the input–output properties of graded transmission between stomatogastric neurones and the role these interactions play during normal behaviour (Graubard, 1978; Raper, 1979; Graubard, Raper & Hartline, 1980).

The stomatogastric ganglion of the lobster comprises a network of about 30 identified and well-characterised neurones, most of which are motoneurones innervating the muscles of the stomach. These neurones are organised into two functionally independent subsystems which generate cyclic motor discharge to the muscles of the pyloric-filters and gastric-mill respectively. The motor patterns underlying both rhythmic behaviours persist in the isolated, deafferented ganglion, and by recording intrasomatically from two or three neurones at a time (e.g. Fig. 6*a*), it has been possible to determine functional monosynaptic pathways within each subsystem (see review, Selverston *et al.*,

1976). A diagram of some major synaptic connections within the pyloric pattern-generator network is shown in Fig. 6*b*. The central interactions between stomatogastric neurones are predominantly inhibitory and mostly reciprocal between cells innervating antagonistic muscles.

Non-spiking synaptic transmission

Although discrete impulse-evoked postsynaptic potentials are a prominent feature of neuronal communication within the stomatogastric

Fig. 6. Synaptic connections between motoneurones within the pyloric subsystem of the stomatogastric ganglion of the lobster. *a*. Spontaneous rhythmic activity recorded intracellularly from the cell bodies of a lateral pyloric (LP) and a pyloric dilator (PD) neurone and extracellularly from the lateral ventricular nerve (LVN). The two cells are reciprocally inhibitory and show initary, fixed-latency IPSPs. *b*. Circuit diagram showing some major pyloric connections. Open circles enclose cells of the same functional type. There are eight pyloric (PY) neurones. The anterior burster (AB) and two PD cells are electrically coupled to one another and are endogenous bursters. All synaptic pathways indicated are chemical and inhibitory. *a*, from Maynard & Selverston, 1975: *b*, adapted from Selverston *et al*., 1976.

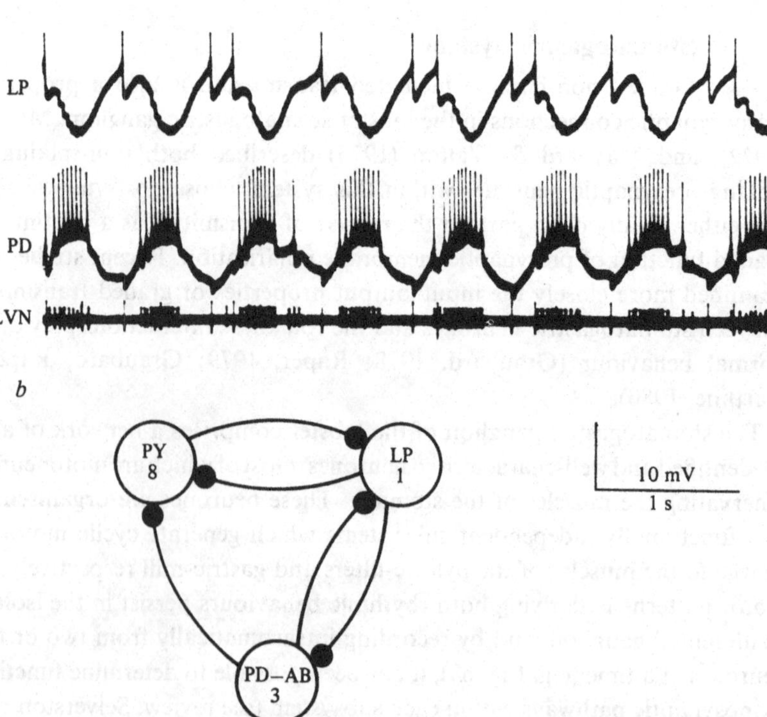

ganglion, there are other chemical synapses which operate without presynaptic action potentials. One example is the EX1–GM 'non-spike inhibition' (NSI) synapse. This responds to soma depolarisation of the non-spiking presynaptic EX1 neurone with a maintained hyperpolarisation and impulse suppression in the postsynaptic gastric mill (GM) motoneurone (Fig. 7*a*; see also Maynard, 1972; Maynard & Walton, 1975). The EX1 cells do not produce

Fig. 7. Graded synaptic transmission from a non-spiking neurone in the stomatogastric ganglion of the lobster. *a*. A depolarising current pulse injected via a separate electrode into the soma of a non-spiking EX1 cell evokes a sustained hyperpolarising response in the postsynaptic gastric mill (GM) motoneurone. *b*. Grading the presynaptic voltage steps causes increasing postsynaptic hyperpolarisation. *c*. Plot of the peak postsynaptic response as a function of presynaptic voltage. Data from the same experiment as illustrated in *b*. For this cell pair, at least 15 mV depolarisation of EX1 from its resting potential (RP) was required to elicit non-spiking inhibition of the GM neurone. Adapted from Graubard, 1978.

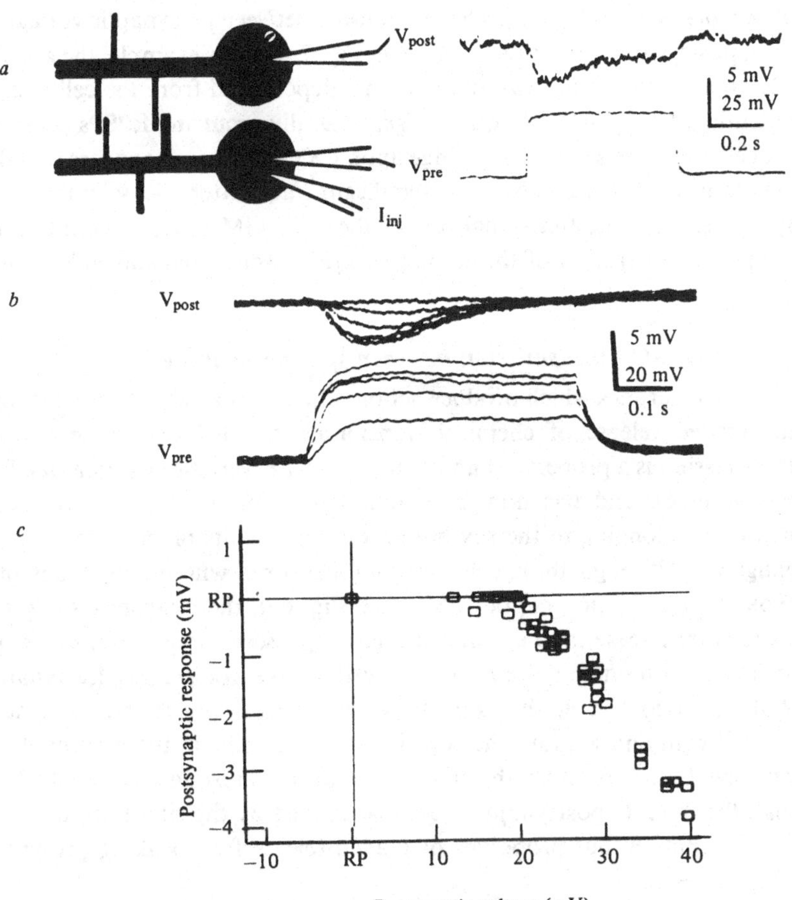

action potentials in response to soma depolarisation or antidromic stimulation of the peripheral axon. Furthermore, EX1–GM NSI remains unaffected by treatment with 10^{-7} M tetrodotoxin (TTX), which is known to abolish action potentials in all stomatogastric neurones (Graubard, 1978). Synaptic transmission is chemically mediated since NSI is blocked by 10^{-4} M picrotoxin (Maynard & Walton, 1975) and by low-calcium high-magnesium saline. In addition, the postsynaptic GM response can be reversed with injection of hyperpolarising current (Graubard, 1978).

The input–output properties of the EX1–GM synapse have been measured by depolarising the presynaptic cell with steps of current injected into the soma while recording the voltage change produced in both the pre-and post-synaptic cell bodies (Graubard, 1978). The typical postsynaptic response follows a synaptic delay with an initial peak of hyperpolarisation and then a decay to the plateau which is maintained for the duration of the presynaptic depolarisation (Fig. 7*a*). There is a threshold EX1 voltage for transmitter release, beyond which increasing synaptic depolarisation causes increasing GM hyperpolarisation (Fig. 7*b*). The relationship between presynaptic voltage and peak postsynaptic response is shown in Fig. 7*c*. In this example, the apparent EX1 release threshold was about 15 mV depolarised from the cell's resting potential of –60 mV. Although large, naturally occurring EPSPs (as seen at the cell body) are known to impinge upon EX1 neurones, it appears that these depolarising inputs do not cause significant transmitter release at the output synapses. Thus the functional role of the EX1–GM synaptic connection in the normal operation of the stomatogastric ganglion remains unknown.

Graded transmission between spiking neurones

Neurones which produce action potentials can also interact through the graded release of chemical transmitter. Graded and spike-mediated transmission is a property of amacrine cells in the vertebrate retina (see Fain, this volume), and has now been found to occur between many spiking neurones belonging to the rhythm-generating circuits of the stomatogastric ganglion. Although these cells are motoneurones whose action potentials evoke typical monosynaptic IPSPs (see Fig. 6*a*), their capacity for graded transmitter release can be demonstrated in quiescent preparations by depolarising the presynaptic neurone below threshold for spiking (Maynard & Walton, 1975). Fig. 8*a* shows the hyperpolarising response of a postsynaptic PD cell with and without the contribution of a spike in the presynaptic LP neurone. In this example, the graded subthreshold response is actually larger than the discrete postsynaptic potential evoked by the single impulse.

The input–output properties of graded release from spiking presynaptic

Fig. 8. Graded synaptic transmission between spiking neurones of the stomatogastric ganglion of the lobster. *a*. Subthreshold depolarisation of a presynaptic LP neurone causes a slow hyperpolarisation of the postsynaptic PD cell. A second sweep shows the increase to this response caused by a single spike-evoked IPSP. *b*, *c*. Recordings from two different PD–PY cell pairs in which spiking activity had been blocked with 2×10^{-7} M TTX. *b*, A step depolarisation of the presynaptic PD neurone causes a peak-plateau hyperpolarising waveform in the postsynaptic PY. *c*. The PY response increases in amplitude with increasing PD depolarisation. For this cell pair, hyperpolarising PD results in a depolarisation of PY, indicating that at rest the postsynaptic cell is held continuously hyperpolarised by the presynaptic neurone. *d*. Peak postsynaptic response plotted as a function of presynaptic voltage. Data from the same experiment as illustated in *c*. In this example, the release threshold of the presynaptic PD cell is about 5 mV below its resting potential (RP) in TTX. *a*, *b*, *d*, from Graubard, Raper & Hartline, 1980; *c*, from Graubard & Calvin, 1979.

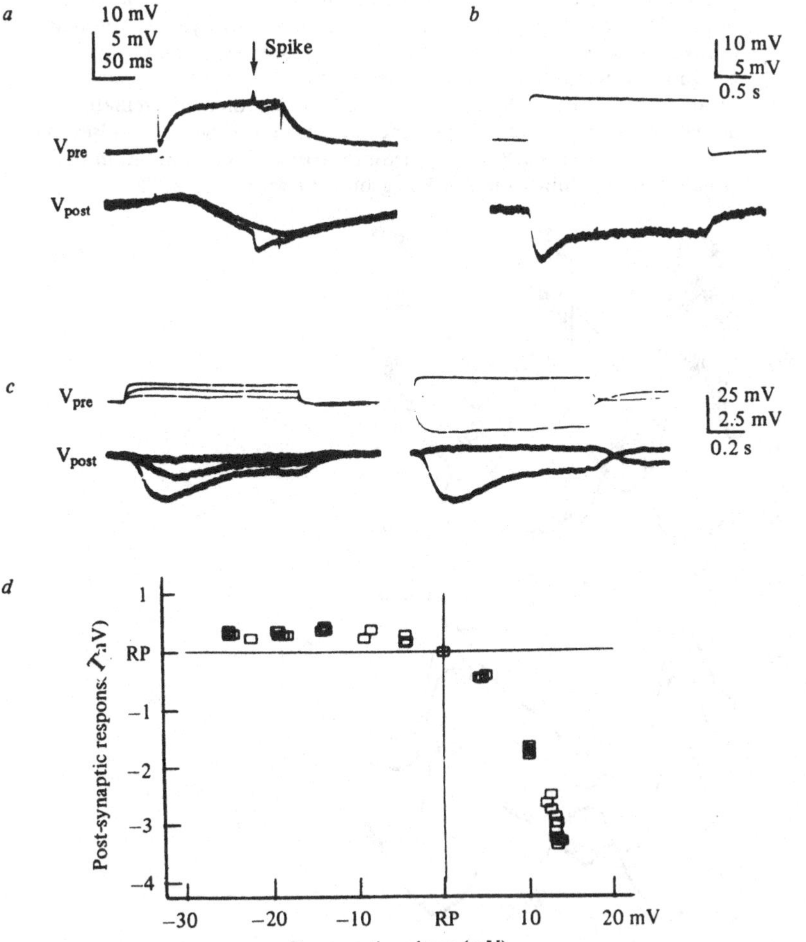

neurones were studied by eliminating all pre- and post-synaptic impulses with TTX (Graubard *et al.*, 1980). Fig. 8*b* shows the postsynaptic PY cell response to a suprathreshold voltage step in the presynaptic PD neurone in which the spikes had been blocked by 2×10^{-7} M TTX. The typical waveform displays both peak and plateau hyperpolarising components (Fig. 8*b*), which increase in amplitude with increasing presynaptic depolarisation (Fig. 8*c*, *d*). These properties are characteristic of the non-spiking EX1–GM synaptic connection described above (compare Fig. 8*b*–*d* with Fig. 7*a*–*c*). A difference of functional importance, however, is that for the pairs of spiking neurones examined, the threshold for transmitter release was invariably closer to and sometimes more negative than the resting potential of the presynaptic neurone. In the case of the cell pair of Fig. 8*c*, the presynaptic PD neurone continuously released

Fig. 9. Evidence that graded transmission is a part of motor pattern generation in the stomatogastric ganglion of the lobster. *a*. Simultaneous intracellular recordings from the cell bodies of four pyloric motoneurones during spontaneous rhythmic activity in an isolated stomatogastric ganglion. *b*. Recordings from the same cells 10 min after the start of localised perfusion with 1.7×10^{-5} M TTX. Coordinated oscillatory activity could be evoked despite the suppression of nerve impulses. The slow waves in the AB/PD cells mediate strong graded inhibition of LP which in turn inhibits the AB/PD group. From Raper, 1979.

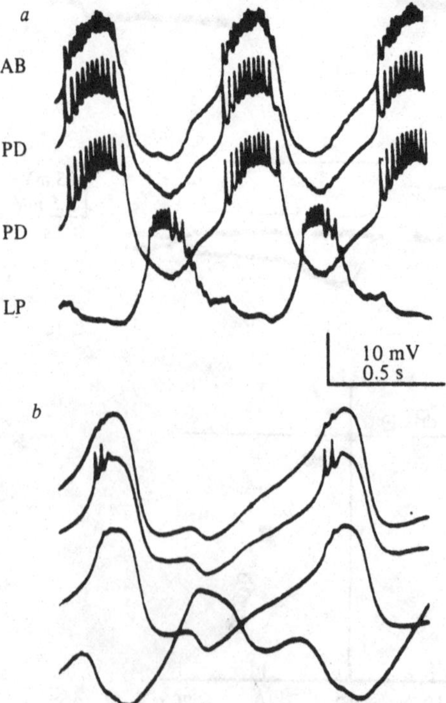

transmitter at rest, since presynaptic hyperpolarisation caused a slight depolarisation (by disinhibition) of the postsynaptic PY cell. The corresponding input–output curve (Fig. 8*d*), indicates a release threshold about 5 mV below the PD resting potential in TTX.

During spontaneous cyclic activity, the motoneurones of the pyloric pattern generator are characterised by large amplitude oscillations in membrane potential (15–25 mV) underlying their rhythmic bursts of action potentials (Figs. 6*a*; 9*a*). Cyclic activity results from the endogenous pacemaker properties of three cells (AB/PD group) which interact synaptically with all other neurones (PY, LP, etc.) in the network (Fig. 6*b*; see Selverston *et al.*, 1976). The spontaneous oscillations observed in pyloric follower cells are now known to result from the graded modulation of inhibitory transmitter release (Raper, 1979). Localised perfusion of TTX suppresses impulse production in a rhythmically active preparation, but the underlying slow waves persist with little change in either amplitude or relative timing (Fig. 9*b*). Under these conditions, chemically-mediated synaptic drive can be inferred by the reversal of postsynaptic waveforms with injected hyperpolarising current. This includes the hyperpolarising responses of any one of the follower cells as well as the reciprocal LP-mediated inhibition of the AB/PD group. There is therefore, direct evidence that graded synaptic interactions play an important role in the normal function of this network of spiking neurones.

This conclusion derived from electrophysiological data is also supported by anatomical evidence. Serial section reconstructions of stomatogastric motoneurones have revealed both input and output synapses distributed over the distal regions of each dendritic process (King, 1976). The close proximity of pre- and post-synaptic contact provides, therefore, a morphological substrate for graded local circuit function within the stomatogastric system.

Conclusions

It has been known for some time that graded transmitter release can be induced artificially from chemical synapses that function normally with presynaptic spikes. Examples include the giant synapses of the squid (Katz & Miledi, 1967), and the lamprey (Martin & Ringham, 1975). After treatment with TTX to block action-potential activity, substantial presynaptic depolarisation is required to reach threshold for synaptic release. Beyond this threshold however, transmitter is released as a continuously graded function of the presynaptic voltage. The high release threshold is indicative of an output synapse which, under normal physiological conditions, relies solely on spike-mediated activation. In contrast, synapses which operate normally without presynaptic impulses would be expected to have a much lower release threshold, thereby maximising the sensitivity of the postsynaptic response to presynaptic voltage fluctuation (Burrows & Siegler, 1978). Graded synaptic transmission is not confined to neurones completely lacking impulses. If the release threshold of a spiking cell is also close to its resting potential, it is able to use both action-potential evoked and graded release at its output synapses. Mixed impulse and graded transmission is a property of amacrine cells in the vertebrate retina (Werblin & Dowling, 1969) and mitral cells of the olfactory bulb (Rall & Shepherd, 1968), and has now been found to occur between spiking neurones belonging to the rhythm generating circuitry of the stomatogastric ganglion.

What is the functional significance of graded release from a spiking neurone? With the discovery of dendro-dendritic and graded synaptic interactions, it has become recognised that while the long-distance projection of neuronal information relies on the propagation of all-or-none action potentials, communication within the localised regions of dendritic networks need only require the electrotonic spread of graded non-spike potentials (e.g. Rakic, 1976). A corresponding diversification of function seems to exist within single neurones of the stomatogastric ganglion. Pyloric neurones not only participate centrally in a localised pattern-generating network in which graded interactions play an important role, they also function as motoneurones, relaying impulse-encoded information to stomach muscles distant from the ganglion.

Similar considerations suggest that neuronal function is more clearly defined in the swimmeret and gill-ventilation systems of Crustacea. In both these systems, it appears that rhythm generation is predominantly the role of neural elements presynaptic to the motoneurones. Although this does not exclude the possibility of central graded interactions at the motor level (Heitler, 1978), it may reflect a greater functional specialisation of cells comprising purely non-spiking local circuit interneurones on one hand, and long distance projection motoneurones on the other.

Are there functional advantages for utilising graded synaptic interactions in motor control? Implicit in the production of coordinated movements underlying many rhythmic behaviours is the need for continuous regulation of motoneuronal discharge over a wide range of frequencies. Smoothly variable control of patterned motor activity could be derived from the summed inputs of many asynchronously occurring unitary postsynaptic potentials impinging upon motoneurones from a large number of spiking premotor neurones. Alternatively, a system with fewer driver components, but which remains no less capable of controlling precisely the discharge rate of its follower cells, is one in which the motoneurones are influenced directly by the graded release of transmitter from presynaptic neurones, without the intervention of action potentials. Support for this scheme comes from a theoretical study on the cockroach walking system (R. A. DiCaprio & K. G. Pearson, personal communication). In a model system where only one premotor element drives a population of motoneurones, a graded non-spiking input can produce the complete experimentally-determined output pattern. A single *spiking* input to this system, on the other hand, could not duplicate the motor pattern. Given the relative economy of neural components within Arthropod nervous systems, it is perhaps not surprising, therefore, that graded interactions have now been discovered within several rhythm generating circuits of this group of animals.

References

Burrows, M. & Siegler, M. V. S. (1978). Graded synaptic transmission between local interneurones and motor neurones in the metathoracic ganglion of the locust. *J. Physiol., Lond.* **285**, 231–55.

Davis, W. J. (1973). Neuronal organisation and ontogeny in the lobster swimmeret system. In *Control of Posture and Locomotion*, ed. R. B. Stein, K. G. Pearson, R. S. Smith & J. B. Redford, pp. 437–55. New York: Plenum Press.

Davis, W. J. & Kennedy, D. (1972). Command interneurones controlling swimmeret movements in the lobster. II. Interaction of effects on motoneurons. *J. Neurophysiol.* **35**, 13–19.

Fourtner, C. (1976). The central nervous system control of insect walking. In *Neural Control of Locomotion*, ed. R. H. Herman, S. Grillner, P. G. S. Stein & D. G. Stuart, pp. 401–18. New York: Plenum Press.

Graubard, K. (1978). Synaptic transmission without action potentials: input– output properties of a non-spiking presynaptic neurone. *J. Neurophysiol.* **41**, 1014–25.

Graubard, K. & Calvin, W. H. (1979). Presynaptic dendrites: implications of spikeless synaptic transmission and dendritic geometry. In *The Neurosciences: Fourth Study Program*, ed. F. O. Schmitt & F. G. Worden, pp. 317–31. Cambridge, Mass: MIT Press.

Graubard, K., Raper, J. A. & Hartline, D. K. (1980). Graded synaptic transmission between spiking neurons. *Proc. Nat. Acad. Sci.* **77**, 3733–5.

Heitler, W. J. (1978). Coupled motoneurones are part of the crayfish swimmeret central oscillator. *Nature, Lond.* **275**, 231–4.

Heitler, W. J. & Pearson, K. G. (1980). Non-spiking interactions and local interneurons in the central pattern generator of the crayfish swimmeret system. *Brain Res.* **187**, 206–211.

Hughes, G. M. & Wiersma, C. A. G. (1960). The co-ordination of swimmeret movements in the crayfish, *Procambarus clarkii* (Girard). *J. Exp. Biol.* **37**, 657–70.

Ikeda, K. & Wiersma, C. A. G. (1964). Autogenic rhythmicity in the abdominal ganglia of the crayfish: the control of swimmeret movements. *Comp. Biochem. Physiol.* **12**, 107–115.

Katz, B. & Miledi, R. (1967). Study of synaptic transmission in the absence of nerve impulses. *J. Physiol., Lond.* **192**, 407–36.

King, D. G. (1976). Organization of crustacean neuropil. II. Distribution of synaptic contacts on identified motor neurones in lobster stomatogastric ganglion. *J. Neurocytol.* **5**, 239–66.

Martin, A. R. & Ringham, B. L. (1975). Synaptic transfer at a vertebrate central nervous system synapse. *J. Physiol., Lond.* **251**, 409–26.

Maynard, D. M. (1972). Simpler networks. *Ann. N. Y. Acad. Sci.* **193**, 59–72.

Maynard, D. M. & Selverston, A. I. (1975). Organisation of the stomatogastric ganglion of the spiny lobster. IV. The pyloric system. *J. comp. Physiol.* **100**, 161–82.

Maynard, D. M. & Walton, K. D. (1975). Effects of maintained depolarization of presynaptic neurons on inhibitory transmission in lobster neuropil. *J. comp. Physiol.* **97**, 215–43.

Mendelson, M. (1971). Oscillator neurones in crustacean ganglion. *Science*, **171**, 1170–3.

Pasztor, V. M. (1968). The neurophysiology of respiration in decapod Crustacea. I. The motor system. *Can. J. Zool.* **46**, 585–96.

Pearson, K. G. (1976). Nerve cells without action potentials. In *Simpler Networks*, ed. J. Fentress. Sunderland, Mass: Sinauer Associates.

Pearson, K. G. & Fourtner, C. R. (1975). Non-spiking interneurones in walking system of the cockroach. *J. Neurophysiol.* **38**, 33–52.

Pilkington, J. B. & Simmers, A. J. (1973). An analysis of bailer movements responsible for gill ventilation in the crab, *Cancer novae-zelandiae. Mar. Behav. Physiol.* **2**, 73–95.

Rakic, P. (Ed.). (1976). *Local Circuit Neurons.* Cambridge, Mass: MIT Press.

Rall, W. & Shepherd, G. M. (1968). Theoretical reconstruction of field potentials and dendro-dendritic synaptic interactions in olfactory bulb. *J. Neurophysiol.* **31**, 884–915.

Raper, J. A. (1979). Non-impulse mediated synaptic transmission during the generation of a cyclic motor program. *Science*, **205**, 304–6.

Selverston, A. I., Russell, D. F., Miller, J. P. & King, D. G. (1976). The stomatogastric nervous system: structure and function of a small neural network. *Prog. Neurobiol.* **7**, 215–89.

Shepherd, G. M. (1972). The neuron doctrine: a revision of functional concepts. *Yale J. Biol. Med.* **45**, 584–99.

Simmers, A. J. (1979). Oscillatory potentials in crab ventilatory neurones. *J. Physiol., Lond.* **287**, 39–40*P*.

Simmers, A. J. & Bush, B. M. H. (1980). Non-spiking neurones controlling ventilation in crabs. *Brain Res.* **197**, 247–52.

Stein, P. S. G. (1971). Intersegmental coordination of swimmeret motoneurone activity in crayfish. *J. Neurophysiol.* **34**, 310–18.

Stewart, W. W. (1978). Functional connections between cells as revealed by dye-coupling with a highly fluorescent naphthalimide tracer. *Cell*, **14**, 741–59.

Werblin, F. S. & Dowling, J. E. (1969). Organisation of the retina of the mudpuppy, *Necturus maculosus*. II. Intracellular recordings, *J. Neurophysiol.* **32**, 339–55.

Wiersma, C. A. G. & Ikeda, K. (1964). Interneurones commanding swimmeret movements in the crayfish, *Procambarus Clarkii* (*Girard*). *Comp. Biochem. Physiol.* **12**, 509–25.

Young, R. E. (1975). Neuromuscular control of ventilation in the crab *Carcinus maenas. J. comp. Physiol.* **101**, 1–37.

MALCOLM BURROWS

Local interneurones in insects

Local interneurones have been known to exist in the segmental ganglia of insects for more than half a century (Zawarzin, 1924). The morphological description of these neurones, based on methylene-blue staining, laid largely dormant until a few years ago. Then Pearson & Fourtner (1975) made intracellular recordings from cockroach neuropile and showed that there were neurones in the thoracic ganglia that normally did not produce action potentials (spikes). Morphologically these neurones proved to be local interneurones, by which is meant neurones with all their branches confined to one ganglion. Manipulation of the membrane potential of these interneurones, without the production of action potentials, could influence the output of motor neurones innervating muscles of a leg. This evidence when coupled with similar evidence derived earlier in crustacea (Mendelson, 1971, and see Simmers this volume), and with evidence in the retina and olfactory bulb of vertebrates for local interactions between neurones not mediated by action potentials, has added considerable impetus to our changing concepts about the functioning of the nervous system (Shepherd, 1972, 1978; Rakic, 1976). Clearly the spike is no longer to be regarded as the universal intracellular signal of neurones. If we now have to contemplate a nervous system which contains interneurones that can process information without spiking, three classes of question must be addressed. *Firstly*, if neurones do not generate spikes, how do they transmit information to other neurones? Are the synaptic potentials generated in these apparently non-spiking neurones as a result of signals from other presynaptic neurones, sufficient in turn to effect communication with postsynaptic neurones? *Secondly*, if individual synaptic potentials can effect intercellular communication, does this imply that one part of the neurone can act with some independence from the other parts? Were these sorts of interactions to be demonstrated they would form a sound physiological base for the belief that local interactions between neurones are more important for integration than has previously been recognised. *Thirdly*, to what extent is the processing of information and the shaping of motor patterns dependent upon graded interactions between neurones that are independent of spikes? Are non-spiking interneurones a rare oddity, or are they numerous and essential elements in the production of behaviour?

In an attempt to shed light on some of these questions I will consider the properties and connectivity of the apparently non-spiking local interneurones that participate in the production of leg movements in the locust (Fig. 1). The advantage of considering the production of motor patterns in the locust are many, but one of prime importance to the above questions is that it is possible to relate the intracellularly recorded activity in local interneurones and in motor neurones, to the movements for which they are the causal elements. Moreover, many of the motor patterns which occur in a locust prepared for intracellular recording are the same as those which are seen in a freely moving locust. Therefore, it is reasonable to assume that the mechanisms of integration observed while recording are those which occur during normal life. In particular the functional state of the neurones, for example the non-spiking ones, can be expected to be normal and not induced by the experimental procedure.

General features of the locust central nervous system

The central nervous system of the locust (*Schistocerca americana gregaria* (Dirsh)) consists of a series of ganglia each joined by a pair of connectives. Typically there is one ganglion for each body segment which lies just inside the ventral body wall, but the brain, metathoracic and last abdominal ganglia are formed by the fusion of ganglia. A blood–brain interface around the central nervous system ensures the maintenance of an internal ionic environment different from the surrounding haemolymph. Gaseous exchange occurs through a complex arrangement of tracheae and tracheoles (Burrows, 1980*a*).

The ganglion chosen for study is the last thoracic segment and is involved with the control of the movements of the hind wings and the large hind legs. It contains the cell bodies of some 2000 neurones which are symmetrically arranged in the left and right halves of the ganglion. There are about 50 motor neurones which innervate the muscles of one hind leg and about 20 that innervate muscles of one hind wing. Most of these motor neurones are identified individuals that have been characterised physiologically and morphologically. More than half of the cell bodies appear to belong to

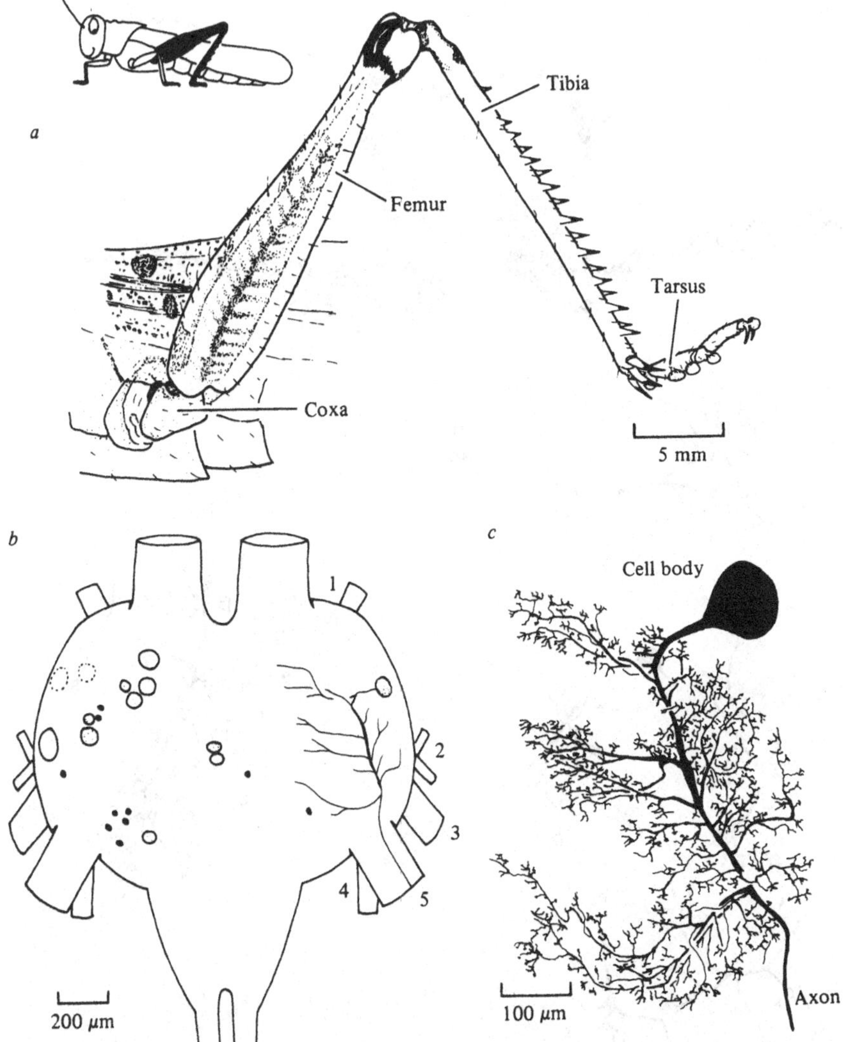

Fig. 1. Some morphological features of the locust that pertain to this study of its local interneurones. *a*. A side view of a locust, and a more detailed drawing of the left hind leg. The tarsus and trochanter can be levated or depressed, the tibia flexed or extended and the coxa adducted, abducted or rotated. *b*. A diagram of the metathoracic ganglion as viewed ventrally; anterior is at the top, and nerves posterior to nerve 5 (the main leg nerve) are omitted. The cell bodies of some of the local interneurones (filled circles) and identified motor neurones that control the movement of the tibia about the femur of the right hind leg are shown. The cell bodies of the three extensor tibiae motor neurones are indicated by open profiles, those of some of the flexor tibiae motor neurones by stippled profiles. The structure of one identified flexor motor neurone (the lateral fast) is outlined on the right of the ganglion and is drawn in more detail in *c*. Not all of the small neuropilar branches have been drawn from this intracellular cobalt-stained preparation.

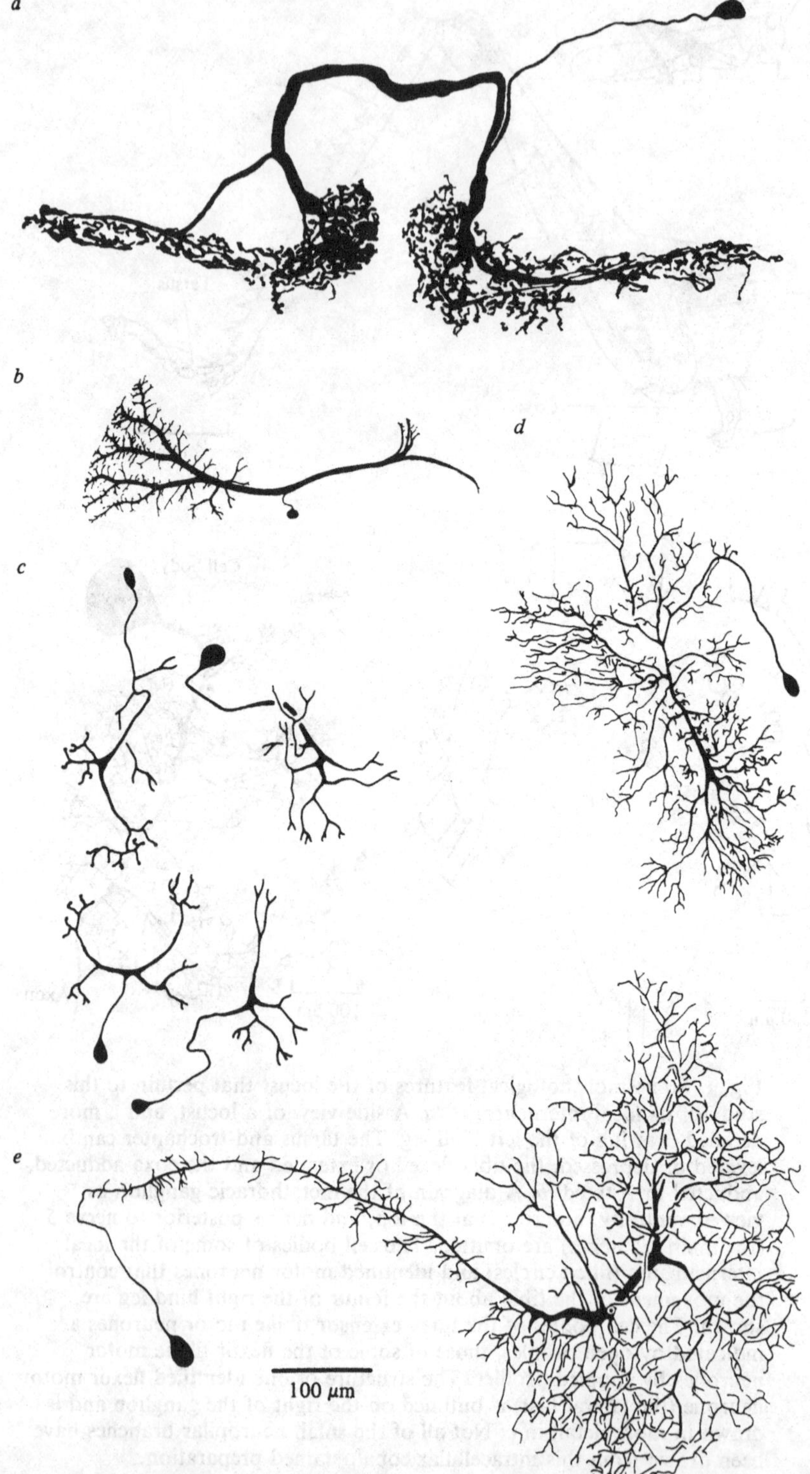
a
b
c
d
e
100 μm

local interneurones, though only a small percentage of them have been characterised so far.

The cell bodies of the interneurones and the motor neurones are intermingled (Fig. 1*b*) in a cortex that surrounds a central core of neuropile and tracts. The motor neurones have cell bodies of larger diameter (30–90 μm) than those of the local interneurones (10–30 μm). Both types of neurone are unipolar (Fig. 1*c*, 2*d*, *e*) and do not appear to have synapses upon their cell bodies. Instead, synaptic interactions take place between the plethora of branches that arise within the neuropile from both motor neurones (Fig. 1*c*) and local interneurones (Fig. 2*c*–*e*). Therefore, if an electrode is to sample events close to the synaptic sites it must be placed in the fine neuropilar processes of the neurones. In order to make such recordings the metathoracic ganglion is exposed dorsally and stabilised within the thorax. The locust is alert and able to move its head, abdomen and some joints of its hind legs. The tracheae in the neuropile regions of the ganglion are ventilated by the pumping movements of the abdomen. An adequate oxygen supply to small neurones deep within a neuropile is probably essential for their normal functioning, and any interrruption of the supply might well suppress spikes in what would normally be spiking neurones. The muscles of one hind leg have wires implanted in them so that motor neurones recorded in the ganglion can be identified and the local interneurones characterised, in part, by their effects upon these known motor neurones.

Morphological and physiological properties of local non-spiking interneurones

Morphology

The definition of the neurones to be described is based upon physiology, which characterises them in a negative way as non-spiking

---

Fig. 2. Drawings of spiking (*a*, *b*) and non-spiking (*c*(?), *d*, *e*) local interneurones in insects. *a*. Cobalt stained preparation of a spiking 'omega' neurone in the prothoracic ganglion of the cricket *Gryllus bimaculatus*, viewed ventrally (from Wohlers & Huber, 1978). *b*. A spiking 'south horizontal neurone' stained with Procion Yellow from the left lobula plate of the fly *Phaenicia sericata* (from Dvorak, Bishop & Hendrick, 1975). *c*. Four local interneurones stained with methylene blue in the mesothoracic ganglion of the dragonfly (*Aeshna*) larva (from Zawarzin, 1924). *d*, *e*. Non-spiking interneurones from the right half of the metathoracic ganglion of the locust viewed dorsally (from Siegler & Burrows, 1979). The neurone in *d* excites the slow extensor tibiae motor neurone, that in *e* excites the flexor tibiae motor neurones. The drawings in *a*, *b* were made from reconstructions of serial sections, *c* was traced from Zawarzin's drawings which show many neurones in one ganglion and those in *d*, *e* were drawn from whole mounts with a camera lucida.

neurones, and upon morphology which characterises them as local, intraganglionic interneurones. A positive physiological designation has not been forthcoming so far, although it is much needed. For the sake of brevity the neurones will often be referred to simply as local interneurones, and can for the present be considered as a distinct class. They are separable from local neurones that do spike in insects (Fig. 2*a*) (Casaday & Hoy, 1977; Popov, Markovich & Andjan, 1978; Wohlers & Huber, 1978), and from parts of spiking neurones that may be distant from the spike-initiating sites and invaded only by small electronic remnants of spikes that resemble excitatory post-synaptic potentials (EPSPs) (cf. Vedei & Moulins, 1977; Graubard, 1978). There are, however, no clear morphological criteria that only can separate spiking from non-spiking local interneurones (Fig. 2).

The morphology of the neurones is revealed by the intracellular injection of cobalt following physiological characterisation. What is revealed are complex and beautiful neurones which despite their diversity of shape have some features in common (Fig. 2*d*, *e*; see also Fig. 5 in the chapter by Rall). Each neurone has a single process which emerges from the cell body and which then breaks up into a profusion of branches. The majority of branches are in the more dorsal areas of the neuropile, where they are intermingled with those of the motor neurones, but other branches extend throughout the depth of the ganglion and into the most ventral neuropilar areas. The volume of neuropile in which the interneurones have branches varies considerably even for those neurones which have effects on motor neurones to the same muscle. None of the interneurones have processes that could be termed an axon and which leave the segmental ganglion. Few have processes that are more than 8–10 μm in diameter, although this measurement is necessarily dependent upon the fixation and staining procedure that is used and may need to be revised in the future. Interneurones with the same shape and with the same physiological effects can be recognised in different locusts (Siegler & Burrows, 1979). This may indicate that some at least of these small interneurones are unique individuals. Interneurones are also known with different shapes and yet with apparently similar physiological effects (Siegler & Burrows, 1979). Different interneurones may have similar actions as far as is revealed by present physiological tests. Estimates of the numbers of these interneurones can at present be little more than personal bias or guesswork. A large population is suggested by the wide diversity of shapes so far revealed, coupled with the fact that further staining tends to reveal still more types. In this regard, we have not yet reached the point where increased effort is rewarded by diminished returns. Counts of cell bodies in a locust abdominal ganglion (Sbrenna, 1971; Lewis, Miller & Mills, 1973), which may give a misleading impression because of the inherent difficulties and consequent inaccuracies of such methods, suggest that as many as 65% of cell bodies belong to local neurones. This *does not* imply that all are non-spiking interneurones. If such figures are true then the metathoracic ganglion may contain as many as 1000 local interneurones.

Physiology

A characteristic feature of many neurones that are recorded in the neuropile is that they apparently do not spike when they participate in the production of behaviour. Such neurones have always proved to have the morphological features just described. This does not mean that these neurones *cannot* spike, or that they *may not* do so under certain conditions. So ingrained in our thinking about neurones is the belief that all spike, that it is necessary to establish that these local interneurones *do not* spike under conditions which are thought to be close to those of real life. This is made all the more necessary following the demonstration that some interneurones in the brains of flies (Fig. 2*b*) that were at one time thought to be non-spiking, may be spiking neurones after all (Hengstenberg, 1977; Eckert, 1978). The problem must be squarely faced that the experimental procedure necessary to study the properties of small interneurones may in fact alter their properties.

The following observations suggest that some locust interneurones are normally non-spiking. First, penetration of an interneurone with a micro-electrode does not inactivate the spike mechanism. No effect of penetration is seen in a post-synaptic neurone which might indicate that a sudden depolarisation has occurred in the interneurone. Second, the lower membrane potential of local interneurones (mean ± S.E. -47.7 ± 0.5 mV) compared with those of spiking neurones (mean ± S.E. -63.5 ± 0.7 mV) recorded in the same region of the ganglion, does not inactivate the spike mechanism. Increasing the potential to the same value as that of the spiking neurones by the application of a steady hyperpolarising current, does not reveal spikes when the interneurones are depolarised during spontaneous or evoked movements. Third, recordings from different places in the same neurone (Pearson & Fourtner, 1975) or from two places simultaneously in the same neurone (Burrows & Siegler, 1978) do not reveal spikes. Current can be applied through one electrode to depolarise the interneurone but no spikes are recorded by either of the two electrodes, even though effects are evoked in post-synaptic neurones. Furthermore, no spikes are observed at either recording site during evoked or spontaneous movements to which the interneurone contributes. Fourth, the motor output produced by the ganglion in which the non-spiking interneurones are recorded, appears to be close to that produced by an intact locust. This observation would only be expected if the local interneurones were functioning in the same way on both occasions. The conclusion is therefore that some local interneurones can function and generate normal behaviour without themselves producing spikes. It is not the intention to show that the neurones *never* spike, but to accumulate evidence which suggests that they *normally do not* spike.

Synaptic transmission between local interneurones and motor neurones

A spiking interneurone evokes discrete synaptic potentials upon its post-synaptic neurones. By contrast a non-spiking interneurone evokes changes in its post-synaptic neurones that are continuously graded with respect to its own membrane potential (Burrows & Siegler, 1976, 1978). These

Fig. 3. Effects of one spiking interneurone, *a*, and two non-spiking interneurones, *b*, *c*, upon an identified flexor tibiae motor neurone recorded in one locust. *a*. A spiking interneurone that evokes EPSPs in the motor neurone. When depolarised with progressively increasing current, the frequency of spikes increases and the evoked EPSPs summate so that they eventually elicit a sequence of motor spikes. *b*. A non-spiking, local interneurone. When depolarised the interneurone evokes a sustained depolarisation of the motor neurone. With increasing currents the amplitude of the evoked depolarisation increases in a graded fashion until a sequence of motor spikes is elicited. *c*. A second non-spiking local interneurone. When depolarised, the interneurone evokes a sustained hyperpolarisation of the motor neurone whose amplitude is graded according to the presynaptic current. Calibration: interneurone (Int) *a*, 16 mV, *b*, *c*, 40 mV; motor neurone (mn) 7 mV; current 7 nA.

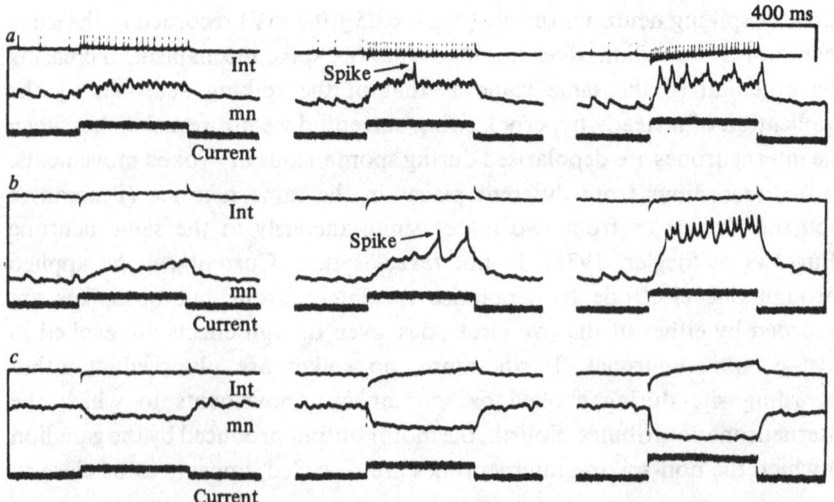

contrasting effects are most readily demonstrated by comparing the effects of a spiking neurone with those of non-spiking interneurones upon the same identified motor neurone recorded in one locust (Fig. 3). The recordings are made simultaneously from the cell body of the motor neurone and the neuropilar processes of the interneurones. Each spike in the spiking neurone is followed by a discrete EPSP of about 1.4 mV with a delay of 1.5 ms that suggests the connection is direct. Injecting a pulse of current into this interneurone evokes a sequence of spikes each of which is followed by an EPSP, so that the motor neurone is depolarised in an uneven fashion (Fig. 3*a*). More current evokes higher frequencies of spikes which in turn evoke summating sequences of EPSPs and eventually spikes in the motor neurone. It is only with the higher frequencies of presynaptic spikes that the depolarisation of the motor neurone becomes even.

Depolarisation of one non-spiking interneurone with a pulse of current produces a more slowly rising depolarisation of the motor neurone that is first detected after a delay of some 2 ms (Fig. 3*b*). When there is no presynaptic spike as a reference point, assessment of the synaptic delay is more difficult, but a variety of physiological tests and observations suggest that the connections between non-spiking interneurones and motor neurones may be direct. When more current is applied to the interneurone, the depolarisation of the motor neurone increases. The depolarisation is relatively smooth, bearing in mind that it is superimposed upon a background of other synaptic potentials. The depolarisation is sustained for the duration of the pulse and its amplitude is graded continuously with respect to the presynaptic current until the motor neurone begins to spike.

Depolarisation of another non-spiking interneurone produces effects of opposite polarity, namely a graded and sustained hyperpolarisation of the motor neurone (Fig. 3*c*). Again the voltage changes in the motor neurone are free from the sorts of discontinuities that a spiking neurone would cause. The amplitude of the hyperpolarisation is graded continuously with respect to the presynaptic current, until with larger currents a plateau is reached. From these sorts of observations there emerge two characteristic and salient features of transmission from the non-spiking interneurones to the motor neurones: the process is graded and can be sustained. These features will now be considered in turn.

Graded effects of local interneurones

Graded transmission should offer greater flexibility and more precise control over a greater range of presynaptic voltages than is possible if the presynaptic signals are digitised as spikes. The effectiveness of graded control can be illustrated by considering the way non-spiking interneurones influence the frequency of spikes in tonic motor neurones and the force that is generated in the muscles.

Injection of sinusoidally changing current into one interneurone alters its membrane potential in a sinusoidal way (Fig. 4). As a consequence, the frequency of spikes in an identified slow flexor tibiae motor neurone, upon which this interneurone synapses, also rises and falls in a sinusoidal way (Fig. 4*a*). Increasing the frequency of the sine waves continues to modulate the

Fig. 4. Graded control by local interneurones of the frequency of spikes in motor neurones and of the force generated by muscles. *a, b.* Sinusoidally varying current passing above resting potential is applied to an interneurone that excites an identified slow flexor tibiae motor neurone. The membrane potential of the interneurone is smoothly modulated by the current. Consequently the frequency of spikes in the motor neurone also varies sinusoidally. With a lower frequency of current (0.3 Hz in *a*), the frequency of motor spikes is greater than when the frequency of stimulating current is increased (1.0 Hz in *b*). *c, d.* An interneurone (not displayed) that excites the slow extensor tibiae motor neurone. Sinusoidally varying current injected into this interneurone modulates the frequency of extensor motor spikes in a sinusoidal fashion. The force developed by the extensor also fluctuates sinusoidally. *d.* With more current, the number of motor spikes is increased, but their frequency is still modulated sinusoidally. As a consequence the muscular force increases in a graded fashion. The interval between each spike is represented by the dots above each spike. Calibrations: voltage, interneurone (int) 40 mV, motor neurone (mn) 7 mV; force 0.2 g; current 15 nA; time, *a*, 800 ms; *b, d*, 400 ms.

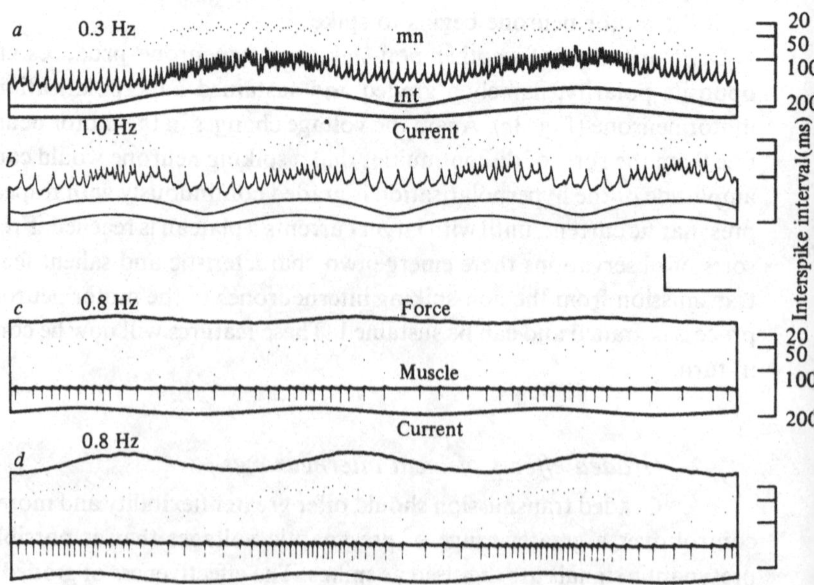

frequency of spikes sinusoidally although the maximum frequency attained in one cycle is lower (Fig. 4*b*). Therefore the frequency of spikes in a motor neurone can be controlled by one interneurone in a precise and graded way.

The next question is: what effect does stimulation of an interneurone have upon the muscular force? This can most easily be answered by considering the extensor tibiae muscle because its excitatory innervation, one slow and one fast motor neurone, is much simpler than that of the flexor. By selection of an appropriate interneurone the force generated by the slow motor neurone alone can be examined. Depolarisation of such an interneurone with sinusoidally varying current modulates the frequency of the spikes in the slow motor neurone, and as a consequence the force generated by the muscle also fluctuates sinusoidally (Fig. 4*c*). If the current is now increased, the frequency of spikes is still modulated sinusoidally, but more spikes at the higher frequencies are evoked. Therefore, more force is generated by the muscle, but again it is modulated sinusoidally (Fig. 4*d*). By adjusting the amount of current that is injected into this interneurone, the amount of force generated by the muscle can be controlled in a graded way. Such delicate and graded control of movement is precisely what is required of the nervous system.

Sustained effects of local interneurones

The pulses of current used in Fig. 3 produce no decrement in the response of the post-synaptic neurone with time. This is in contrast to transmission at spiking synapses when spikes have been artificially abolished (Katz & Miledi, 1967) or at some apparently non-spiking synapses in crustacea (Graubard, 1978). Even pulses of current lasting several seconds injected into a locust interneurone produce effects in motor neurones that persist without decrement for the duration of the stimulus. For example, a pulse of current lasting almost 5 s in one interneurone excited the levator tarsus motor neurones, some flexor tibiae motor neurones and caused the force developed by the flexor tibiae muscle to rise throughout the stimulation (Fig. 5*a*). Continuous depolarisation of an interneurone produces a gradual decline in the post-synaptic response, but an effect is still detectable after 7 min stimulation (Burrows & Siegler, 1978). It is the normal action of some of the interneurones to have continuous effects upon their post-synaptic neurones. These effects are achieved by the tonic release of transmitter indicating that these interneurones act in much the same way as rods or cones in the retina. A depolarisation of one of these interneurones increases, whilst a hyperpolarisation decreases the release of transmitter. Great care is necessary in making the observations on these particular interneurones, because for all the interneurones only small voltage changes are necessary to effect the release of transmitter. Therefore, damage inflicted by the electrode as it penetrates an interneurone could be sufficient to cause a tonic depolarisation and thereby lead to the erroneous conclusion that the interneurone is releasing large amounts of transmitter tonically. Nevertheless, such interneurones can be shown to shown to exist, and perhaps all the local interneurones will behave in this way in the appropriate behavioural context. Consider the interneurone illustrated in Fig. 5*b*. When depolarised, this interneurone evokes a hyperpolarisation of a fast flexor tibiae motor neurone; when the interneurone is hyperpolarised, the motor neurone is depolarised. This can be explained by supposing that the motor neurone is normally held tonically hyperpolarised by the transmitter released from this local interneurone. In addition to modulating the membrane potential of this flexor motor neurone, the interneurone has more widespread effects. When it is depolarised, tonically spiking flexor tibiae motor neurones are inhibited and at the same time the frequency of spikes in the slow extensor tibiae motor

Fig. 5. Sustained effects of local non-spiking interneurones upon motor neurones. *a*. A pulse of current lasting approximately 5 s is injected into a local interneurone that excites flexor tibiae and levator tarsi motor neurones. The frequency of spikes in both sets of motor neurones is sustained throughout, and the force developed by the flexor tibiae muscle gradually increases. There is a reflexive feedback on to the fast extensor tibiae motor neurone as a result of the developing flexion force. *b*. An interneurone that releases transmitter tonically upon the lateral fast flexor tibiae motor neurone. One cycle of sinusoidally varying current (passing about zero which is indicated by the dotted line on the bottom trace), is injected into the interneurone. When depolarised, the interneurone hyperpolarises the flexor and excites the slow extensor tibiae motor neurones (arrow, fourth trace). When hyperpolarised, the interneurone depolarises the impaled flexor and excites other flexor tibiae motor neurones (third trace). Calibration: voltage, interneurones (int) 40 mV, motor neurones (mn) *a*, 8 mV, *b*, 4 mV; force 0.4 g; current *a*, 30 nA, *b*, 6 nA; time *a*, 400 ms, *b*, 200 ms.

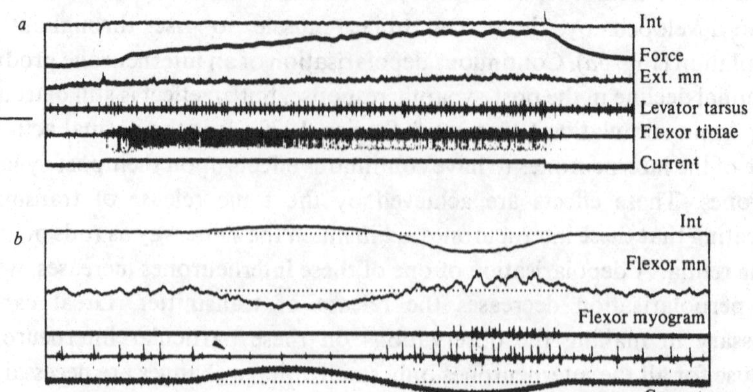

neurone increases (Fig. 5*b*). Conversely when the interneurone is hyperpolarised many flexor tibiae motor neurones begin to spike (Fig. 5*b*). The result is that when the interneurone is depolarised, the tibia extends and when hyperpolarised the tibia flexes. Thus voltage changes of either polarity in the interneurone cause movements of the tibia because of the delicate balance between the excitatory and inhibitory effects upon antagonistic motor neurones that is permitted by the tonic release of transmitter.

Synaptic transmission is chemically mediated

All the evidence accumulated so far supports the interpretation that synaptic transmission between the local non-spiking interneurones and motor neurones is mediated by chemical synapses and not by electrical junctions. Of course, this does not exclude the possibility that some connections are mediated by electrical junctions. First, current of either polarity injected into a motor neurone has no effect on a presynaptic local interneurone, as would be expected if an electrical junction were to exist. Second, the observed synaptic delays are longer than those expected of electrical junctions. Third, reversal potentials can be demonstrated for evoked hyperpolarising effects in motor neurones. Fourth, depolarising responses are associated with an increase in the number of voltage fluctuations in the motor neurones (Burrows & Siegler, 1978), which are probably associated with quantal release of transmitter.

Synaptic potentials and the release of transmitter

By placing two electrodes in one interneurone it is possible to obtain an estimate of the change in voltage needed to evoke the release of transmitter. The results so obtained indicate that a depolarisation of 1–2 mV evoked at one point by a pulse of current applied some 200 μm away, is sufficient to produce effects in a post-synaptic motor neurone (Burrows & Siegler, 1978). Interpretation of this experiment is complicated because we do not know the spatial distribution of the output synapses of the interneurones, or the spatial attenuation of voltage through neurones of such labyrinthine structure. This is where the contribution of Rall may prove invaluable (next chapter, this volume). Nevertheless it would seem possible that discrete synaptic potentials, which may exceed 2 mV amplitude, could themselves mediate transmission to the post-synaptic neurones.

Direct evidence that single synaptic potentials effect the release of transmitter can be obtained in the following way (Burrows, 1979*a*). An interneurone is depolarised that evokes an hyperpolarisation of an identified flexor tibiae motor neurone, recorded in one of its neuropilar processes, and at the same time excites the slow extensor tibiae motor neurone (Fig. 6*a*). The connection between the interneurone and the flexor motor neurone appears to be direct and chemically mediated. When the tibia of a hind leg is flexed, the interneurone is depolarised by a sequence of EPSPs, the flexor motor neurone is hyperpolarised and the extensor is excited. A resistance reflex is thus activated which opposes the forced flexion. Each EPSP in the interneurone is followed by an inhibitory post-synaptic potential (IPSP) in the motor neurone (Fig. 6*b*). The relationship of the PSPs is more clearly seen when they are averaged (Fig. 6*c*). The IPSP follows the EPSP in the interneurone with a delay of 1.5 ms which is the same as that observed when a hyperpolarisation is evoked by stimulating the interneurone directly. This indicates either that

Fig. 6. A single synaptic potential can cause the release of transmitter from a non-spiking local interneurone. *a*. Depolarisation of the interneurone evokes a sustained hyperpolarisation of an identified flexor tibiae motor neurone and spikes in the slow extensor tibiae motor neurone, as recorded in the extensor muscle. *b*. A forced and sustained flexion of the tibia evokes a sequence of EPSPs in the interneurone which is accompanied by a sequence of IPSPs in the flexor motor neurone. Spikes also occur in the slow extensor motor neurone. *c*, *d*. The EPSP in the interneurone triggers a signal averager and 32 occurrences are averaged. *c*. The interneurone is at its normal resting potential and the EPSP is followed by a prominent IPSP in the flexor motor neurone. *d*. The interneurone is hyperpolarised with 3 nA of current with a consequent increase in the amplitude of the EPSP. However, the IPSP in the flexor motor neurone is now much reduced. Calibration: voltage, *a*, interneurone (int) 16 mV, motor neurone (mn) 4 mV, *b*, 4 mV; *c*, *d*, 0.8 mV; current 15 nA; time *a*, *b*, 200 ms, *c*, *d*, 20 ms.

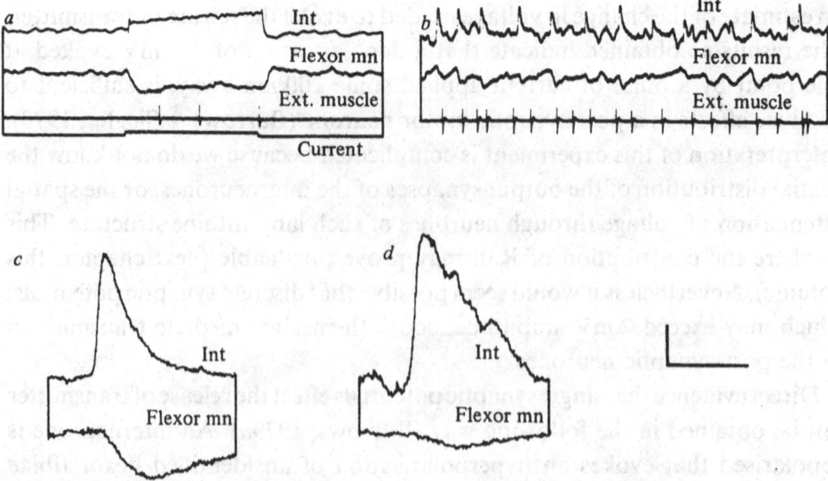

the EPSP directly causes the IPSP, or that both PSPs are derived from the same presynaptic neurone.

To show that the EPSP in the interneurone causes the IPSP in the motor neurone, the interneurone is tonically hyperpolarised. The expectation is that the EPSP, although now larger, should not reach the same absolute level of depolarisation and should therefore release less transmitter. When the hyperpolarisation is applied, the IPSP in the motor neurone is greatly reduced and its latency is increased (Fig. 6*d*). This is consistent with the expectation that only the peak of the EPSP is now able to release transmitter. The amplitude of the IPSP decreases in a graded way depending upon the level of hyperpolarisation in the interneurone. Taken together, the interpretation of these experiments is consistent with the fact that single, discrete EPSPs in an interneurone can evoke transmitter release on to a motor neurone.

There is a behavioural correlate of this method of information transfer. If the interneurone is hyperpolarised and the tibia is forcibly flexed, the resistance reflex evoked is less effective; the hyperpolarisation in the flexor motor neurone is reduced and there are fewer extensor spikes. Therefore, single synaptic potentials can transmit information of behavioural importance from one neurone to the next. The local interneurones are, however, not simply acting as a relay, or inverting mechanism for these particular synaptic potentials. The interneurones allow the effectiveness of the EPSPs to be modulated by other synaptic potentials that are present.

Contribution of local interneurones to reflexive actions

The above experiments suggest that the local interneurones have causal effects during reflex actions. During voluntary or forced movements of a hind leg, the membrane potentials of local interneurones that affect leg motor neurones are modulated by synaptic inputs. The interneurones are either hyperpolarised or depolarised by a barrage of synaptic potentials, the source of which has yet to be located in the locust. In the cockroach some local interneurones receive direct inputs from certain sensory neurones (Pearson, Wong & Fourtner, 1976). The local interneurones of the locust participate in reflexive actions in ways that are in accord with their experimentally demonstrated connections with particular motor neurones. This is well illustrated by an interneurone that excites the two extensor tibiae motor neurones and which is depolarised when the tibia of a hind leg is forcibly flexed (Fig. 7*a*). The depolarisation in the interneurone precedes that in the fast extensor tibiae motor neurone and subsequently there are many parallel fluctuations of voltage of similar sign in the two neurones. Another interneurone, when depolarised, hyperpolarises a flexor tibiae motor neurone.

When the tibia is flexed the interneurone is depolarised and this is followed by a hyperpolarisation of the flexor motor neurone, that interrupts its spikes, and an excitation of the slow extensor tibiae motor neurone (Fig. 7*b*). In this pair of neurones the voltage changes are of opposite polarity. For both pairs of neurones, the changes in the interneurones are appropriate for them to play a casual role in controlling the motor neurones. However, during these vigorous and complex movements it is difficult to demonstrate that there are causal effects. Undoubtedly the action of one interneurone does influence the output of a motor neurone, for hyperpolarising it can diminish the response. The conclusion must be that the action of a motor neurone depends upon the combined effects of a group of local interneurones, and of course other neurones. For example, a group of twelve local interneurones have already been characterised that influence the excitability of the slow extensor tibiae motor neurone (Burrows, 1980*b*). Therefore in order to understand the role played by the local interneurones in generating motor patterns, an essential prerequisite is to examine possible interconnections between them. Only then can an attempt to made to understand the distribution of the output connections of an interneurone with particular sets of motor neurones.

Fig. 7. Reflexive actions of a hind leg and the participation of local interneurones. *a*. An interneurone which when depolarised excites both the slow and the fast extensor tibiae motor neurones. When the tibia is forcibly flexed, the depolarisation of the interneurone precedes and then parallels that of the fast extensor motor neurone. *b*. An interneurone that hyperpolarises an identified flexor tibiae motor neurone. When the tibia is forcibly flexed the interneurone depolarises and the motor neurone is hyperpolarised and stops spiking. The spikes in the flexor and extensor muscle alternate as the tibia is moved rhythmically. Calibration: voltage, interneurones (int) 8 mV, motor neurones (mn) 3.5 mV; time, *a*, 100 ms, *b*, 200 ms.

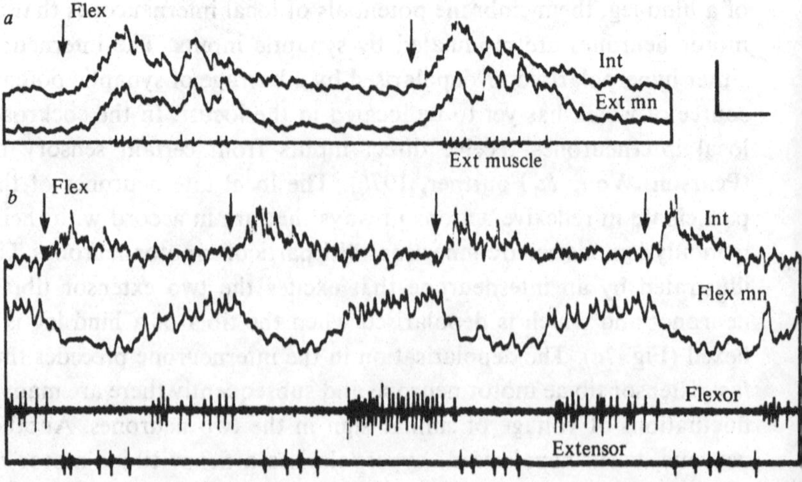

Interconnections between local interneurones

To demonstrate connections between two local interneurones requires that simultaneous recordings be made from the fine neuropilar processes of both. Against my original expectation, whenever an interaction was found between two local interneurones it was always inhibitory and one-way (Burrows, 1979*b*). The interactions are mediated by the graded release of chemical transmitter. Of 90 paired recordings from interneurones some 38% have displayed connections that are one-way and inhibitory. No connections have been revealed between the other pairs of interneurones. The interconnections occur between interneurones which have the following properties.

1. Control of opposing movements of one joint: for example some of the interneurones that excite flexor tibiae motor neurones are inhibited by some of the interneurones that excite extensor tibiae motor neurones and vice versa.

2. Control of movements of different joints; for example some interneurones that excite depressor tarsi motor neurones inhibit interneurones that excite adductor coxae interneurones.

3. Control of the output of the same motor neurone; for example two interneurones that excite the slow extensor motor neurone are connected by a one-way inhibitory synapse (Fig. 8). Depolarisation of the first interneurone produces a hyperpolarisation in the second interneurone that is graded with respect to the presynaptic current (Fig. 8*a*). Current injected into the second interneurone has no effect upon the first interneurone, but nevertheless excites the slow extensor tibiae motor neurone (Fig. 8*b*). When each interneurone is depolarised separately, their excitatory effects upon the slow extensor motor neurone are similar (Fig. 8*c*, *d*). However, when both interneurones are depolarised together, their combined effects are little more than their individual effects (Fig. 8*e*). A possible explanation is that whenever the first interneurone is depolarised it inhibits the action of the second interneurone.

Inhibitory interactions between the local interneurones imply that the action of a particular interneurone must be dependent upon the activity of, probably, numerous other local interneurones. Each interneurone contributes to the patterning of the motor output only in the context of the activities of the others. The changes in behavioural context which can affect the output evoked by an interneurone can be extremely subtle. For example some interneurones undergo tonic shifts in their membrane potential of several mV when the femoral–tibial angle of one hind leg is altered by a few degrees. Given this dependence upon the prevailing behavioural context, an inhibitory connection between two interneurones affecting the same motor neurone can be interpreted to indicate that the two are used in different behavioural contexts, in some of which it is necessary to exclude the action of one interneurone. Implicit in this assumption is that each of the two interneurones makes widespread but slightly different connections with various motor neurones, or receive inputs from different neurones. Either of these interneurones acting alone is incapable of eliciting the full range of spike frequencies from the slow extensor motor neurone, or the maximal amount of force from the muscle. Therefore the effects of one of these interneurones must sum with those of other interneurones in order to produce the movements characteristic of normal behaviour. Summation of effects can be shown to occur from interneurones that are apparently not interconnected directly (Fig. 8*f*). Consider two interneurones that excite flexor tibiae motor

Fig. 8. Interactions of local interneurones. *a–e*. Two local interneurones that excite the slow extensor tibiae motor neurone. *a*. Sinusoidally varying current is injected into the first interneurone. As a result the second interneurone is hyperpolarised and the extensor motor neurone is excited whenever the current is depolarising. *b*. A pulse of current injected into the second interneurone is without effect on the first, but does excite the extensor motor neurone. The interneurones are therefore connected by a one-way inhibitory synapse. *c, d*. Pulses of current injected into either interneurone increase the frequency of extensor motor spikes to about the same level. *e*. When the two interneurones are depolarised together, they are no more effective at exciting the extensor motor neurone than when activated separately. *f*. Two interneurones that do not synapse upon each other produce an increased motor output when they are depolarised at the same time. Calibration: voltage, *a*, 8 mV, *b*, int 2 40 mV, int 1, 8 mV, *c–f*, 40 mV; current 15 nA. int = interneurone.

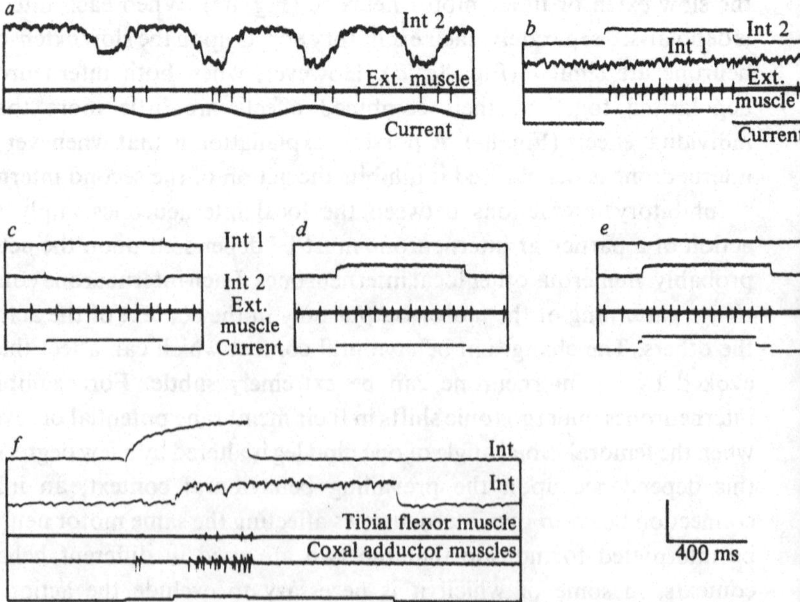

neurones and some coxal motor neurones. Both interneurones when depolarised individually had weak effects only upon coxal motor neurones. When the depolarisation of the two was arranged to coincide, there was then a much greater effect on the coxal motor neurones, and an effect upon flexor neurones became apparent (Fig. 8*f*).

Control of groups of motor neurones

A movement results from the coordinated output of a group of motor neurones. The local interneurones have clear effects on all the leg motor neurones examined so far, making them likely candidates for an organisational role in coordinated movements. To assess the effects of one interneurone upon a group of motor neurones requires that must be satisfied with less than the acceptable amount of evidence concerning the directness of connections. It has been feasible until now to record intracellularly from no more than three neurones in one ganglion. Therefore rigorous evidence for the directness of only two of the apparently many connections of an interneurone can be obtained. The problem is complicated still further by the interconnections between the interneurones; current injected into one interneurone will spread 'laterally' to other interneurones as well as 'forwards' to the motor neurones. With these limitations of the method in mind, it is still possible to gain a considerable insight into the way the motor neurones are organised by the local interneurones. The following generalisations can be made.

The motor neurones that innervate one muscle are subdivided into overlapping sets of various sizes by the local interneurones. For example, some interneurones excite only the fast, others only the slow motor neurones while others have effects upon both. This type of organisation allows an individual motor neurone to be activated on its own, or in various combinations with its fellow motor neurones that are appropriate for a particular behaviour. Few, if any, interneurones have effects which are limited to the motor neurones of one muscle. Instead they affect movements about more than one joint in a hind leg.

The motor neurones to muscles that move one joint are organised by the local interneurones in at least three ways. First, where only two muscles are involved there are interneurones with effects upon motor neurones of one muscle, but which have no effects on those to the other muscle. This permits the independent inhibition or excitation of one muscle in combination with muscles of other joints. Second, the muscles can be divided into antagonistic groups and controlled reciprocally. One interneurone can have excitatory or inhibitory effects upon different motor neurones (Burrows, 1980*b*). Third, the motor neurones can be activated at the same time so that the muscles co-contract. Considerable flexibility and control is permitted by this sort of organisation in that the balance can be shifted between both the number and the particular types of interneurones and hence the sets of motor neurones that are used.

Some of the local interneurones produce dramatic movements of a hind leg that are coordinated both spatially and temporally. In fact the movements so evoked are components of the normal behaviour of the locust. One interneurone causes movement about three joints of the hind leg; a rotation of the coxa, a flexion of the tibia and a levation of the tarsus. The extent of the movement about a particular joint depends on the current injected into the interneurone. With low current the tibia may flex weakly with no accompanying movement of the tarsus. With more current the tibia flexes more vigorously, there is a concomitant rotation of the coxa and a delayed levation of the tarsus. Undoubtedly reflexes add to the effects on the motor neurones once a movement has been initiated but it would appear that the interneurone has direct effects upon motor neurones to muscles of each joint.

The organisation of a population of local interneurones

What has emerged from this study of one ganglion in the locust is that local interneurones and graded synaptic transmission are responsible for a considerable amount of the integration that occurs. The normal functional state of these local interneurones studied so far appears to be non-spiking. This does not imply that the neurones cannot produce spikes under appropriate natural or experimentally induced conditions. The point to be emphasised about these neurones is their ability to release transmitter in a graded fashion in response to small changes in their membrane potentials. All observations of ganglia that are producing motor outputs not discernibly different from those of the normal locust suggest that the interneurones transmit information to other neurones without themselves producing spikes. Indeed individual synaptic potentials can cause the release of transmitter. This finding when coupled with the anatomical observation that the branches of one interneurone can be separated by as much as 1200 μm (Siegler & Burrows, 1979) and with the calculations of Rall concerning attenuation within a local interneurone (this volume) suggests that local processing of information occurs. Consider for example, a synaptic potential of 0.3 mV at one site in a neurone (e.g. the one in Fig. 5 of Rall, in the next chapter of this volume) that can effect the release of transmitter on to a post-synaptic neurone. From Rall's calculation this potential can be attenuated by a factor of at least 3.2 (and probably by much more) depending upon the geometry of the neurone, so that at another site in the neurone it may only be 0.9 mV in amplitude. At this second site it may well be too small to effect the release of transmitter from the output synapses there. It is possible therefore, that processing of information can occur independently at the two sites in the neurone. This opens the intriguing possibility, which must be established experimentally, that a neurone can perform several simultaneous, but largely independent computations. Such an interpretation is dependent upon the determination, in the future, of the membrane properties of the interneurones, the spatial distribution of input and output synapses and whether these synapses are intermingled. Nevertheless there is a precedent for the isolation of parts of a neurone in the B-type horizontal cells of the cat retina (Nelson, Lützow, Kolb & Gouras, 1975; Boycott, Peichl & Wässle, 1978). In addition, the concept of neurones performing local computations, which is considerably strengthened by the observations on the locust interneurones, is at the heart of current thinking about nervous systems (Shepherd, 1972, 1978, and this volume; Rakic, 1976; Schmitt & Worden, 1979).

The dominance of inhibitory connections amongst the local interneurones suggests that their organisation will not prove to be too dissimilar from that of other neurones involved in the generation of motor patterns. For example, inhibitory connections predominate amongst the motor neurones that generate gut movements in the lobster (Selverston, Russell, Miller & King, 1976) and between interneurones that are responsible for the heart-beat (Thompson & Stent, 1976) or swimming movements (Stent *et al.*, 1978) of the leech. In both the lobster and the leech graded synaptic transmission occurs even between neurones that may normally spike (Graubard, 1978; Nicholls & Wallace, 1978). The expectation is therefore that the non-spiking local interneurones and graded synaptic transmission observed in the locust, are not simply a quirk in the organisation of arthropod nervous systems, but general features of all nervous systems.

This work was supported by a grant from the S.R.C. Some of the experiments, and many of the ideas were developed with Dr M. V. S. Siegler, to whom I owe much.

References

Boycott, B. B., Peichl, L. & Wässle, H. (1978). Morphological types of horizontal cell in the retina of the domestic cat. *Proc. R. Soc. Lond., Series B*, **203**, 229–45.

Burrows, M. (1979*a*). Synaptic potentials effect the release of transmitter from locust non-spiking interneurons. *Science*, **204**, 81–3.

Burrows, M. (1979*b*). Graded synaptic interactions between local pre-motor interneurones of the locust. *J. Neurophysiol.* **42**, 1108–1123.

Burrows, M. (1980*a*). The tracheal supply to the central nervous system of the locust. *Proc. R. Soc. Lond. Series B*, **207**, 63–78.

Burrows, M. (1980*b*). The control of sets of motor neurones by local interneurones in the locust. *J. Physiol., Lond.* **298**, 213–33.

Burrows, M. & Siegler, M. V. S. (1976). Transmission without spikes between locust interneurones and motoneurones. *Nature, Lond.* **262**, 222–4.

Burrows, M. & Siegler, M. V. S. (1978). Graded synaptic transmission between local interneurones and motor neurones in the metathoracic ganglion of the locust. *J. Physiol., Lond.* **285**, 231–55.

Casaday, G. B. & Hoy, R. R. (1977). Auditory interneurons in the cricket *Teleogryllus oceanicus*: physiological and anatomical properties. *J. comp. Physiol.* **121**, 1–13.

Dvorak, D. R., Bishop, L. G. & Eckert, H. E. (1975). On the identification of movement detectors in the fly optic lobe. *J. comp. Physiol.* **100**, 5–23.

Eckert, H. E. (1978). Response properties of dipteran giant visual interneurones involved in the control of optomotor behaviour. *Nature, Lond.* **271**, 358–60.

Graubard, K. (1978). Synaptic transmission without action potentials: input-output properties of a non-spiking presynaptic neuron. *J. Neurophysiol.* **41**, 1014–25.

Hengstenberg, R. (1977). Spike responses of 'non-spiking' visual interneurone. *Nature, Lond.* **270**, 338–40.

Katz, B. & Miledi, R. (1967). A study of synaptic transmission in the absence of nerve impulses. *J. Physiol., Lond.* **192**, 407–36.

Lewis, G. W., Miller, P. L. & Mills, P. S. (1973). Neuromuscular mechanisms of abdominal pumping in the locust. *J. exp. Biol.* **59**, 149–68.

Mendelson, M. (1971). Oscillator neurons in crustacean ganglia. *Science*, **171**, 1170–3.

Nelson, R., v. Lützow, A., Kolb, H. & Gouras, P. (1975). Horizontal cells in the cat retina with independent dendritic systems. *Science*, **189**, 137–9.

Nicholls, J. & Wallace, B. G. (1978). Modulation of transmission at an inhibitory synapse in the central nervous system of the leech. *J. Physiol., Lond.* **281**, 157–70.

Pearson, K. G. & Fourtner, C. R. (1975). Nonspiking interneurons in walking system of the cockroach. *J. Neurophysiol.* **38**, 33–52.

Pearson, K. G., Wong, R. K. S. & Fourtner, C. R. (1976). Connexions between hair-plate afferents and motoneurones in the cockroach leg. *J. exp. Biol.* **64**, 251–66.

Popov, A. V., Markovich, A. M. & Andjan, A. S. (1978). Auditory interneurons in the prothoracic ganglion of the cricket, *Gryllus bimaculatus* deGeer. 1. The large segmental auditory neuron (LSAN). *J. comp. Physiol.* **126**, 183–92.

Rakic, P. (Ed.). (1976). *Local Circuit Neurons*. Cambridge, Mass: MIT Press.

Sbrenna, G. (1971). Postembryonic growth of the ventral nerve cord in *Schistocerca gregaria* Forsk. (Orthroptera: Acrididae). *Boll. Zool.* **38**, 49–74.

Schmitt, F. O. & Worden, F. G. (Ed.) (1979). *The Neurosciences: Fourth Study Program*. Cambridge, Mass: MIT Press.

Selverston, A. I., Russell, D. F., Miller, J. P. & King, D. G. (1976). The stomatogastric nervous system: structure and function of a small neural network. *Prog. Neurobiol.* **6**, 1–75.

Shepherd, G. M. (1972). The neuron doctrine: a revision of functional concepts. *Yale J. Biol. Med.* **45**, 584–99.
Shepherd, G. M. (1978). Microcircuits in the nervous system. *Sci. Amer.* **238**, 92–103.
Siegler, M. V. S. & Burrows, M. (1979). The morphology of local non-spiking interneurones in the metathoracic ganglion of the locust. *J. comp. Neurol.* **183**, 121–48.
Stent, G. S., Kristan, W. B. Jr., Friesen, W. O., Ort, C. A., Poon, M. & Calabrese, R. L. (1978). Neuronal generation of the leech swimming movement. *Science*, **200**, 1348–57.
Thompson, W. J. & Stent, G. S. (1976). Neuronal control of heartbeat in the medicinal leech. III. Synaptic relations of heart interneurons. *J. comp. Physiol.* **111**, 309–33.
Vedel, J.-P. & Moulins, M. (1977). Functional properties of interganglionic motor neurons in the stomatogastric nervous system of the rock lobster. *J. comp. Physiol.* **118**, 307–25.
Wohlers, D. W. & Huber, F. (1978). Intracellular recording and staining of cricket auditory interneurones (*Gryllus campestris* L., *Gryllus bimaculatus* DeGeer). *J. comp. Physiol.* **127**, 11–28.
Zawarzin, A. (1924). Zur Morphologie der Nervenzentren. Das Bauchmark der Insecten. Ein Beitrag zur vergleichenden Histologie (Histologische Studien über Insekten VI). *Z. wiss. Zool.* **122**, 323–424.

WILFRID RALL

Functional aspects of neuronal geometry

Introduction

This chapter will review some implications of cable theory for the role of neurones without impulses and present the results of some very recent computations. Exposition and review of neuronal cable properties can be found in the book by Jack, Noble & Tsien (1975) and in a chapter in a recent handbook (Rall, 1977). Here I disclaim any pretence at scholarly review; in addition to the usual difficulties in keeping abreast of an expanding literature, I have had problems with my vision (including disastrous cataract surgery on one eye) which interfered with my reading for much of the past two years. I apologise to those authors who should have been cited, and welcome corrective personal communications.

Perspective

Many modes of neural activity

There is much we still don't know about neuronal interactions, and it is wise to keep our minds open to many possibilities. Given basic membrane properties together with one of several basic branching patterns, it is apparent that the same neuronal machinery can be adjusted to function in several different modes: phasic or tonic, spontaneously rhythmic or not, synaptic input or output brief or prolonged, synaptic output graded in response to graded input and/or impulsive all-or-none output, etc. Somewhat similar statements can be made about assemblies of neurones and systems of neurones. (While the above statements may well be self evident to biologists, the neural modeller is constantly confronted with the fact that different selections of parameter values in the same mathematical model will result in significantly different modes of behaviour.)

I assume that some neurones (and neural systems) have parameter values that ensure robustness of one activity mode, while others (having different parameter values) are poised for ready shift between different activity modes. There is understandable survival value in the evolution of some units and systems characterised by robust inflexibility of mode, and in the evolution of other units and systems characterised by inherent adaptability.

Neurones without impulses

Neurones without impulses can be expected to exist in both of the above categories: the robust and the adaptable. The robust type, which never generates an impulse, would interact synaptically (in both presynaptic and

Fig. 1. Olfactory bulb, showing histological layers (*a*), and field potentials (*b*), as described and discussed in the text. *c* illustrates the postulated dendro-dendritic synaptic activation sequence in a single granule cell gemmule (hatched) in contact with a mitral secondary dendrite (cylinder) at synaptic contacts indicated by thickenings. Time epochs (I–III) are as in *b*. *D* is depolarisation, $\mathscr{E}$ is synaptic excitation, $\mathscr{I}$ is synaptic inhibition and *H* is hyperpolarisation.

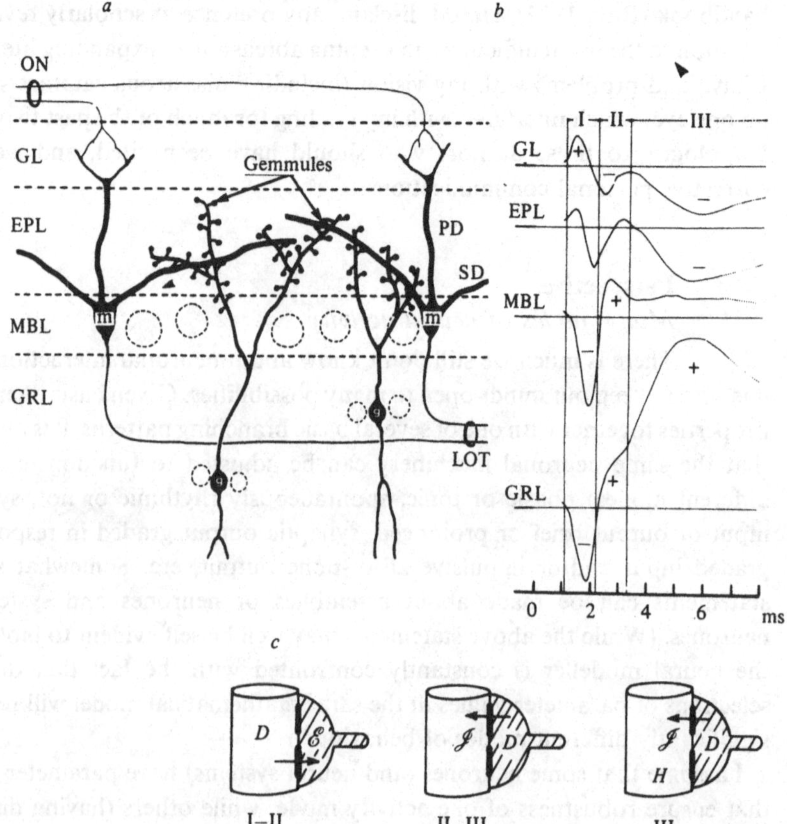

post-synaptic roles) with other neurones that it contacts. It might also have electrotonic junctions with some of its neighbours, and possibly also be involved in neurohumoral interactions. The latter, however, would seem to be of more importance for the adaptable category, whose shifting between different activity modes might be controlled in this way.

I will not attempt to survey the many examples of neurones without impulses that have been discovered in the last few years; this book as a whole will provide that kind of perspective. Thus I comment only briefly upon the earliest example to come to my attention, namely the granule cell population of mammalian olfactory bulb (Rall, Shepherd, Reese & Brightman, 1966; and Shepherd, this volume). These neurones possess no axons and few, if any impulses, yet they are present in very large numbers and are now understood to function as interneurones.

The schematic diagram (Fig. 1*a*) shows only a few of the hundreds of thousands of neurones that are found in the indicated layers of mammalian olfactory bulb. The output neurones are the mitral cells (m) with cell bodies in the mitral body layer (MBL) and whose axons exit in the lateral olfactory tract (LOT). These mitral cells extend their primary dendrites (PD) outward into the glomerular layer (GL), where they receive synaptic input from many fine axons of the olfactory nerve (ON). The mitral cells also extend several secondary dendrites (SD) laterally outward into the external plexiform layer (EPL), where they are in intimate association with the dendritic spines (or gemmules) belonging to the dendrites of the granule cell population, whose cell bodies (g) lie deeper in the granular layer (GRL) of the bulb. Because these granule cells have no axons, their physiological function had been obscure. However, it should be noted that the careful studies of Phillips, Powell & Shepherd (1963) and Shepherd (1963) did identify these granule cells as interneurones which could be presumed to deliver recurrent synaptic inhibition to the mitral cells; their studies also demonstrated the correlation of highly reproducible field potentials in response to LOT stimulation (Fig. 1*b*), as a function of depth in the bulb, with four distinct histological layers of the bulb, as is indicated by the juxtaposition of parts *a* and *b* in Fig. 1. It was this correlation, together with a recognition of the importance of anatomical symmetry and electrophysiological synchrony, that provided a basis for a theoretical reconstruction of such field potentials (Rall & Shepherd, 1965, 1968).

Here I do not intend to explain that theoretical reconstruction; it was presented carefully in the 1968 reference above and has been discussed further elsewhere (Rall, 1970; Shepherd, 1974; Klee & Rall, 1977). What this reconstruction did was to convince us that the field potentials during time periods I and II (in Fig. 1*b*) could be largely accounted for by extracellular current flow between soma and dendrites of the synchronously activated mitral cell population, and that (for quantitative reasons) the field potentials during time period III could not be due to mitral cell activity. Because only the granule cell population could account for the size and depth of the potentials of period III, we became convinced that there must be significant and extensive depolarisation of granule cell dendrites in the EPL during period III. We recognised that this could result from synaptic excitation delivered to those dendrites in the EPL at a time (late period I and early period II) when the mitral secondary dendrites would be depolarised and hence in suitable condition to activate hypothetical mitral-to-granule dendro-dendritic excitatory synapses (Fig. 1*c*: I–II). Also, we recognised that granule cell membrane depolarisation in the EPL would be well timed and well placed to activate hypothetical granule-to-mitral dendro-dendritic inhibitory synapses (Fig. 1*c*: II–III and III). Fortunately for us, we did not have to wait long to learn from independent research by our colleagues, Reese & Brightman, that electron microscopy reveals reciprocal dendro-dendritic synapses which we found to be consistent with those postulated synaptic actions (Hirata, 1964; Andres, 1965; Rall *et al.*, 1966); these findings were further confirmed and amplified by Reese & Brightman (1970), Price & Powell (1970) and by Jackowski, Parnevalas & Lieberman (1978).

This fortunate agreement between experiment and theory meant that we had an explanation for how the granule cells (devoid of axons and with few, if any, impulses) could function as inhibitory interneurones. At that time there was no precedent for these notions and they were met with considerable scepticism. Dogmatically inclined critics objected to our designating as 'dendritic' a process which can function presynaptically; our answer was that 'dendritic' is a descriptive morphological term, and that function is a separate question to be determined by observations rather than dogma; furthermore, these processes function both presynaptically and post-synaptically which was even harder for dogmatists to swallow. Others were startled at the suggestion that synapses could be activated without a presynaptic impulse, but we could point to already existing evidence for graded release of synaptic transmitter for graded depolarisation of presynaptic membrane, both at the neuromuscular junction (Castillo & Katz, 1954; Liley, 1956; Katz & Miledi, 1965, 1967), and at squid synapses (Katz & Miledi, 1966; Bloedel, Gage, Llinás & Quastel, 1966; Kusano, Livengood & Werman, 1967). In vertebrates and arthropods (see other chapters) we now have direct experimental evidence, obtained by intracellular recording, for neurones without impulses.

In the chapter after this one, Dr Shepherd includes documentation of the fact that granule cells of the olfactory bulb receive and integrate synaptic input from several different afferent sources: thus these particular neurones incorporate an important integrative function that is common to many other types of interneurones. In our original presentation, we also pointed out that the inhibitory action of the granule cell population could account for lateral inhibition that might contribute to olfactory discrimination, and for rhythmic activity of the mitral cell population, a well-known phenomenon in olfactory bulb.

Returning to the general case of neurones without impulses, it should be noted that such neurones need not necessarily be sensory neurones or interneurones. Such a neurone could be an effector neurone provided that it is located near enough to the effector organ to make direct synaptic contact with it. Such an arrangement would be particularly appropriate for effector organs that give graded responses to graded synaptic input. It should also be pointed out that neurones without axons are not entirely restricted to local interactions within one restricted region of a nervous system, because they could be presynaptic and/or post-synaptic to contacting processes (whether these be axons, collaterals, long dendrites or non-commitally called neurites) that belong to neurones which are primarily located outside the local region in question.

Returning to the adaptable category of neurones without impulses, we can find more possibilities. Some may possess axons and function in several modes, some of which would involve no impulses, while other modes might include impulses. The means of shifting between different modes might be neuro-hormonal or electrophysiological, and may provide fascinating research opportunities. It should also be noted that those neurones which have several distinct dendritic arbors separated by significant electrotonic distance could participate in local interactions involving one arbor that would be almost independent of other local interactions involving another arbor. Different degrees of coupling between such arbors could serve to provide different amounts of coupling between two or more multineuronal activities (or between neuronal subsystems).

This suggestion of different degrees of coupling between arbors, and hence between associated subsystems, can be augmented by a further suggestion that the degree of coupling might be modified by changes of membrane resistance or of core resistance. Then we have an interesting preliminary model for one possible aspect of nervous system plasticity. I hope to say more about this and about dendritic spine-stem resistance changes, later.

Neurone models and experiments

Mathematical modelling of dendritic neurones was begun because of a need to combine three different kinds of knowledge into a coherent theory. There are anatomical measurements of dendritic branching and of synaptic arrangements. Then there are well-established membrane models for passive

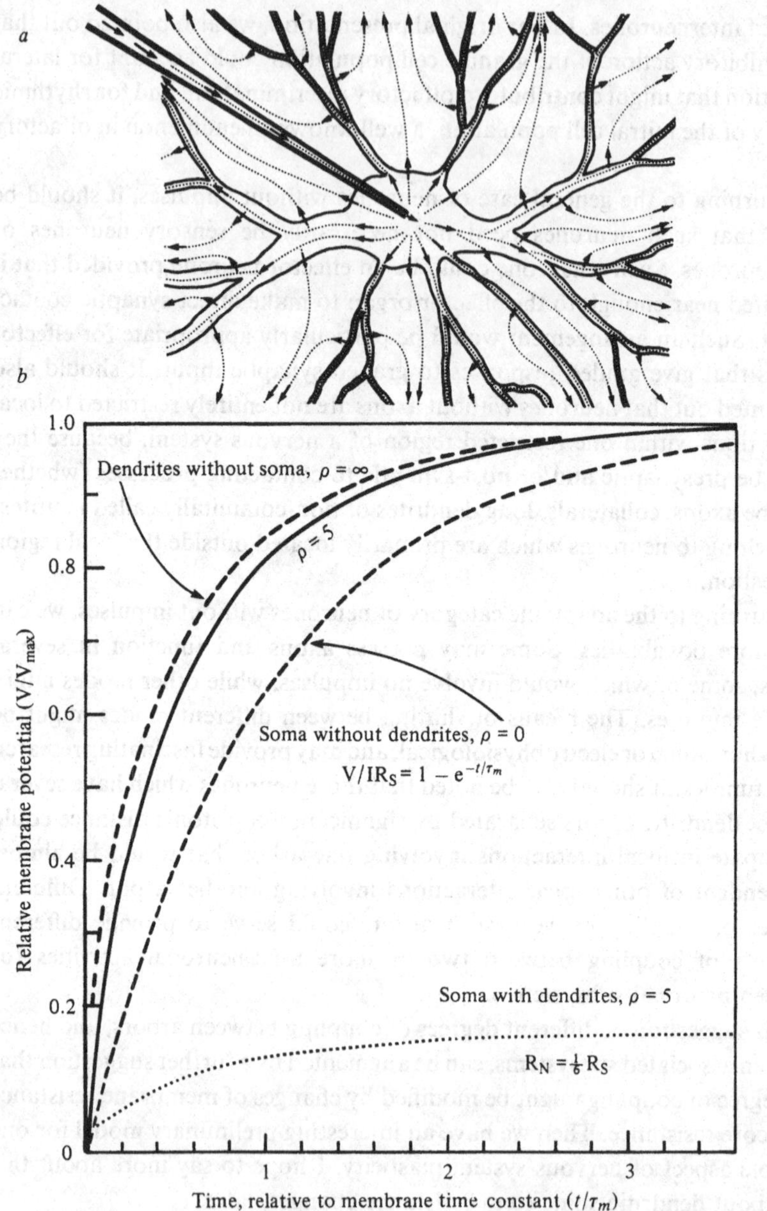

Fig. 2. *a.* Diagram showing flow of electric current from a microelectrode whose tip penetrates a neurone soma. Full extent of dendrites not shown; external electrode to which current flows is at a distance far beyond limits of this diagram; adapted from Rall (1959). *b.* Transient response at soma or origin of dendrites when a current step is suddenly turned on at $t = 0$; adapted from Rall (1957). The upper limiting case ($\rho = \infty$) represents $V/V_{max} = \text{erfc}(\sqrt{t/\tau})$, corresponding to dendritic cylinders of semi-infinite

membrane, synaptic membrane and for impulse-generating membrane. Also we have detailed quantitative electrophysiological information that has been obtained from individual neurones by means of intracellular and extracellular microelectrodes. The need to interpret experiments correctly and to take all relevant data into account has led to a family of mathematical models; different choices of simplifying assumptions are appropriate to different experimental situations. A few equations will be included in the Appendix; further details of these mathematical models can be found in the original papers; some details and discussion can also be found in Jack *et al.* (1975) and in Rall (1977).

Electrode inside soma

When an intracellular electrode enables us to measure the input resistance of a neurone, and to record the transient response to the application of a current step or brief current pulse, how do we interpret these measurements? Fig. 2 illustrates this problem for an electrode inside the soma; how much current flows into the dendrites, how far does it spread, and how does this dendritic current influence the observed transients? Here we note briefly that the answer lies between two limiting cases. One of these can be designated dendritic dominance (uppermost curve of Fig. 2*b*, labelled 'Dendrites without soma, $\rho = \infty$', and determined by the cable properties of dendrites represented as cylinders of semi-infinite length); the other limiting case can be designated somatic dominance (curve labelled 'Soma without dendrites, $\rho = 0$', and determined by a single exponential approach to a steady maximum, which corresponds to lumped somatic membrane, uncomplicated by any attached cables). The effect of finite dendritic length was taken into account later (Rall, 1962, 1964, 1967) and it was shown how the effective electrotonic length of the whole neurone (as represented by the most nearly equivalent cylinder) can be estimated from the ratio of two time constants found by peeling the sum of exponential decays present in experimental transient data (Rall, 1969; Nelson & Lux, 1970; Lux, Schubert & Kreutzberg, 1970; Burke & ten Bruggencate, 1971); see Jack & Redman (1971) or Jack *et al.* (1975) for a different but related method of estimating electrotonic length.

Before leaving Fig. 2, I wish to comment about an apparent paradox that

---

length. The other limiting case ($\rho = 0$) represents a single exponential, as labelled. Both the solid curve and the dotted curve are for the case ($\rho = 5$), as discussed in text. The symbol, ρ, represents the ratio of steady current into all dendrites to the steady current across the soma membrane; it is dendritic input conductance divided by some membrane conductance. See Rall (1960) for additional details.

sometimes bothers people; why does the transient appear to rise faster when dendritic current drain is added to the soma? This can be resolved by considering the lowest (dotted) transient in Fig. 2*b*, for the case ($\rho = 5$) where the final steady current into the dendrites is five times that which flows across the soma membrane. For this case, the neuronal input resistance, R_N, equals one-sixth of the soma membrane resistance, R_S. Thus, when the same constant current is applied both to the soma without dendrites, and to this soma with dendrites, the final steady voltage, IR_N, will be only one-sixth as great as the steady voltage, IR_S, for the soma alone; however the initial slope, dV/dt as $t \to 0$, depends upon the soma membrane capacity and is exactly the

Fig. 3. Branching diagram and graph of steady-state distribution of voltage as a function of x/λ in all branches and trees of this neurone model, for two cases, where x is the distance along the branch and λ is the length constant of each branch. Solid lines show result for current injection to one branch terminal, labelled I, with voltage also shown in sibling branch (S), first cousin branches (C-1), second cousin branches (C-2) and along the main line leading to the soma and then (leftward) into all of the other five dendritic trees. Dashed lines show result for injection of same current at soma, which is here represented merely as the common original of all six trees. Details of model and the equations can be found in the paper (Rall & Rinzel, 1973) from which this figure was adapted.

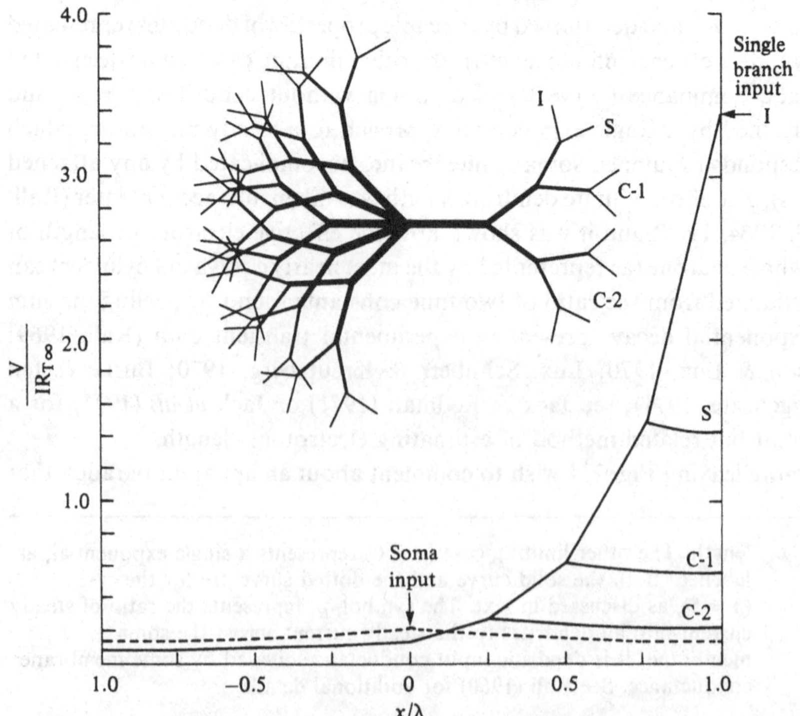

same in both of these cases. It is only when the dotted transient is rescaled relative to its maximum (i.e. scaled up by a factor of 6) that the early rate of rise of the curve becomes more steep for $\rho = 5$ than for $\rho = 0$. (Incidentally; the limiting case ($\rho = \infty$) has an infinitely steep initial slope which can be understood in terms of a zero lumped capacity at $x = 0$.)

Steady voltage attentuation in dendritic neurones

Fig. 3 illustrates several points I wish to make about extensively branched neurones. It is convenient, but not necessary, to represent this neurone as composedof six identical dendritic trees, and to assume symmetrical branching that satisfies the constraint ($\Sigma d^{3/2}$ = constant, where d = branch diameter) which underlies mathematical transformation between a dendritic tree and an equivalent cylinder (Rall, 1962). One tree receives input current at a single branch terminal; five other trees (grouped to the left) receive no input to any of their branches. The computed graph shows a steep voltage decrement in the input branch, with successively less steep decrement as the current flows toward the soma and other trees through branches of successively larger calibre. These voltage decrements may be compared with the dashed curve, obtained for the same amount of current injected at the soma (i.e. at the locus of common origin of the six equal dendritic trees). The explanation for the larger voltage at the soma in this second example is that here each tree receives an equal, one-sixth portion of the total current, while in the first example, more than one-sixth of the total current is dissipated across the membrane resistance of the input tree, leaving less than five-sixths of the total current to flow into the five other trees. The input resistance at the single branch terminal is 15.5 times that at the soma, for this particular example; many other examples, as well as a mathematical solution for such input resistance ratios can be found in Rall & Rinzel (1973).

It is noteworthy that the amount of atttenuation in each branch does not depend merely upon branch calibre, as can be verified by comparing attenuation in the input branch with that in its sibling branch, which is of equal calibre. The explanation is that attenuation depends also upon the boundary condition at the other end of each branch. In the case of the sibling branch the far end is regarded as sealed, allowing no current leakage and implying a boundary condition, $dV/dx = 0$ at this terminal. This boundary condition can also be expressed as a conductance ratio, $G_k/G_\infty = 0$, where G_k represents the conductance met by current as it approaches the end, while G_∞ represents what the conductance at that point would be if the same cylinder were extended to infinite length. In the case of the input branch, for current flow from terminal to proximal branch point, its 'other' end is not

Fig. 4

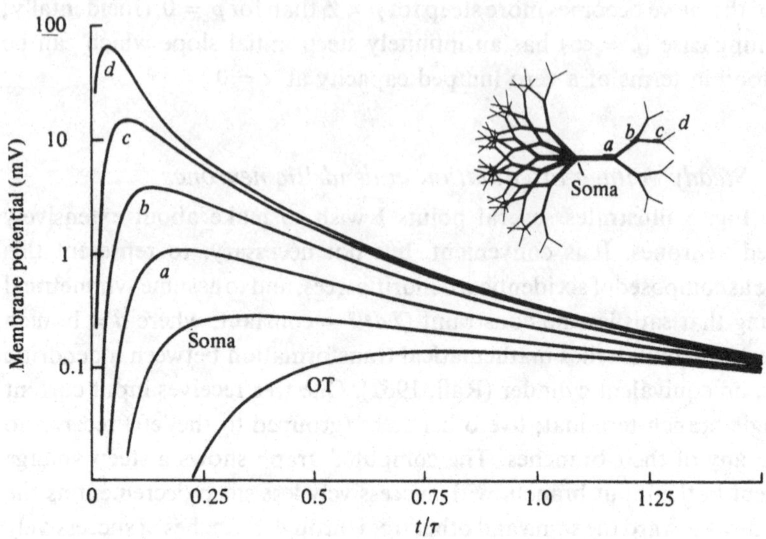

Transient responses at left, for six locations (as labelled) in the same neurone model. These transients and the attentuation factors tabulated at lower left were computed (Rinzel & Rall, 1974) for a brief synaptic current whose rise time was 0.02 of τ; when peak current is 5 nA and the input resistance at the soma is 2 MΩ, the amplitude values shown at left are in mV. As explained in the text, the input resistance values shown are those which correspond to the resistivities and the branch dimensions shown at bottom. It can be seen (by inspecting the values for $d^{3/2}$) that these branch dimensions yield a tree that can be transformed to an equivalent cylinder; also, each order of branching corresponds to the same surface area (minor discrepancy results from listing rounded values).

	Attenuation factors		Input resistances (MΩ)		Branch dimensions				Surface area/ branch
	Peak	Steady state	at soma	2		d (μm)	$d^{3/2}$ ($\mu m^{3/2}$)	Δx (μm)	μm^2
V_a/V_{soma}	3.8	2	(a)	3.3	trunk	10	32	250	2500π
V_b/V_{soma}	13.6	4.5	(b)	6.6	1st (ab)	6.3	16	200	1260π
V_c/V_{soma}	52.5	10.4	(c)	14.0	2nd (bc)	4.0	8	160	640π
V_d/V_{soma}	235	24	(d)	31.0	3rd (cd)	2.5	4	128	320π

These values imply: $R_m/R_i = 40$ cm; $R_m \simeq 2871\Omega\,cm^2$; $R_i \simeq 71.8\Omega\,cm$.

sealed, but very open; in fact the value of $d^{3/2}$ increases threefold (two for the parent branch plus one for the sibling branch). However the conductance ratio, G_k/G_∞, is not given merely by the ratio of $d^{3/2}$ values, because the value of G_k depends also upon the boundary conditions one step away, in both directions. For this input branch, calculation yields a value very close to 5.0 for the conductance ratio, G_k/G_∞. For the electrotonic distance, $\Delta x/\lambda = 0.25$, this conductance ratio yields (using eqn (A5) of the Appendix) an attenuation factor of 2.3 for the input branch, while the sibling branch, with $G_k/G_\infty = 0$, yields 1.03 as its attentuation factor.

In spite of its importance, this effect of the boundary condition upon steady-state voltage decrement is frequently overlooked by those accustomed to think only of simple exponential decrement with distance in long, unbranched cylinders. Both the equations needed and a diagram (Fig. A1) illustrating eleven different examples of steady-state decrement are included in the Appendix. Here, we go on to consider transient attenuation factors as well as input resistance values computed for this same neurone model.

Transient peak attenuation compared with steady state

When the injected current has a brief time course (rise time = 0.02 of τ and peak amplitude of 5 nA) and the site of injection is a single branch terminal (d, in Fig 4) of the same neurone model as in Fig. 3, computations yield the voltage transients shown in Fig. 4 for six different locations: the soma, three branch points (labelled a, b and c), the input terminal (d), and finally, any or all branch terminals of the five other trees, designated (OT). The mathematical expressions and the computational scheme for these results can be found in Rinzel & Rall (1974). Here we note both attenuation and delay of the transient peaks, as we move inward from the input terminal. Values of peak attenuation factors, from points a, b, c and d to the soma, are tabulated at lower left, where they may be compared with the corresponding values for steady-state attenuation. As might be expected, transient peak attenuation is greater than for the steady state; it can also be stated that for briefer synaptic current, attenuation would be slightly greater, while for slower synaptic potentials, the attenuation factors would lie between the two tabulated sets.

This kind of information is clearly relevant to neurones without impulses because we wish to estimate the effect of synaptic input in one region of a neurone upon the membrane potential elsewhere in the neurone. However, one may ask whether this model resembles real neurones. Do realistic values of resistance and of dendritic dimensions lead to consistent results?

Resistance values and branch dimensions

Suppose, for example, that we choose the branch dimensions shown at lower right in Fig. 4. These dimensions imply that each dendritic tree is equivalent to a cylinder of 10 μm diameter and 1000 μm length, and that the whole neurone surface area can be expressed, $A_N = 6\pi \times 10^4\ \mu m^2$. Let us see if the originally assumed electrotonic length, $L = 1.0$ per tree is viable; this implies that equivalent cylinders of 10 μm diameter have a length constant, $\lambda = 1000\ \mu m$, and together, these values imply a value of 40 cm for R_m/R_i, because λ equals the square root of $[(R_m/R_i)(d/4)]$. Also, if $L = 1.0$ is correct, we can make use of the relation

$$R_m/R_N = (A_N \tanh L_N)/L_N$$

where one assumes a uniform membrane resistivity, R_m, and where R_N, A_N and L_N represent the input resistance, the surface area and the electrotonic length of that cylinder (or those parallel six cylinders), which is equivalent to the whole neurone. For the values of A_N and L_N already stated, this relation implies a value of 14.36×10^{-4} cm² for R_m/R_N.

If we choose a reasonable candidate value for whole neurone input resistance, such as $R_N = 2\ M\Omega$, the ratio above for R_m/R_N implies that $R_m = 2870\ \Omega cm^2$, and then the earlier value for R_m/R_i implies that $R_i = 71.8\ \Omega cm$, which is also a reasonable value. Using this set of values for R_m, R_i and branch dimensions, we can calculate the input resistance values corresponding to microelectrode penetrations at previously noted branch locations (labelled *a*, *b*, *c* and *d* in Fig. 4).

Before leaving this example completely, suppose we suddenly discover that we previously overlooked four more orders of branching, with successive diameters of 1.6, 1.0, 0.63 and 0.4 μm, and successive lengths of 100, 80, 63 and 50 μm. Addition of these branches will double the values of both A_N and L_N. For the same membrane, one would expect the input resistance to be reduced, or if the input resistance is to be preserved, then the values of R_m and R_i must be increased. In the above expression for R_m/R_N, the value of $\tanh L_N$ increases from 0.762 to 0.964, which increases the value for R_m/R_N to 18.2×10^{-4} cm². Thus, if we preserve $R_N = 2\ M\Omega$, we obtain an increase of R_m from 2870 to 3640 Ωcm² and an increase of R_i from 71.8 to 91 Ωcm both of which seem still to be reasonable values. If we preserve the previous values for R_m and R_i, we obtain a new input resistance value of about 1.6 MΩ.

I hope these numerical results for this hypothetical example seem both helpful and realistic. They bear upon the interrelation of several measureable quantities, namely voltage attenuation factors, input resistance values and dendritic dimensions. My purpose is to interest experimental investigators who make such measurements to consider carrying out such calculations.

Non-spiking interneurone of locust

Very recent research provides elegant anatomical and electrophysiological information about non-spiking interneurones in the metathoracic ganglion of locust (Burrows & Siegler, 1978; Siegler & Burrows, 1979). I could not resist attempting calculations based upon these data. The left side of Fig. 5 shows two drawings of a neurone which Burrows & Siegler had injected with cobalt ions; drawing *a* shows processes revealed by cobalt sulphide precipitation alone, while drawing *b* shows changes revealed by intensification with silver. The increased richness of the branching is apparent, and Burrows & Siegler describe the changes in diameter of the major processes, as well as the additional branching, in some detail. In *c* and *d* of Fig. 5, I show my explicit schematic abstractions which incorporate the quantitative details; thus (15) indicates the 15 μm diameter of the cell body while (3 × 550) indicates a 3 μm diameter process of 550 μm length, extending from branch point b to branch point e. Also, five major branches extending from g toward h, which were reported as being approximately 2.5 μm in diameter, are here assigned 200 μm length and designated (2.5 × 200) × 5. These five branches give rise to many finer branches of 0.5 to 1.0 μm diameter, which have been arbitrarily represented by two orders of branching designated (1 × 50) × 2, followed by ($\frac{1}{2}$ × 50) × 2; this two-ordered tuft implies (for convenience) that each 2.5 μm diameter branch ends in two 1 μm diameter branches (hi), and that each of these ends in two 0.5 μm diameter branches (it). The same tuft was used at f and h in case *c*; in case *d* it was used unchanged at f and h, and added at d and along be and ef.

To those who find such details rather finicky, I comment that in order to calculate input resistance values, electrotonic lengths, and attenuation factors, one must be completely explicit about the simplified geometry; one can, of course, work with several different versions of simplified geometry. Here, I will present the results of calculations for two versions, case *d* corresponding to intensification with silver, and case *c*, corresponding to the less complete information obtained using cobalt without intensification. Details of the method of calculation can be found in the Appendix. Here I summarise a few results with the help of Table 1.

Using trial resistivity values of $R_m = 3000\ \Omega cm^2$ and $R_i = 83\ \Omega cm$ (see Appendix), the dependence of the length constant, λ, upon branch diameter reduces to

$$\lambda = 300(d)^{1/2}\ \mu m, \quad (\text{for } d \text{ in } \mu m)$$

while the dependence of G_∞ (input conductance for semi-infinte length) upon branch diameter reduces to

$$G_\infty \approx 3(d)^{3/2}\,nS, \quad (\text{for } d \text{ in } \mu m)$$

Fig. 5. Left side from Siegler & Burrows (1979) where *a* shows branches seen and drawn with cobalt stain before intensification with silver, and *b* shows new drawing of same neurone as seen after intensification with silver. Diagrams *c* and *d* show schematic abstractions based upon *a* and *b* and the descriptive text of Siegler & Burrows (1979, pp. 128–32). The use of diagrams *c* and *d* as the basis for calculations is described in the text and in the Appendix. Figures in brackets are branch dimensions.

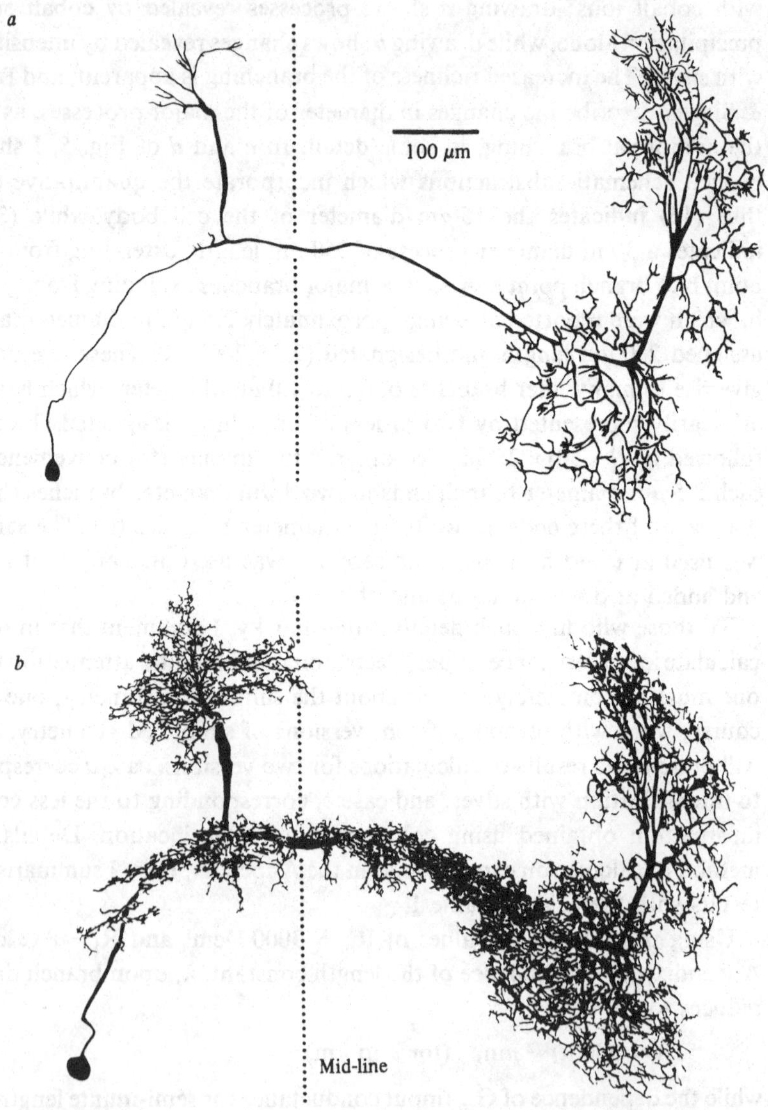

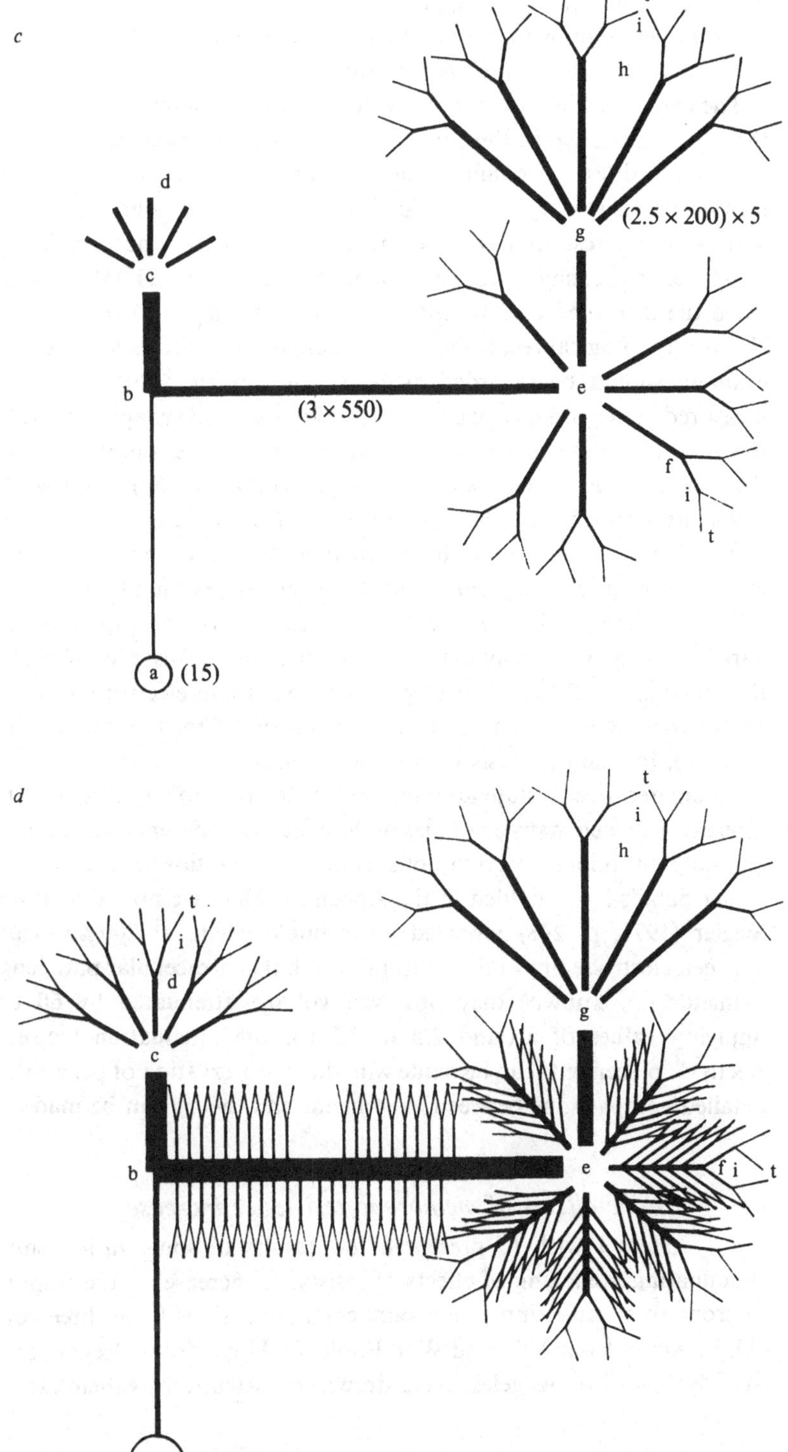
c
t
i
h
d
g
(2.5 × 200) × 5
c
b
e
(3 × 550)
f
i
t
a
(15)
d
t
i
h
t
i
d
g
c
b
e
f i
t
a
(23)

where nanoSiemens (nS) equals ohm^{-9}. Without specific evidence to the contrary, we begin with the simplifying assumption that R_m and R_i have the same values in all parts of the neurone.

The calculated input resistance values in Table 1 show (in both cases) that the input resistance at the soma is about ten times that near location e. If experiment does not confirm this, one may be forced to assume different membrane resistivity values for different parts of the neurone.

It is of interest that for the intensified geometry the calculated input resistance in the major processes ranges between 6 and 11 MΩ; this is about twice the 3 to 5 MΩ estimated by Burrows & Siegler (1978, p. 248) from electrophysiological recordings with a similar neurone; they noted that their estimate should be regarded as below the resting value because it was measured in the presence of considerable background synaptic input. It seems that the trial resistivity values used here have placed the computed results near the correct values when the intensified geometry (Fig. 5*d*) was used, but not when the geometry before intensification (Fig. 5*c*) was used. However, it would be very interesting to have experimental values for several controlled levels of synaptic background, including zero background.

Those interested in the calculated electronic distances between different parts of this neurone, may note, for example, that the distance, cbeg, between the arbor at c and the arbor at g corresponds to an electrotonic distance of 1.0 for the intensified diameters and a value of 1.5 for the smaller diameters (Fig. 5*c*), for the trial resistivity values cited above.

Calculated steady-state attenuation factors are shown at right in Table 1. There can be seen many examples of the effects of different electrotonic length and different boundary conditions (conductance ratios G_j/G_∞ and G_k/G_∞). Fuller details are provided in the Appendix. Here we note that Burrows & Siegler (1978, p. 248) reported an example where at 200 μm separation between electrode tips in the neuropile (with the intracellular path length and diameter not known) they observed voltage attenuated by 60 to 70%, implying values of around 2.5 to 3.3 for the attenuation factor. When electrode placements can be made with direct observation of penetration loci, detailed comparisons between experiment and theory can be made.

Some effects of membrane resistance increase

Here I wish to present a few results of some rather simple but illuminating modelling of effects of resistance increase on the response of a neurone to synaptic input. Personal communications from three colleagues (D. L. Alkon from NIH and Woodshole, G. Hoyle from Oregon, and C. D. Woody from Los Angeles) have drawn my attention to their examples of

Table 1. *Locust interneurone*

Calculated input resistance values (MΩ) at different locations				Calculated steady-state attenuation factors	
Location	Fig. 5*c* (Table A1 in Appendix)	Fig. 5*d* (Table A2 in Appendix)		Fig. 5*c* (Table A1 in Appendix)	Fig. 5*d* (Table A2 in Appendix)
a	170	59	V_a/V_b, from a to b	17.1	9.6
b	29	9.7	V_b/V_a, b to a	2.9	1.9
c	31	10.9			
e	16	6.1	V_b/V_c, from b to c	1.05	1.05
g	17	8.5	V_c/V_b, c to b	1.1	1.2
			V_b/V_e, from b to e	3.7	1.9
			V_e/V_b, e to b	3.1	1.6
			V_e/V_g, from e to g	1.6	1.16
			V_g/V_e, g to e	1.8	1.6

neurones which exhibit significantly increased responses under physiological conditions where they can also measure a significant increase in the input resistance of the neurone.

The upper part of Fig. 6 reproduces a sketch (Alkon, 1973) based on serial sections of photoreceptors and optic ganglian cells of the nudibranch mollusc, *Hermissenda crassicornis*. There are only five photoreceptors in each eye; each photoreceptor has a fairly large soma with a rather short process which ends in an arbor of fine terminal branchings, where synaptic contacts are made. Modelling was done in collaboration with J. Rinzel and R. Miller and in consultation with D. L. Alkon, in an effort to explore symbiosis between experiment and theory for this system. At first we used a compartmental model with a chain of three compartments for the coupling cylinder and from four to twelve compartments representing branching and electrotonic distance in the arbor. Then we realised that the coupling cylinder is electrotonically short and can, at least for most purposes, be modelled simply as a lumped core resistance coupling the soma to the dendritic arbor. Also, we decided, in spite of a dedication to dendritic branching, that sometimes it can be useful to collect all dendritic branches into a single lump, as shown in Fig. 6. The resistance values shown in Table 2 were chosen for their simplicity and because they provide input resistance values and attenuation factors that are in general agreement with a large number of experiments.

The major variable in our computational experiments was the soma membrane resistance, R_S, which was doubled and then redoubled. This necessarily results in an increase of input resistance values; however the factors of input resistance are seen to be closer to 1.5 than to the factor of 2 in the value of R_S. Also, assuming that membrane capacities remain unchanged, the time constant of the soma membrane was doubled and redoubled. This resulted in an increase of the slower system time constant, τ_0, also by factors of about 1.5; the equalising time constant, τ_1, hardly changes. (The slow time constant governs decay of total charge in the system; the faster equalising timeconstant is important when there is a large membrane potential difference between soma and dendrites, because then current flow between these regions tends to equalise their potentials rapidly (Rall, 1969). When τ_S equals τ_D, the value of τ_0 is the same (40 ms in Table 2). When τ_S exceeds τ_D, the valueof τ_0 must lie between them; a mathematical expression for this is included in the Appendix, but an intuitive explanation is that the inherently slower self decay of the soma membrane is made more rapid during the final slow decay phase because of sustained current flow between soma and dendrites, which also slows the inherently faster self decay of the dendritic membrane. See Appendix for equations defինining these values.)

What changes should we expect in the response to applied current or to

Fig. 6. Upper sketch from Alkon (1973) showing two photoreceptors (soma labelled B) belonging to both eyes of *Hermissenda crossicornis*. Each photoreceptor has a process, about 120 μm long, ending in a fine dendritic arbor where synaptic contacts are made. One diagram represents such a soma with its process and arbor; the next abstracts the arbor to a single lump. The values assumed in model computations are shown in Table 2, together with some of the resulting attenuation factors, input resistance values and time constants.

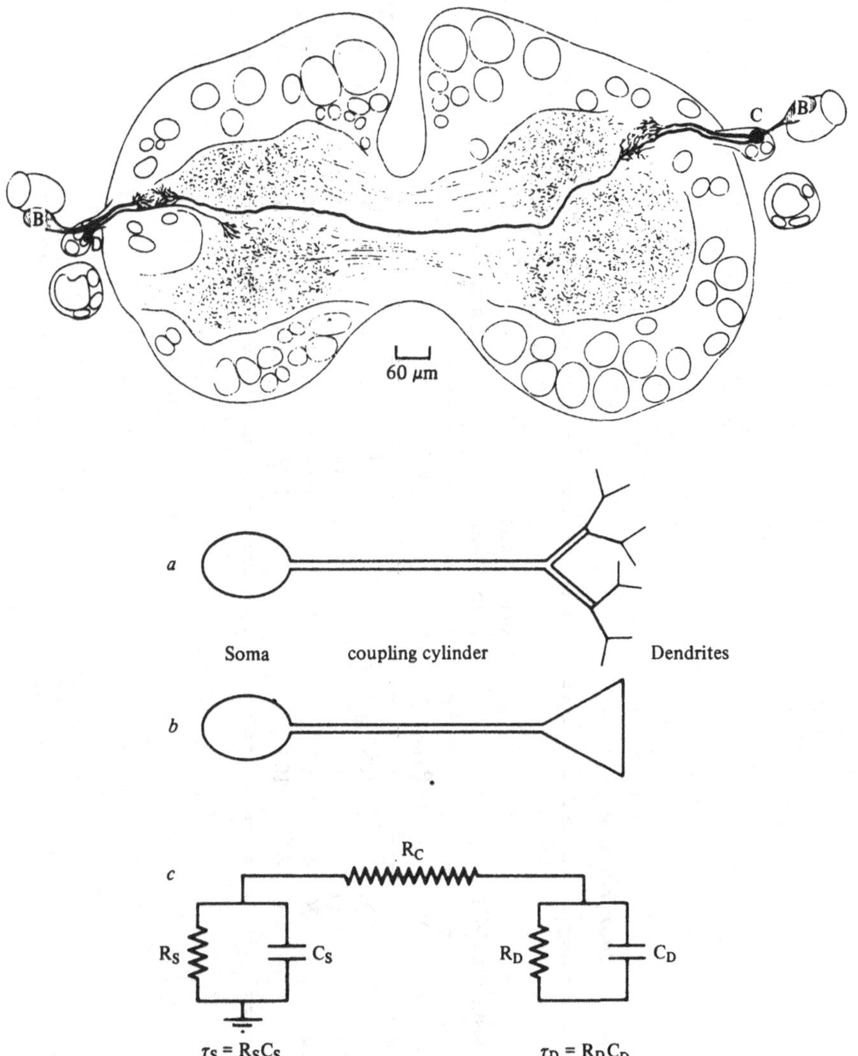

Table 2. *Photoreceptor neurone*

Resistances (MΩ)			Steady-state attenuation		Time constants (ms)			
					Membrane		System	
R_S	R_C	R_D	V_S/V_D	V_D/V_S	τ_S	τ_D	τ_0	τ_1
30	20	100	1.2	1.67	40	40	40	5.3
60	20	100	1.2	1.33	80	40	66	5.4
120	20	100	1.2	1.17	160	40	98	5.5
120	100	100	2	1.83	160	40	108	18

Fig. 7. Results obtained with the model shown in Fig. 6, with values shown in Table 2. *a*. Effect of doubling and redoubling soma resistance (above) upon EPSPs and IPSPs generated in the dendrites. Each synaptic potential was generated by a synaptic conductance transient (Rall, 1967) whose rise time was 10 ms relative to $\tau_D = 40$ ms (i.e. $T_p = 0.25$) and whose peak conductance was 20% of the resting dendritic membrane conductance. *b*. Effect of LLD (long lasting depolarisation) at the soma upon a stochastic sequence of EPSPs generated at the dendrites. Each EPSP was generated by the same synaptic conductance time course as *a*. The nine synaptic inputs occurred during 1000 ms, at the times indicated by the arrowheads along the lowermost time axis; these times were generated by making 1000 successive trials, each with independent 0.01 probability of success. The control shows limited temporal summation when there is neither depolarisation nor soma resistance increase. The response with LLD shows significantly greater temporal summation seen both in the soma and the lumped dendrites. Normalised voltage is expressed relative to the excitatory equilibrium potential; thus, for $V_\epsilon = 60$ mV, the half scale value of -0.08 would correspond to 4.8 mV of hyperpolarisation.

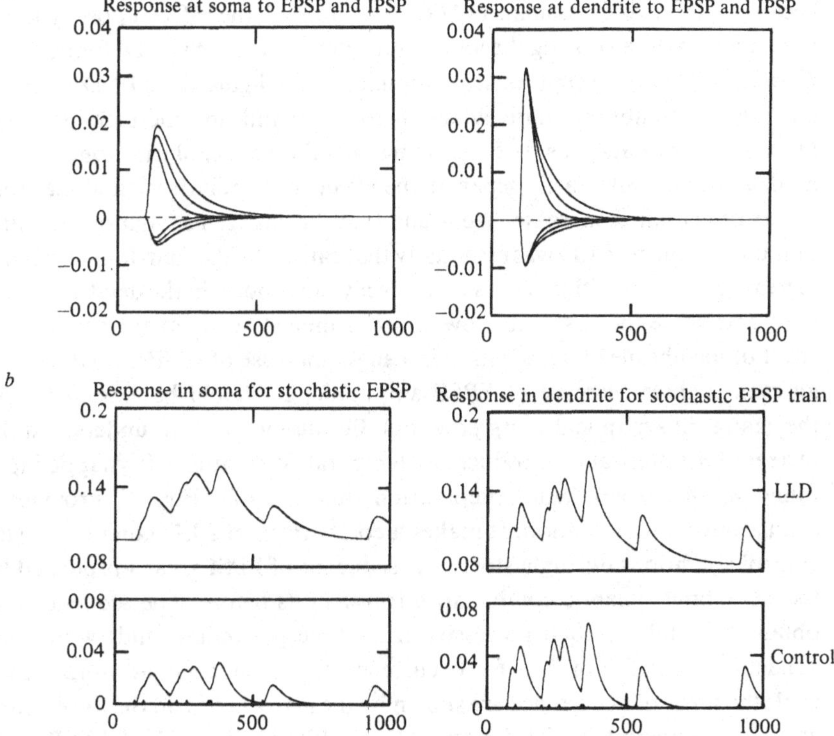

synaptic input in the lumped dendrites? For the same applied current, the increased input resistance should result in increased response amplitude and the increased time constant, τ_0, should result in a slower approach to steady state. Both of these effects were seen in computed results.

Here I wish to show what we found for brief synaptic input, where the rise time of synaptic conductance was set at 10 ms, being one-fourth of the resting dendritic membrane time constant. Then we obtained the results shown in the upper frames of Fig. 7, where we show three EPSP and three IPSP for the same three values of R_S in Table 2. It is noteworthv that at the input location in the dendrites, the peak amplitude is not significantly increased by the resistance increase; the major effect is that synaptic potential decay has become slower because of the increased value of τ_0. (It should be remarked that change in peak amplitude at the dendrites was seen with very slow synaptic potentials; the absence of change here can be attributed to the short rise time.)

We were interested in these synaptic potentials with changed R_S values, because Alkon & Grossman (1978) have noticed the effect upon synaptic potentials observed during the long-lasting depolarisation (LLD) found in the *Hermissenda* photoreceptors after stimulation by light; their LLD refers to an altered membrane state which features significant increase of soma membrane resistance as well as soma membrane depolarisation. In our modelling, we could easily separate the effects of depolarisation alone from effects of resistance increase. We found that the change in synaptic potential time course during LLD was essentially that caused by the four-fold resistance increase at the soma; this is the slowed decay time course in the dendrites, with both amplitude increase and slowed decay time showing at the soma. (The effect of membrane depolarisation is a slight increase of IPSP amplitude and an even slighter decrease of EPSP amplitude, as should be expected from the change in synaptic driving potential. Because we wish to understand the increased depolarisation producedby temporal summation of synaptic input found in the experimental preparation during LLD, Fig. 7*b* provides a simulation of this situation. It makes a comparison of LLD conditions with control conditions during a stochastic sequence of EPSPs, each produced by the same brief synaptic conductance transient as before. The sequence was obtained by taking 1000 successive trials (one per millisecond) with independent probability of 0.01 for occurrence of a synaptic input during each trial; the times of occurrence are shown by the arrowheads on the lowest time axis. It is apparent in the dendrites that, although individual EPSP peak amplitude remained unchanged, the amount of temporal summation was significantly increased, and we can attribute this to the slower decay of the EPSP, or equivalently, the increased time constant, τ_0, of the system. In the soma, we see the effect of both increased EPSP amplitude and increased temporal summation, written on top of the LLD depolarisation. The effect is to augment the LLD depolarisation of the soma by about 50%. This contributes to the positive, cross-augmenting effect between LLD and EPSP activity of type B cells in the observations of Alkon & Grossman (1978). I cannot elaborate here but merely remark that this effect is further enhanced by an associative behavioural change (Crow & Alkon, 1978) and that underlying regenerative feedback can be attributed to the voltage dependence of Ca-ion conductance, together with the dependence of K-ion conductance upon intracellular Ca-ion concentration (Alkon, 1979).

The role of an increased membrane resistance as important for changed responses observed following conditioning has also been emphasised by Woolacott & Hoyle (1976) and by C. D. Woody and his colleagues (personal communication). Note three phyla: molluscs, insects and mammals!

Effect of increased coupling resistance

Before leaving this model, I wish to mention that we also looked at the effect of increasing the coupling resistance. As expected, there was an increase of input resistance, of both attenuation factors, and of the equalizing time constant, τ_1. It is noteworthy that the slow time constant, τ_0, remains unchanged for the simple case where $\tau_S = \tau_D$; however, when τ_S differs from τ_D, even τ_0 shows a small factor of increase; see Table 2. It was predicted, and found that, in this case, the synaptic potential amplitude in the dendrites is increased (70% above control) because although the dendritic capacity and self-decay resistance remain unchanged, the equalising current, $(V_D - V_S)/R_C$, that flows from dendrites to soma must be decreased by the increased value of R_C. The attenuation factor, V_D/V_S, was greatly increased, as expected. This provides an example of a decreased degree of coupling between two regions of a neurone. It may be noted that when R_S and R_C are both increased, we have all of the previous effects combined: increased synaptic potential amplitude in the dendrites, increased decay timeconstant, and increased attentuation factor from dendrites to soma.

Dendritic spines, constrictions and plasticity

If I had more space I would include the results of computations made for dendritic spines and for the constrictions that occur between dendritic varicosities. Briefly, I can say that when a dendritic spine stem or when one or more dendritic constrictions introduce very large core resistance between a synaptic region and another region, the effect is to increase the input resistance and the response amplitude in the spine head or in that part of the arbor which is upstream from the constrictions, and to reduce (because of a large I × R drop) the response observed in regions on the other side (downstream) from the constriction. Which effect is more important? The answer need not always be the same. For a neurone whose arbor can function presynaptically as well as post-synaptically, the local augmentation of synaptic potential amplitudes (either by a constriction downstream or by locally increased membrane resistivity) could be very significant for the level of presynaptic activity that this arbor puts out to its recipient neurones. In addition, other regions of this neurone would be less affected, and this could be important as a decoupling between different neurone pathways or subsystems. Also, this arbor would be less influenced by what goes on elsewhere in the neurone, but I add the caution that the coupling should not be assumed to be equal in both directions. (Remember the effect of G_k/G_∞ upon steady-state decrement in the input branch and its sibling branch in Figs. 3 and 4, and also in the Appendix.)

In the consideration of dendritic spine stem resistance (Rall & Rinzel abstracts for the XXV International Physiological Congress and for the First Annual Meeting of the Society for Neuroscience, both in 1971; see also Rall, 1974, 1978) it was found that the value of a resistance ratio, R_{SS}/R_{BI} (spine-stem resistance divided by the input resistance at the dendritic branch where the spine is attached) is particularly important. When this ratio is reduced below 0.1, there is no increase in the depolarisation delivered to the dendrite; however, when this ratio is increased from 0.1 to 1, the potency is halved and further increase of R_{SS} further reduces the potency. This suggested the idea of a useful operating range, if indeed, spine-stem resistance is used by the organism to change the relative potency of the many synapses from different afferent sources ending upon dendritic spines; some preliminary estimates based upon available anatomy and physiology did not rule out this tentative hypothesis (Rall, 1974, 1978). There is also other suggestive experimental support for this idea (e.g. Bliss & Lømo, 1973; Purpura, 1974; Fifkova & Van Harreveld, 1977; Trubatch, Loud & Van Harreveld, 1977; and Coss & Globus, 1978), and there are exciting prospects for further testing of this hypothesis. I already mentioned the interesting results of Hoyle and collaborators, of Alkon and collaborators, of Woody and collaborators and of Burrows & Siegler. I close by commenting that I believe the prospects for study of local interactions (as well as degrees of coupling between regions) in identified neurones without impulses, for both robust function and adaptable function, are especially exciting. As a theoretical modeller, I welcome new data which provide both a challenge and an opportunity for interpretation and testing.

Appendix

This brief catalogue of the principal equations is intended to be useful, but for more complete presentation, definitions and discussion the reader is referred to a handbook chapter (Rall, 1977).

The cable equation for a core conductor of constant diameter and constant resistivities is usually written

$$\lambda^2 \partial^2 V/\partial x^2 - V - \tau \partial V/\partial t = 0 \tag{A1}$$

Steady-state solutions (for $\partial V/\partial t = 0$) can all be expressed in the form

$$V(X) = A_1 \exp(x/\lambda) + A_2 \exp(-x/\lambda) \tag{A2}$$

but in the analysis of branched core conductors it is advantageous to use two alternative expressions for a uniform finite length extending from x_j to x_k.

Suppose that current injection at $x = x_j$ causes a steady voltage designated V_j at x_j. It makes a difference whether the voltage at $x = x_k$ is left free to be determined by the conductances of additional branches or of a termination, or whether a voltage clamp is applied at $x = x_k$. For voltage clamping, where both V_j and V_k are specified, the solution for values between $X_j = x_j/\lambda$ and $X_k = x_k/\lambda$ can be expressed

$$V(X) = \frac{V_j \sinh(X_k - X) + V_k \sinh(X - X_j)}{\sinh(X_k - X_j)} \tag{A3}$$

This equation was used to compute the upper two dashed curves in Fig. 8; it is important to remember that $V = V_m - E_r$ implies that $V_m = E_r$ when $V = 0$.

Fig. A1. Voltage decrement for different boundary conditions. The numbers inserted along the vertical dotted line represent the value of the conductance ratio, G_1/G_∞. Adapted from Rall, 1959.

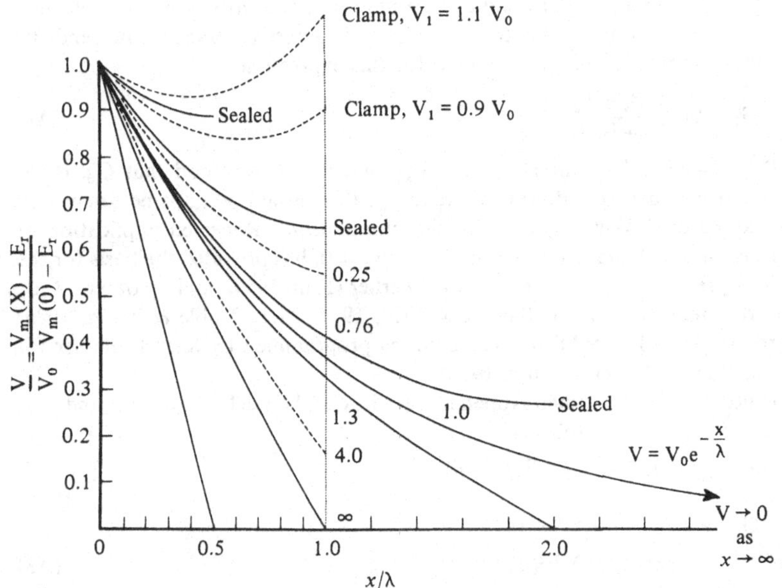

For calculations in dendritic trees, the most useful expression for X between X_j and X_k is

$$V(X) = [\cosh (X_k - X) + (G_k/G_\infty) \sinh (X_k - X)] V_k \tag{A4}$$

where G_k is the input conductance onward into branches or termination at $x = x_k$, and G_∞ is its reference input conductance for the case of an extension of this cylinder from $x = x_k$ to ∞.

The curves shown as solid lines in Fig. A1 were published before (Rall, 1959); of these, only one (the longest) shows single exponential decrement with distance; its boundary condition can be expressed either as $G_k/G_\infty = 1$, or as $V \to 0$ as $x \to \infty$. Three pairs of curves (solid lines) show for three different electronic lengths ($x/\lambda = 0.5$, 1.0 and 2.0) the effect upon voltage decrement of choosing two different boundary conditions: the upper curve of each pair corresponds to a sealed end boundary condition ($G_k/G_\infty = 0$); the lower curve of each pair corresponds to a voltage clamped boundary condition, where a conductance ratio, $G_k/G_\infty = \infty$, is implied by the electronic clamp which holds $V = 0$, meaning that the transmembrane potential, V_m, is clamped to its resting value, E_r.

The two upper dashed curves illustrate voltage clamping at $x/\lambda = 1.0$ to values different from rest; the upper curve clamps V_m to the value $(E_r + 1.1V_0)$ at $x/\lambda = 1.0$; if the amount of depolarising current applied at $x = 0$ had been just sufficient to make V_0 represent 90.9% depolarisation at $x = 0$, then this case would correspond to a perfect local short circuit (of the resting membrane battery) holding $V_m = 0$ at $x/\lambda = 1$.

The other two dashed curves illustrate the effects of fourfold conductance decrease or increase at $x/\lambda = 1.0$; also, the G_k/G_∞ values of 0.76 and 1.3 represent the apparent boundary conditions at $x/\lambda = 1.0$ that result from the two different terminations at $x/\lambda = 2$.

By setting $X = X_j$ in eqn (A4) one immediately obtains an expression for the attenuation factor

$$V_j/V_k = \cosh (X_k - X_j) + (G_k/G_\infty) \sinh (X_k - X_j) \tag{A5}$$

The expression for the input conductance at $x = x_j$ is a little more complicated. Because it is used repeatedly for the treatment of successive branches in dendritic calculations, a shorthand notation is used for this expression

$$G_j = G_\infty \left(\frac{B_k + T_k}{1 + B_k T_k} \right) \tag{A6}$$

Where $B_k = G_k/G_\infty$, $T_k = \tanh [(x_k - x_j)/\lambda_{jk}]$, and it is understood that G_∞ is for the particular cylinder extending from x_j to x_k; this should perhaps be designated $G_{\infty jk}$, to avoid confusion and error in thecomputations. Repeated application of eqn(A6), starting with branch terminals where $B_k = 0$, has provided the basis for the calculations presented here, those presented earlier (Rall, 1959), and by others (Lux, Schubert & Kreutzberg, 1970; Barrett & Crill, 1974; Jack, Noble & Tsien, 1975; Graubard, 1975; and unpublished calculations programmed by Sheryl Glasser for Selverston, Russell, Miller & King, 1976).

At this point, a self-contained Appendix requires explicit definitions of λ and G_∞, which can be expressed as follows

$$\lambda = (r_m/r_i)^{1/2} = [(R_m/R_i)(d/4)]^{1/2} \tag{A7}$$

and

$$G_\infty = \lambda/r_m = (\lambda r_i)^{-1} = (r_m r_i)^{-1/2}$$
$$= (\pi/2)(R_m R_i)^{-1/2}(d)^{3/2} \tag{A8}$$

where d represents cylinder diameter, R_m is the membrane resistivity usually given in Ωcm^2 and R_i is the volume resistivity of the core usually given in Ωcm; also, $r_m = R_m/(\pi d)$ is membrane resistance for unit length of cylinder in Ωcm, and $r_i = R_i/(\pi d^2/4)$ is the core resistance per unit length in Ωcm^{-1}.

To my knowledge, experimental values of R_m range from 1000 to 10000 Ωcm^2 while those of R_i range from 30 to 200 Ωcm. Because these two ranges have positive correlation, the resistance ratio, R_m/R_i seems to range between 20 and 100 cm; I usually begin with a trial value of 36 to 40 cm which makes λ equal $300(d)^{1/2}$ or $316\,(d)^{1/2}$ when d and λ are both expressed as μm. The product, $R_m R_i$, seems to range from $3 \times 10^4\ \Omega^2 cm^3$ for squid up to as much as $10^6\ \Omega^2 cm^3$ at the upper range for cat spinal motoneurones; I usually begin with a trial value between 22 and 40 times $10^4\ \Omega^2 cm^3$, which makes G_∞ equal between $3(d)^{3/2}$ and $4(d)^{3/2}$ in nS when d is in μm; nanoSiemens (nS) means $(10^9\ \text{ohm})^{-1}$.

To gain entry into Table A1, we comment that the terminal branchlets (indexed *it*), with dimensions $(0.5 \times 50) \times 2$ have a terminal conductance ratio, $B_k = 0$. Each branchlet has an input conductance, $G_{jk} = G_\infty \tanh(l/\lambda) = 0.23$ nS; a tuft of two of these has an input conductance, $2G_{jk} = 0.46$ nS. From this we can obtain the B_k value for each of the two branches (indexed *hi*) with dimensions $(1 \times 50) \times 2$, as $B_k = 0.46/3 = 0.15$, and this with $T_k = 0.17$ yields $G_{hi} = 0.93$ for each branch, or $2G_{hi} = 1.9$ nS for the tuft. The entire Table can be filled out in this way.

The same method applies for Table A2, where several diameters are increased and many units composed of (1×50) ending in $(0.5 \times 50) \times 2$, corresponding to $G_{hi} = 0.93$ nS, are added at *d*, along *be*, and along *ef*. If ten of these are spaced along each *ef* branch, they add 9.3 nS per 130 μm, or 0.07 nS μm^{-1} to g_m, the membrane conductance per μm, which was 0.02 nS μm^{-1} before; this 4.5-fold increase of g_m (from 0.02 to 0.09) is responsible for increasing G_∞ from 8.5 to 18 nS and decreasing the effective λ from 424 to 200 μm for *ef* in Table A2. For the larger cylinder, *be*, an 8 μm diameter implies $g_m = 0.83$ nS μm^{-1}, which would become increased to almost 0.9 nS μm^{-1} by adding the same linear density of side branches used above for *ef*; here, this represents only an 8.4% increase of g_m, causing only a 4% increase of G_∞ (from 68 to 71 nS) and a 4% decrease of λ (from 850 to 816 μm).

For the simplified model of Fig. 6 and Table 2, it is not difficult to derive the following expressions for steady-state attentuation factors:

$$V_S/V_D = (R_C + R_D)/R_D = 1 + R_C/R_D, \quad \text{[from S to D]}$$

$$V_D/V_S = (R_C + R_S)/R_S = 1 + R_C/R_S, \quad \text{[from D to S]}$$

and for input resistances

$$R_{NS} = R_S(R_C + R_D)/(R_S + R_C + R_D), \quad \text{[at soma]}$$

$$R_{ND} = R_D(R_C + R_D)/(R_S + R_C + R_D), \quad \text{[at dendritic origin].}$$

It is a little more difficult to solve the systemfor its two time constants. That result can be expressed most simply in terms of four reciprocal time constants, $\mu_S = (R_S C_S)^{-1}$, $\mu_{CS} = (R_C C_S)^{-1}$, $\mu_D = (R_D C_D)^{-1}$ and $\mu_{CD} = (R_C C_D)^{-1}$, and the sum of these four quantities, designated $\Sigma\mu$. Then τ_0 and τ_1 are reciprocals of the quadratic roots defined by

$$\text{roots} = \Sigma\mu/2 \mp [(\Sigma\mu/2)^2 - \mu_S\mu_D - \mu_S\mu_{CD} - \mu_D\mu_{CS}]^{1/2}.$$

Table A1. *Values for Fig. 5c, calculated sequentially*

Branch indices and values				For current from *j* to *k*				For current from *k* to *j*		
jk	$(d \times l)(\mu m) \times n$	G_∞(nS)	tanh (*l*/a)	B_k	G_{jk} (nS)	nG_{jk}	V_j/V_k	B_j	G_{kj} (nS)	V_k/V_j
it	$(0.5 \times 50) \times 2$	1.0	.23	0·0	0.23	0.46	1.03	10.2	3.1	3.5
hi	$(1 \times 50) \times 2$	3.0	0.17	0.15	0.93	1.9	1.04	7.2	10.0	2.2
gh	$(2.5 \times 200) \times 5$	11.9	0.40	0.16	6.3	31.5	1.16	4.3	20.6	2.9
eg	(3×150)	15.6	0.28	2.0	22.9	—	1.64	2.6	26.0	1.8
[*fi* and *it* dimensions same as for *hi* and *it*]										
ef	$(2 \times 130) \times 6$	8.5	0.30	0.22	4.2	25.2	1.12	7.0	20.0	3.3
be	(3×550)	15.6	0.79	1.62	16.5	—	3.7	1.15	15.8	3.1
cd	$(2 \times 75) \times 5$	8.5	0.18	0.0	1.53	7.65	1.02	3.7	19.7	1.7
bc	(6×130)	44.1	0.18	0.17	15.0	—	1.05	0.44	25.0	1.1
ab	(1×350)	3.0	0.82	10.5	3.5	—	17.1	0.80	2.9	2.9

Resulting input conductance ($nS = ohm^{-9}$) and input resistance ($M\Omega = ohm^{6}$)

$G_a = G_s + G_{ab} = 5.9$	$R_a = 170$
$G_b = G_{ba} + G_{bc} + G_{be} = 34.0$	$R_b = 29$
$G_c = G_{cb} + 5G_{cd} = 32.7$	$R_c = 31$
$G_e = G_{eb} + 6G_{ef} + G_{eg} = 64.0$	$R_e = 16$
$G_g = G_{ge} + 5G_{gh} = 58.0$	$R_g = 17$

Table A2. *Values for Fig. 5d, calculated sequentially*

Branch indices and dimensions				For current from *j* to *k*				For current from *k* to *j*		
jk	$(d \times l)(\mu m) \times n$	G_∞ (nS)	$\tanh(l/\lambda)$	B_k	G_{jk} (nS)	nG_{jk}	V_j/V_k	B_j	G_{kj} (nS)	V_k/V_j
it	$(0.5 \times 50) \times 2$	1.0	0.23	0.0	0.23	0.46	1.03	10.8	3.1	3.6
hi	$(1 \times 50) \times 2$	3.0	0.17	0.15	0.93	1.9	1.04	8.5	10.6	2.5
gh	$(2.5 \times 200) \times 5$	11.9	0.40	0.16	6.3	31.5	1.16	9.4	24.5	5.2
eg	(6×150)	44.1	0.20	0.71	35.0	—	1.16	2.9	86.5	1.6
	[*fi* and *it* dimensions same as for *hi* and *it*]									
ef	2×130 with 10 sidebranches	18.0	0.57	0.11	11.5	69.0	1.3	8.4	27.9	7.1
be	$\overline{8 \times 550}$ with 42 sidebranches	71.0	0.59	0.97	70.0	—	1.93	0.47	59.0	1.6
	[*di* and *it* dimensions same as for *hi* and *it*]									
cd	$(2 \times 75) \times 5$	8.5	0.18	0.22	3.3	16.5	1.06	10.4	31.3	2.9
bc	(8×130)	68.0	0.15	0.24	25.6		1.05	1.14	75.0	1.2
ab	(2×350)	8.5	0.68	8.9	11.7		9.6	0.62	7.8	1.9

Resulting input conductance values ($nS = ohm^{-9}$), and input resistance ($M\Omega = ohm^{6}$)

$G_a = G_s + G_{ab} = 17.0$	$R_a = 59.0$
$G_b = G_{ba} + G_{bc} + G_{be} = 103.0$	$R_b = 9.7$
$G_c = G_{cb} + 5G_{cd} = 91.5$	$R_c = 10.9$
$G_e = G_{eb} + 6G_{ef} + G_{eg} = 163.0$	$R_e = 6.1$
$G_g = G_{ge} + 5G_{gh} = 118.0$	$R_g = 8.5$

References

Alkon, D. L. (1973). Neural organization of a molluscan visual system. *J. Gen. Physiol.* **61**, 444–61.

Alkon, D. L. (1979). Voltage dependence of calcium and potassium photo-receptor currents during and after light steps. *Biophys. J.* **25**, 268*a*.

Alkon, D. L. & Grossman, Y. (1978). Long-lasting depolarization and hyperpolarization in the eye of *Hermissenda*. *J. Neurophysiol.* **41**, 1328–42.

Andres, K. H. (1965). Der Feinbau des Bulbus olfactorius der Ratte unter besonderer Berücksichtigung der synaptischen Verbindungen. *Z. Zellforsch. Mikroskop. Anat.* **65**, 530–61.

Barrett, J. N. & Crill, W. E. (1974). Specific membrane properties of cat motoneurones. *J. Physiol., Lond.* **239**, 301–24.

Bliss, T. V. P. & Lømo, T. (1973). Long-lasting potentiation of synaptic transmission in the dentate area of the anaesthetized rabbit, following stimulation of the perforant path. *J. Physiol., Lond.* **232**, 331–56.

Bloedel, J., Gage, P. W., Llinás, R. & Quastel, D. M. J. (1966). Transmitter release at the squid giant synapse in the presence of tetrodotoxin. *Nature, Lond.* **212**, 49–50.

Burke, R. E. & ten Bruggencate, G. (1971). Electrotonic characteristics of alpha motoneurones of varying size. *J. Physiol., Lond.* **211**, 1–20.

Burrows, M. & Siegler, M. V. S. (1978). Graded synaptic transmission between local interneurons and motor neurones in the metathoracic ganglion of the locust. *J. Physiol., Lond.* **285**, 231–55.

Castillo, J. del & Katz, B. (1954). Changes in end-plate activity produced by pre-synaptic polarization. *J. Physiol., Lond.* **124**, 586–604.

Coss, R. G. & Globus, A. (1978). Spine stems on tectal interneurons in jewel fish are shortened by social stimulation. *Science*, **200**, 787–90.

Crow, T. & Alkon, D. L. (1978). Neural correlates of an associative behavioral change in *Hermissenda crassicornis*. *Soc. Neurosci. Abstr.* **4**, 191.

Fifkova, E. & Van Harreveld, A. (1977). Long-lasting morphological changes in dendritic spines of denate granular cells following stimulation of the entorhinal area. *J. Neurocytol.* **6**, 211–30.

Graubard, K. (1975). Voltage attenuation with *Aplysia* neurons: the effect of branching pattern. *Brain Research*, **88**, 325–32.

Grossman, Y., Schmidt, J. A. & Alkon, D. L. (1979). Calcium-dependent potassium conductance in the photoresponse of a nudibranch mollusk. In press.

Hirata, Y. (1964). Some observations on the fine structure of the synapses in the olfactory bulb of the mouse, with particular reference to the atypical synaptic configurations. *Arch. Histol. Jap.* **24**, 293–302.

Hoyle, G. (1975). Identified neurons and the future of neurotheology. *J. Exp. Zool.* **194**, 51–74.

Jack, J. J. B., Noble, D. & Tsien, R. W. (1975). *Electric Current Flow in Excitable Cells*. Oxford: Clarendon Press.

Jack, J. J. B. & Redman, S. J. (1971). An electrical description of the motoneurone, and its application to the analysis of synaptic potentials. *J. Physiol., Lond.* **215**, 321–52.

Jackowski, A., Parnevalas, J. G. & Lieberman, A. R. (1978). The reciprocal synapse in the external plexiform layer of the mammalian olfactory bulb. *Brain Research*, **159**, 17–28.

Katz, B. & Miledi, R. (1965). Propagation of electric activity in motor nerve terminals. *Proc. R. Soc. Series B*, **161**, 453–82.

Katz, B. & Miledi, R. (1966). Input–output relation of a single synapse. *Nature, Lond.* **212**, 1242–5.

Katz, B. & Miledi, R. (1967). The release of acetylcholine from nerve endings by graded electric pulses. *Proc. R. Soc. Series B*, **167**, 23–38.

Klee, M. & Rall, W. (1977). Computed potentials of cortically arranged populations of neurons. *J. Neurophysiol.* **40**, 647–66.

Kusano, K., Livengood, D. R. & Werman, R. (1967). Tetraethylammonium ions: effect of presynaptic injection on synaptic transmission. *Science*, **155**, 1257–9.

Liley, A. W. (1956). The effects of presynaptic polarization on the spontaneous activity at the mammalian neuromuscular junction. *J. Physiol., Lond.* **134**, 427–43.

Lux, H-D., Schubert, P. & Kreutzberg, G. W. (1970). Direct matching of morphological and electrophysiological data in cat spinal motoneurons. In *Excitatory Synaptic Mechanisms*, ed. P. Andersen & J. K. S. Jansen, pp. 189–98. Oslo: Universitetsforlaget.

Nelson, P. G. & Lux, H-D. (1970). Some electrical measurements of motoneuron parameters. *Biophys. J.* **10**, 55–73.

Phillips, C. G., Powell, T. P. S. & Shepherd, G. M. (1963). Responses of mitral cells to stimulation of the lateral olfactory tract in the rabbit. *J. Physiol., Lond.* **168**, 65–88.

Price, J. L. & Powell, T. P. S. (1970). The synaptology of the granule cells of the olfactory bulb. *J. Cell Sci.* **7**, 125–55.

Purpura, D. P. (1974). Dendritic spine 'dysgenesis' and mental retardation. *Science*, **186**, 1126–8.

Rall, W. (1957). Membrane time constant of motoneurons. *Science*, **126**, 454.

Rall, W. (1959). Branching dendritic trees and motoneuron membrane resistivity. *Exptl. Neurol.* **1**, 491–527.

Rall, W. (1962). Theory of physiological properties of dendrites. *Ann. NY Acad. Sci.* **96**, 1071–92.

Rall, W. (1964). Theoretical significance of dendritic trees for neuronal input-output relations. In *Neural Theory and Modeling*, ed. R. Reiss, pp. 73–97. Stanford: Stanford University Press.

Rall, W. (1967). Distinguishing theoretical synaptic potentials computed for different soma-dendritic distributions of synaptic input. *J. Neurophysiol.* **30**, 1138–68.

Rall, W. (1969). Time constants and electrotonic length of membrane cylinders and neurons. *Biophys. J.* **9**, 1483–1508.

Rall, W. (1970). Dendritic neuron theory and dendrodendritic synapses in a simple cortical system. In *The Neurosciences: Second Study Program*, ed. F. O. Schmitt, pp. 552–65. New York: Rockefeller.

Rall, W. (1974). Dendritic spines, synaptic potency and neuronal plasticity. In *Cellular Mechanisms Subserving Changes in Neuronal Activity*, ed. C. D. Woody, K. A. Brown, T. J. Crow, Jr. & J. D. Knispel, Brain Information Service Research Report No. 3, pp. 13–21. Los Angeles: University of California.

Rall, W. (1977). Core conductor theory and cable properties of neurons. In *Handbook of Physiology*, Sect. 1. *The Nervous System, Vol. 1. Cellular Biology of Neurons*, ed. E. R. Kandel, Vol. 1, Part 1, pp. 39–97. Bethesda: American Physiological Society.

Rall, W. (1978). Dendritic spines and synaptic potency. In *Studies in*

Neurophysiology, presented to A. K. McIntyre, ed. R. Porter, pp. 203–9. Cambridge: Cambridge University Press.

Rall, W. & Rinzel, J. (1973). Branch input resistance and steady attenuation for input to one branch of a dendritic neuron model. *Biophys. J.* **13**, 648–88.

Rall, W. & Shepherd, G. M. (1965). Abstract 943 of XXIII International Congress of Physiological Sciences, Tokyo.

Rall, W. & Shepherd, G. M. (1968). Theoretical reconstruction of field potentials and dendrodendritic synaptic interactions of olfactory bulb. *J. Neurophysiol.* **31**, 884–915.

Rall, W., Shepherd, G. M., Reese, T. S. & Brightman, M. W. (1966). Dendro-dendritic synaptic pathway for inhibition in the olfactory bulb. *Exptl. Neurol.* **14**, 44–56.

Reese, T. S. & Brightman, M. W. (1970). Olfactory surface and central olfactory connections in some vertebrates. In *CIBA Foundation Symposium on Taste and Smell in Vertebrates*, ed. G. E. W. Wolstenholme & J. Knight, p. 115. London: Churchill.

Rinzel, J. & Rall, W. (1974). Transient response in a dendritic neuron model for current injected at one branch. *Biophys. J.* **14**, 759–90.

Selverston, A. I., Russell, D. F., Miller, J. P. & King, D. G. (1976). The stomatogastric nervous system: structure and function of a small neural network. *Prog. Neurobiol.* **7**, 215–90.

Shepherd, G. M. (1963). Neuronal systems controlling mitral cell excitability. *J. Physiol., Lond.* **168**, 101–17.

Shepherd, G. M. (1974). *The Synaptic Organization of the Brain*. London: Oxford.

Siegler, M. V. S. & Burrows, M. (1979). The morphology of local non-spiking interneurones in the metathoracic ganglion of the locust. *J. Comp. Neurol.* **183**, 121–47.

Trubatch, J., Loud, A. V. & Van Harreveld, A. (1977). Quantitative stereological evaluation of K Cl induced ultrastructural changes in frog brain. *Neuroscience*, **2**, 963–74.

Woollacott, M. H. & Hoyle, G. (1976). Membrane resistance changes associated with single, identified neuron learning. *Soc. Neurosci. Abstracts*, **2**, 339.

GORDON M. SHEPHERD

Synaptic and impulse loci in olfactory bulb dendritic circuits

The processing of information in the nervous system involves an interplay of impulse and synaptic activity. In some regions, as in the vertebrate retina, the processing mechanisms may rely heavily on a direct transformation of graded synaptic inputs into graded synaptic outputs within the constituent neurones. In other regions, such as the spinal cord or cerebellum, input–output relations may require the intermediation of impulse activity in most or all of the neurones.

The olfactory bulb is a rather complex region in this regard. Evidence has been adduced for a variety of properties, including dendro-dendritic synapses and interactions, graded input–output functions, dendritic compartments, dendritic spikes, and long-lasting synaptic actions. My aim here will be to review our recent work in several of these areas, particularly as they pertain to synaptic and dendritic properties. The results support the concept of an ensemble of distributed loci for synaptic and impulse activity within the dendritic substrate of the bulb.

Organisation of the olfactory bulb

Like most regions of the nervous system, the vertebrate olfactory bulb is composed of three types of neuronal elements: input fibres, principal cells and intrinsic cells. These are illustrated schematically in Fig. 1. The sensory input arrives in the axons of olfactory receptor cells; there are in addition numerous centrifugal fibres with origins in the midbrain and forebrain. The main type of principal neurone, giving rise to output axons from the olfactory bulb, is the mitral cell, which is also differentiated into smaller varieties of tufted cell. The intrinsic neurones are divided into two main populations, periglomerular cells and internal granule cells; there are also several varieties of short-axon cell.

Synaptic circuits are concentrated at two main levels, as indicated in the insets of Fig. 1. In the olfactory glomeruli, the olfactory axons make synapses on to dendrites of mitral (and tufted) and periglomerular cells; in addition, there are synapses between the dendrites of these cells. These circuits provide for the initial processing of the olfactory input. In the external plexiform layer there are synapses between the mitral cell dendrites and the dendritic spines of granule cells. Nearly all of these are arranged in reciprocal pairs, as indicated in the inset diagram, and they provide for control of mitral cell output to olfactory cortical regions. In addition, there are centrifugal inputs to all levels of the bulb except the intraglomerular neuropile; these provide for driving or biassing of bulbar elements by these central regions.

Fig. 1. Schematic diagram of neuronal elements of mammalian olfactory bulb. ON, olfactory nerves; PG, periglomerular short-axon cell; T_s, superficial tufted cell; T_M, middle tufted cell; M/T_d, displaced mitral or deep tufted cell; M, mitral cell; G, granule cell; SA, short-axon cell of deep layer; C, centrifugal fibre. Histological layers shown at right: GLOM, glomerular layer; EPL, external plexiform layer; MBL, mitral body layer; GRL, granule layer. Insets at left show main types of synaptic connections in glomeruli (above) and external plexiform layer (below). (From Getchell & Shepherd, 1975*a*, *b*).

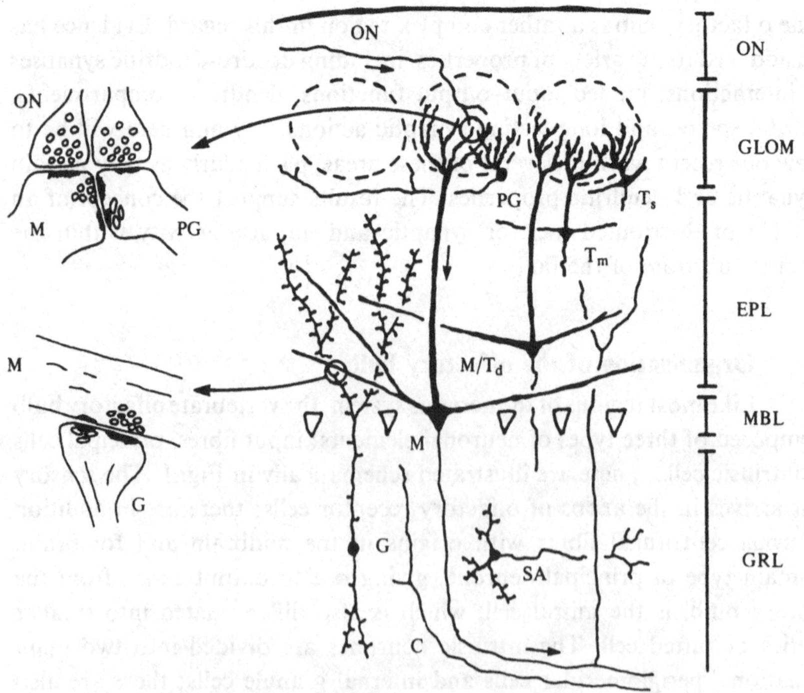

Morphology of dendro-dendritic synapses

There is general agreement that the synapses between dendrites in the olfactory bulb have a morphology that is similar to that seen in other parts of the nervous system. As indicated in the diagram, the mitral-to-granule synapse has asymmetric membrane densities and round presynaptic vesicles, which are characteristic of Gray's type I synapse. The granule-to-mitral synapse, on the other hand, has symmetric membrane densities and flattened or pleomorphic presynaptic vesicles, characteristic of Gray's type II synapse. There is thus a correlation of the type I synapse with mitral-to-granule excitation, and the type II synapses with granule-to-mitral inhibition. Similar correlations between these morphological types and their physiological actions have been found at a number of other sites in the nervous system (reviewed in Shepherd, 1979*b*).

Freeze-fracture studies (Fig. 2*a*) have shown that the mitral-to-granule synapse is associated with an accumulation of intramembranous particles on the inner surface of the outer leaflet of the postsynaptic membrane (Landis, Reese & Raviola, 1974), whereas these particles are absent at the granule-to-mitral synapse. Thus, the intramembranous particles are associated with the type I, presumable excitatory, synapse, whereas they are absent at the type II, presumable inhibitory, synapse. Landis & Reese (1974) found a similar correlation between intramembranous particles, synaptic type and synaptic action in the cerebellum.

In contrast to these studies, Ramon-Moliner (1977) pointed out that in single electron microscope sections the granule-to-mitral synapse is difficult to discern: the membrane densities are often indistinct, and there is often not a clear relation between the junction site and the accumulation of synaptic vesicles. He therefore raised the question of whether the type II synapse exists as a morphological entity, and suggested that, if it did not, the implication is that the inhibition of mitral cells by granule cells must therefore be mediated by a 'non-synaptic' mechanism. Prompted by this report, Jackowski, Parnevalas & Leiberman (1978) have thoroughly reinvestigated the synapses in the external plexiform layer of the olfactory bulb, with a variety of methods: serial sections, goniometry, and EPTA (ethanolic phosphotungstic acid) and BIUS staining. The results (Fig. 2*b*, *c*) fully confirmed again the presence and morphological characteristics of the granule-to-mitral, presumably inhibitory, synapses, as originally described, as well as the nearly 1:1 relation between them and the mitral-to-granule synapses in the reciprocal pairs. In retrospect, it appears that single electron microscope sections were inadequate for the specific criticisms raised by Ramon-Moliner. However, the suggestion of 'non-synaptic' interactions, if taken to mean interactions not mediated

Fig. 2. Fine structure of dendro-dendritic synapses connecting mitral and granule cells. *a*. Freeze-fracture specimen, showing post-synaptic intramembranous particles (arrows) in P face of granule-cell spine at sites of mitral-to-granule synapses (note presynaptic vesicles (V) in mitral (M) cell dendrite). × 10000. From Landis *et al.*, 1974. *b*. Reciprocal synaptic pair, oriented from granule spine (left) to mitral dendrite (right), indicated by arrow, and in reverse direction below. × 65000. *c*. Another reciprocal pair, visualised by the EPTA (ethanolic phosphotungstic acid) method. Mitral-to-granule synapse, from right to left, shown above, and reverse direction below. × 55000. *b*, *c* from Jackowski *et al.*, 1978.

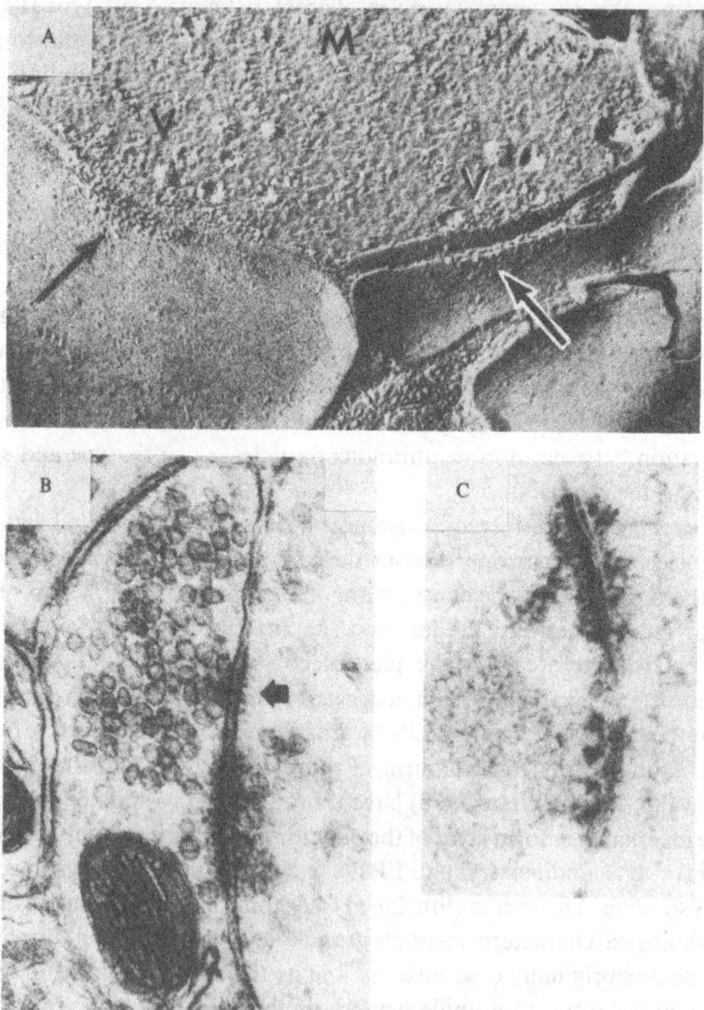

directly through morphological contacts, remains a possibility at these sites in the olfactory bulb, as well as generally throughout the nervous system, particularly in the light of increasing evidence for transneuronal transfer of substances and very slow 'synaptic' or 'neuromodulatory' interactions.

Functional properties of mitral cells

Electrophysiological analysis of the olfactory bulb has taken advantage of the fact that the olfactory nerves and mitral cell axons form separate and distinct pathways. The neuronal elements and circuits can therefore be activated synchronously by orthodromic and antidromic volleys, respectively. Numerous studies in the in vivo mammalian olfactory bulb have shown that an antidromic impulse in a mitral cell is followed by an IPSP lasting 100–200 ms (reviewed in Shepherd, 1972). It has been postulated that the inhibition is mediated by dendro-dendritic synaptic interactions between mitral dendrites and granule-cell spines. The original suggestion arose from a study that combined information about electrophysiological properties and anatomical connections with computer simulations of the mitral and granule cell populations (Rall, Shepherd, Reese & Brightman, 1966; Rall & Shepherd,

Fig. 3. Intracellular responses of mitral cell to olfactory nerve volleys in the isolated olfactory bulb of turtle. *a.* Initial phases of response, at high resting membrane potential (solid trace), and during depolarising current injection (dashed traces). *b.* Long-lasting character of responses, on slow time base; dashed vertical line indicates initial single impulse response; horizontal bracket indicates duration of trace in *a.* Time: 1 s divisions. *c.* Control response to depolarising current injection, showing impulse discharge. Voltage: 20 mV. *d.* Blockage of response to current injection due to long-lasting IPSP. Time: 0.5 s. Voltage calibration: 20 mV (vertical bar in *c*). From Mori & Shepherd, 1979*a*.

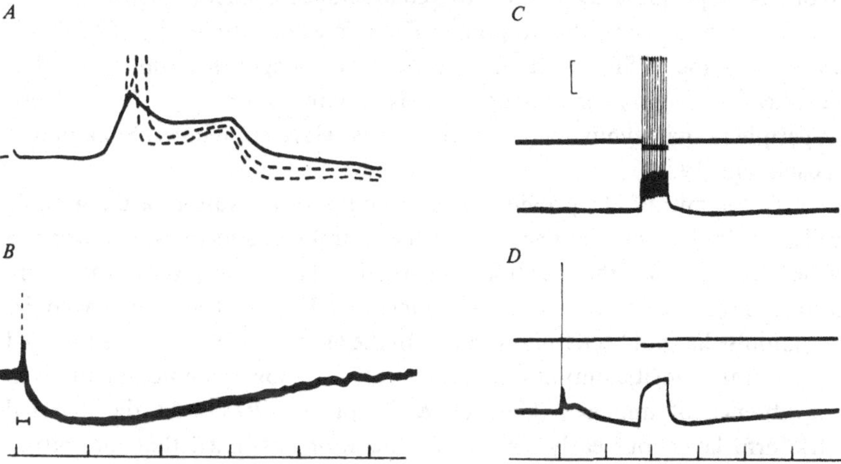

1968; Rall, this volume). The dendro-dendritic synaptic connections have been confirmed in numerous laboratories (see above), and the postulated synaptic interactions have been supported by various studies, most recently by the intracellular analyses of Mori & Takagi (1978*a*, *b*).

Currently we are developing an isolated preparation of the turtle olfactory bulb in order to carry out further intracellular studies of neuronal properties under conditions that are stable and amenable to pharmacological manipulations. Some initial results are illustrated in Fig. 3. In *a* are depicted intracellular recordings from a presumed mitral cell that responded to single orthodromic volleys in the olfactory nerves. Initially the membrane potential was relatively high. The response (continuous trace) consists of an initial depolarisation followed by a brief repolarising wave, a small depolarisation and then a late slow hyperpolarisation. As the membrane was artifically depolarised, single impulse responses occurred at decreasing latencies (superimposed dashed lines). Most of this early sequence represents the synaptic potentials in the glomerular dendritic tufts spreading to the cell body of the mitral cell, where the action potential is set up that propagates into the axon.

The later phase of the response is shown in *b*. On this slow time base the entire trace in *a* is indicated by the small bracket; the later phase can be seen to consist of a large-amplitude, long-lasting hyperpolarisation. The extremely long duration is notable; here it is approximately 4 s; instances as long as 4.7 s have been encountered. As is characteristic of an IPSP, the hyperpolarisation is graded with volley strength; its inhibitory nature is shown by the suppression of impulse responses to injected current (*c*, *d*). IPSPs have been demonstrated in other isolated brain preparations (e.g. hippocampal slice (Schwartzkroin, 1975); olfactory cortex (Scholfield, 1978), but the large amplitude and extremely long duration in the olfactory bulb are unusual. Work is in progress to identify the conductance changes and transmitter substances underlying this response. There is evidence (Mori & Shepherd, 1979) that the IPSP has a voltage-sensitive component, similar to that reported in olfactory cortex neurones (Mori, Satou & Takagi, 1978), and also in peripheral parasympathetic ganglion cells (Hartzell, Kuffler, Stickgold & Yoshikami, 1977).

Since the mitral cells provide the only route for activation of the granule cells, the initial synaptic response of the mitral cell and the single impulse generated by it are the functional properties that set in motion the entire sequence of events seen in the responses in Fig. 3. The role played by the impulse has not been fully clarified. In the glomeruli, it has been suggested that mitral dendritic output may be activated by slow synaptic potentials in the absence of impulses (Getchell & Shepherd, 1975*a*). In the external plexiform layer, our early computer simulations assumed that the mitral-

Fig. 4. Different types of spike activity in the rabbit olfactory bulb. *a*. Extracellular responses of unit in granule layer to volleys in lateral olfactory tract (a) and olfactory nerves (b). Time: ms. From Shepherd, 1963. *b*. Extracellular responses in granule layer to volleys in lateral olfactory tract of increasing intensity (× 1– × 4 threshold). Voltage: 2 mV; time: ms. from Shepherd, 1963. *c*. Intracellular response of mitral cell to olfactory nerve volley, showing fast prepotential (arrow) between synaptic response and impulse. From Mori & Takagi, 1975. *d*. Intracellular responses of unit in granule layer (*b*–*d*). Note larger depolarising waves and spike blockage during current injection (*c*, *d*). Voltage: 10 mV; time: 10 ms. From Mori & Takagi, 1978*a*.

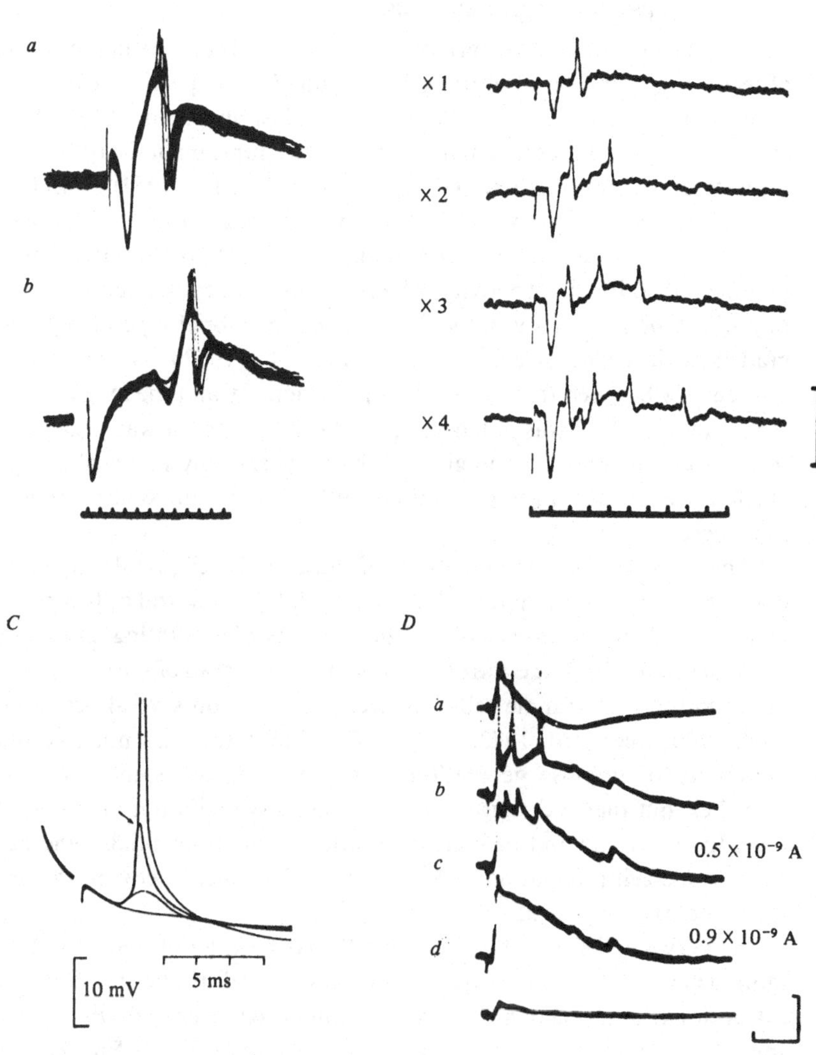

to-granule synapses were activated by impulse spread from the cell body into the dendrites, but the computations could not distinguish between passive spread in large dendrites or active propagation in very thin dendrites; we suggested that the two cases were not mutually exclusive (Rall & Shepherd, 1968). Recently, Mori & Takagi (1975) have obtained evidence that some mitral cells show fast prepotentials (i.e. non-propagating impulse activity) that may be generated in the primary dendrite near the glomerular tuft (see Fig. 4*c*, below).

Properties of granule cells

The position of the granule cell as an interneurone in the mediation of recurrent inhibition of mitral cells suggested a comparison with Renshaw inhibition in the spinal cord (Phillips, Powell & Shepherd, 1963). Renshaw cells discharge impulses throughout most of the duration of the IPSP that they mediate on to motoneurones (Eccles, Fatt & Koketsu, 1954). Search of the granule cell layer with extracellular microelectrodes, however, showed that spike-generating units are uncommon, in contrast to the high density of granule cell bodies. Unit activity, when encountered, consists either of a single impulse response to an input volley (Fig. 4*a*) or a brief repetitive discharge, graded with volley strength, but lasting only 10–20 ms (Fig. 4*b*). The discrepancy between this restricted impulse activity and the long duration of the mitral cell IPSP was noted in the early studies and it was suggested that the granule cell processes might well have special physiological properties, 'such as sustained transmitter actions outlasting the initial impulse activity' (Shepherd, 1963).

Subsequent computer simulations of the granule cell population indicated that the granule cells might mediate the inhibition of mitral cells through the dendro-dendritic synapses without the necessity of generating impulses (Rall & Shepherd, 1968). The electrotonic basis for the spread of synaptic potentials in the mitral and granule cells has been discussed on several occasions (cf. Rall, 1970; Shepherd, 1972, 1979*a*). The simulations did not rule out the possibility of impulse generation in the granule cell spines or dendritic branches, but they did suggest that such activity could not propagate to or from the cell body and be consistent with the field potentials generated by the granule cell population and the potential gradient between the granule layer and external plexiform layer.

The very small size of the granule cells has been an obstacle to obtaining more definitive data on the question of granule cell properties. Some of the extracellular units encountered in the granule cell layer undoubtedly are the somewhat larger short-axon cells located in that layer (cf. Fig. 2). Mori &

Takagi (1978*a*) have obtained intracellular recordings from granule layer neurones in the rabbit. They confirm that impulse activity is restricted and brief (see Fig. 4*d*). Some of the cells could be partially filled with Procion Yellow dye, and appeared to have the morphology of granule cells. It may be noted that these granule layer responses bear some resemblance to the responses of amacrine cells to light stimuli in the retina (cf. Werblin & Dowling, 1969; see Introductory chapter to this volume, Fig. 4).

Recently we have returned to computer simulation studies, constructing more detailed models of individual mitral and granule cell dendrites and the reciprocal synapses connecting them (Shepherd & Brayton, 1979). These studies confirm that recurrent and lateral inhibition of mitral cells can be mediated under the assumption of strictly passive properties of mitral and granule cell dendrites (apart from an initial impulse near the mitral cell body, and appropriate conductance changes at the dendro-dendritic synapses). This, of course, does not rule out the presence of excitable membrane in granule cell dendrites; the fact that we cannot definitely answer this question reflects the tentative stage of our analysis.

Some complexities of the problem are illustrated by considering further the granule cell spines. Like dendritic spines elsewhere, these are small, of the order of 1 μm in diameter and 2–3 μm long, connected to their parent branch by a thin neck that may be only 0.2 μm in diameter and 1–3 μm in length. They lie at a considerable distance, both morphologically and electrotonically, from the granule cell bodies, so that recordings made at the cell body are likely to give only an extremely attenuated version of the actual activity in the spines. Similarly, impulse activity in the cell body (e.g. Fig. 4*d*) may or may not reach the spine, whether by active or passive spread.

Computer simulations of the spines are yielding results which in some respects go beyond one's intuitive predictions. In Fig. 5*a*, simultaneous excitatory input is delivered on two spines arising from a branch of a granule cell dendritic tree. The conductance changes, though small, give rise to relatively large potential transients, which, after a rapid rise, decay very rapidly, giving the transient a spike-like character. The large amplitude reflects in part the extremely high input resistance of the small spine head; the rapid decay however reflects the large conductance load of the attached dendritic branch and the rest of the dendritic tree. Intracellular recordings from dendritic spines are beyond the reach of our present techniques, but these computer simulations indicate that synaptic activity itself could generate spike-like activity in the finest processes of the synaptic neuropile.

A further problem facing analysis of granule cells and similar small neurones is shown in the remainder of Fig. 5 (Fig. 5*b*–*d*). In this simulation experiment we varied the time interval between delivery of synaptic excitation to the two

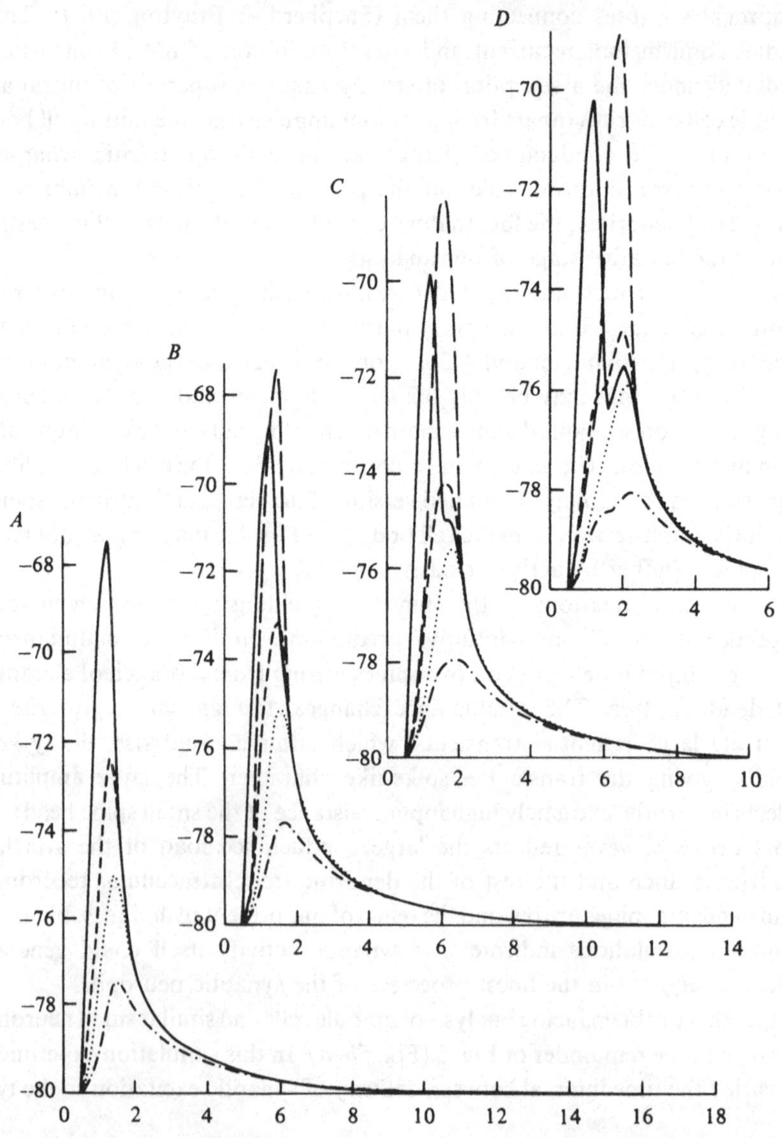

Fig. 5. Computer simulations of voltage transients in granule cell spines. Model consisted of two spines arising from a granule cell dendritic branch; each spine received an excitatory synapse, which caused a brief depolarising conductance change in the spine head. Transients shown are for two spine heads (full lines, long dashed lines), the intervening segment of a dendritic branch (shorter dashed lines), and the distal (dotted lines) and proximal (dotted and dashed lines) lumped parts of the dendritic tree. The excitatory inputs were delivered simultaneously to the two spines (*A*), and at successive 0.25 ms intervals (*B–D*). Note sharp transients in spines, but smaller merged waves in dendritic tree. Abscissa: ms; ordinate, mV. G. M. Shepherd, R. K. Brayton & K. Mori, unpublished observations.

spines. The transients in the two spines retain their spike-like conformation, with the second riding on the electrotonic spread from the first. However, in the dendritic branch the transients are merged into a slower, lower-amplitude wave that gives little indication of its composite nature. Thus, a recording from the branch, or from the rest of the dendritic tree or the cell body, gives little guide to the activity occurring in the spines. Conversely, each spine can act as an individual input–output unit with its own precise timing, but this precision is lost in the summed electrotonic potentials of its branches.

Synaptic and impulse loci: distribution and variability

From the foregoing considerations it can be seen that analysis of impulse and non-impulse (e.g. synaptic) activity in the neurones of the olfactory bulb is still at an early stage, because of the technical limitations in recordings from small cells and because of our ignorance of the functional properties of the finest nerve cell processes. At this stage it is best, therefore, to keep in mind the range of possible properties that may exist. Referring again to the diagram of Fig. 1, one can consider, at one extreme, that impulse activity is limited to the axons of the input fibres and the axon hillocks and axons of mitral cells and periglomerular cells. Activation of dendro-dendritic synapses in this scheme is entirely by electrotonic spread in the dendrites of mitral, granule and periglomerular cells. At the other end of the range, impulse generation could occur at several additional sites, in the dendrites of mitral, granule and periglomerular cells, and the granule cell bodies. The rest of the dendritic membranes of these cells might be weakly or partially excitable. One or another type of active sites may be present in given cells of a given species, and under the particular conditions of a given experiment. A general conclusion in this regard does not seem warranted, and may be quite inappropriate for describing accurately all cells in all species.

The situation in the olfactory bulb thus appears to be more complicated than in the retina, where cells normally generating impulses can be differentiated from those that do not (cf. Fain, this volume). Similarly, non-spiking neurones are being recognised in many invertebrate species, and their functions differentiated from those of spiking neurones (see other contributions to this volume). While it remains possible that the olfactory granule cell is a non-spiking neurone, it seems likely that in general the olfactory bulb provides neuronal substrates which are fractionated into multiple loci for signal intergration and input–output functions. Because of this, parts of a given neurone may operate in a spikeless mode, while other parts may operate with impulses. One must also take account of the likelihood that the presence of excitable foci may vary between different species, during different developmental stages, and under different experimental conditions. Taken together, these various factors may provide a generalised framework for approaching the analysis of synaptic and excitable properties in many other regions of vertebrate and invertebrate central nervous systems.

This work has been supported by research grant NS–07609 of the National Institute of Neurological and Communicative Disorders and Stroke.

References

Eccles, J. C., Fatt, P. & Koketsu, K. (1954). Cholinergic and inhibitory synapses in a pathway from motor collaterals to motoneurones. *J. Physiol., Lond.* **216**, 524–62.

Getchell, T. V. & Shepherd, G. M. (1975*a*). Synaptic actions on mitral and tufted cells elicited by olfactory nerve volleys in the rabbit. *J. Physiol., Lond.* **251**, 497–522.

Getchell, T. V. & Shepherd, G. M. (1975*b*). Short-axon cells in the olfactory bulb: dendro-dendritic synaptic interactions. *J. Physiol., Lond.* **251**, 523–48.

Hartzell, H. C., Kuffler, S. W., Stickgold, R. & Yoshikami, D. (1977). Synaptic excitation and inhibition resulting from direct action of acetylcholine on two types of chemoreceptors on individual amphibian parasympathetic neurones. *J. Physiol., Lond.* **271**, 817–46.

Jackowski, A., Parnevalas, J. G. & Lieberman, A. R. (1978). The reciprocal synapse in the external plexiform layer of the mammalian olfactory bulb. *Brain Research*, **159**, 17–28.

Landis, D. M. D. & Reese, T. S. (1974). Differences in membrane structure between excitatory and inhibitory synapses in the cerebellar cortex. *J. Comp. Neurol.* **155**, 93–110.

Landis, D. M. D., Reese, T. S. & Raviola, E. (1974). Differences in membrane structure between excitatory and inhibitory components of the reciprocal synapse in the olfactory bulb. *J. Comp. Neurol.* **155**, 67–92.

Mori, K., Satou, M. & Takagi, S. F. (1978). Fast and slow inhibitory postsynaptic potentials in the piriform cortex neurons. *Proc. Jap. Acad.* **54B**: 484–9.

Mori, K. & Shepherd, G. M. (1979). Synaptic excitation and long-lasting inhibition of mitral cells in the *in vitro* turtle olfactory bulb. *Brain Res.* **172**, 155–9.

Mori, K. & Takagi, S. F. (1975). Spike generation in the mitral cell dendrite of the rabbit olfactory bulb. *Brain Research*, **100**, 685–9.

Mori, K. & Takagi, S. F. (1978*a*). An intracellular study of dendrodendritic inhibitory synapses on mitral cells in the rabbit olfactory bulb. *J. Physiol., Lond.* **279**, 569–88.

Mori, K. & Takagi, S. F. (1978*b*). Activation and inhibition of olfactory bulb neurones by anterior commissure volleys in the rabbit. *J. Physiol., Lond.* **279**, 589–604.

Phillips, C. G., Powell, T. P. S. & Shepherd, G. M. (1963). Responses of mitral cells to stimulation of the lateral olfactory tract in the rabbit. *J. Physiol., Lond.* **168**, 65–88.

Rall, W. (1970). Dendritic neuron theory and dendrodendritic synapses in a simple-cortical system. In *The Neurosciences: Second Study Program*, ed.-in-chief F. O. Schmitt, pp. 552–65. New York: Rockefeller.

Rall, W. & Shepherd, G. M. (1968). Theoretical reconstruction of field potentials and dendrodendritic synaptic interactions in olfactory bulb. *J. Neurophysiol.* **31**, 884–915.

Rall, W., Shepherd, G. M., Reese, T. S. & Brightman, M. W. (1966). Dendro-dendritic synaptic pathway for inhibition in the olfactory bulb. *Exp. Neurol.* **14**, 44–56.

Ramon-Moliner, E. (1977). The reciprocal synapses of the olfactory bulb: questioning the evidence. *Brain Research*, **128**, 1–20.

Scholfield, C. N. (1978). A depolarising inhibitory potential in neurones of the olfactory cortex *in vitro*. *J. Physiol., Lond.* **275**, 547–57.

Schwartzkroin, P. A. (1975). Characteristics of CA1 neurons recorded intracellularly in the hippocampal *in vitro* slice preparation. *Brain Research*, **85**, 423–36.

Shepherd, G. M. (1963). Neuronal systems controlling mitral cell excitability. *J. Physiol., Lond.* **168**, 101–17.

Shepherd, G. M. (1972). Synaptic organization of the mammalian olfactory bulb. *Physiol. Rev.* **52**, 864–917.

Shepherd, G. M. (1979*a*). Functional analysis of local circuits in the olfactory bulb. In *The Neurosciences: Fourth Study Program*, ed.-in-chief F. O. Schmitt & F. G. Worden, pp. 129–43. Cambridge, Mass: MIT Press.

Shepherd, G. M. (1979*b*). *The Synaptic Organization of the Brain.* Second Edition. New York: Oxford University Press.

Shepherd, G. M. & Brayton, R. K. (1979). Computer simulation of a dendrodendritic synaptic circuit for self and lateral inhibition in the olfactory bulb. *Brain Reseach*, **175**, 377–82.

Werblin, F. S. & Dowling, J. E. (1969). Organization of the retina of the mud puppy, *Necturus maculosus*. II. Intracellular recording. *J. Neurophysiol.* **32**, 339–55.

THEODORE H. BULLOCK

Spikeless neurones: where do we go from here?

This book consolidates and confirms an old idea. In 1957 it was possible to say '...These considerations also lead us to the suggestion that much of normal nervous function *occurs without impulses* [emphasis in original] but mediated by graded activity, not only as response but also as stimulus' (Bullock, 1957). The notion had appealed to me for some time (Bullock, 1945) and in 1947 I wrote in a review 'The far-reaching implications of the assumption that neurons can affect each other by means distinct from classical impulses in synaptic pathways are obvious'. I referred to Bremer (1944) and Gerard (1941) who influenced me the most in this view, which remained for a long time ignored in the conventional orthodoxy. The claim by Parry (1947), based on the insect ocellus, that subthreshold activity in one cell can cause spikes in post-synaptic neurones, was plausible as a mechanism, although the data offered were not definitive. A related idea, called Wedensky inhibition, was still older. This stated that stimulation, even when subthreshold, can exert an influence, at the right repetition rate, by maintaining a depressed state and preventing recovery – in peripheral nerve. I found an equivalent effect at the squid giant synapse in the stellate ganglion, upon stimulating the presynaptic axon close to the ganglion; presumably local potentials in the presynaptic axon terminals were preventing recovery (Bullock, 1948).

The purpose of this chapter is to stand back and attempt an overview, to speculate on areas of future developments. This compels me to begin by re-examining the questions, already treated in the Introduction by Shepherd, of what the set of neurones under consideration includes, how they are defined and how they are recognised.

Whether a neurone spikes or not can sometimes be difficult to decide, even if one has solved the technical problems of access, identification and intracellular recording. As is well-known some crustacean leg muscle fibres only fire all-or-none, regenerative, propagating action potentials under special conditions. Various other cells, such as some electrocytes, show graded local potentials up to overshooting peak amplitude. Abortive spikes of diverse amplitude are familiar in axons. Good spikes may appear tiny because they often do not invade the soma. Retinal rods can give a calcium spike but normally this is swamped by the large potassium conductance (Fain, this volume; Fain, Quandt & Gerschenfeld, 1977). Spike amplitude can be graded as in *Aplysia* bag cells, where it shows facilitation (F. Strumwasser, personal communication, Strumwasser, Kaczmarek & Viele, 1978). In many types of neurones it decreases with high frequency repetition. All this means two things. 1. There can be seemingly transitional or intermediate forms of activity. If they were not so rare we could not make the dichotomy that forms the basis for this symposium. 2. To be sure that a neurone can spike it should be shown to have a sharp threshold for an all-or-none event with absolute refractoriness, since a large action potential may be a graded, non-regenerative response that would spread decrementally even though not entirely passively.

So much for the problems of recognising neurones that function without impulses. One further specification is needed to embrace the full set of neurones under consideration.

My bias is to include, whenever thinking in terms of roles or of assessment, both the permanently spikeless neurones – or those not known to produce spikes under quasi-normal conditions, and also those quite capable of spiking but playing some active role in intercellular communication during epochs when they are not spiking. To neglect this second class of cells would be quite arbitrary and would, I believe, lead to a serious underestimation of the importance of neuronal signalling without spikes. To be sure, the inclusion of this class of cells and events (Graubard *et al.*, 1980) opens up a very large can of worms, as we will see shortly. However it is my strong feeling that it is much more heuristic to have them out in the open than to continue to neglect them. They certainly belong in any assessment. Indeed it may actually turn out that you can sometimes demonstrate a clear influence of a neurone without spikes but can not be sure that this neurone is unable to spike under other conditions, untested.

Now, it is very clear, with the impressive body of evidence that has been reviewed and the beautiful new evidence that has been presented in this book, that spikeless neurones are here to stay, are not experimental artefacts of abnormal preparations, or idiosyncratic anomalies of some exceptional lower animals, but are indeed widespread in both central and peripheral nervous systems, of both invertebrates and vertebrates. With this much established, people have turned to more advanced questions, chiefly concerned with the details of their mechanisms in each case. The neglected questions such as where they are likely to occur, and when nature uses spikeless neurones, will require inductive leaps as to rules or tendencies, and then drastic extrapolations to new situations, necessarily little understood. These questions are chiefly for the future. I will speak to them briefly and gingerly, but will not achieve much of a list of predictions. Then I would like to extrapolate drastically from the knowledge at hand and attempt to paint a new and more realistic picture of what's going on in neuropile and grey matter.

Where may we expect spikeless neurones?

The examples already established and referred to in this book are valuable because they rule out some otherwise plausible hypotheses. Spikeless neurones do not correlate well with axon-less neurones. We have spikeless neurones with good axons in the crab leg (Ripley, Bush & Roberts, 1968), the *Limulus* lateral eye (Waterman & Wiersma, 1954) and barnacle photoreceptor (Gwilliam, 1963), and we have amacrine cells with good spikes in the vertebrate retina (Werblin & Dowling, 1969). Supposedly amacrine cells in the substantia gelatinosa are believed to spike (but the absence of an axon here is doubtful) and some workers (Mori & Takagi, 1975) believe the axon-less granule cells in the olfactory bulb may spike. Spikeless neurones do not correlate well with short axon, Golgi type II or intrinsic neurones; many such cells can spike – in cerebellum, hippocampus and olfactory bulb. Not only do some retinal intrinsic neurones spike but granule cells of the cerebellum and probably basket and Golgi cells as well. Spikes have been recorded in the finest textured neuropile, that of the corpora pedunculata of the insect brain (Maynard, 1966; Vowles, 1964; Erber & Menzel, 1977), although it is not easy to decide whether they can be attributed to the intrinsic neurones. The same examples may be used to rule out a strict correlation with size of neurones. Siegler & Burrows (1979) apparently could not distinguish a morphological class of non-spiking cells among the many local interneurones in the locust ganglion. Judging from the retina, as Fain made clear, we cannot expect a strong correlation with either electrically or chemically transmitting neurones.

However, even though strict correlations appear to be excluded, it may still be supposed that we are more likely to find spikeless neurones among small, short axon and amacrine cells. This consideration suggests especially such places as the octopus optic lobe, which is very rich in amacrines (Bullock & Horridge, 1965; Young, 1971). I expect there are spikeless neurones in the mammalian cortex (Szentagothai, 1978), although evidence of dendrodendritic and reciprocal synapses (Sloper & Powell, 1978*a*) and of gap junctions (Sloper & Powell, 1978*b*) are by themselves not enough and there is not yet any solid physiological evidence. At the moment therefore, the only clear examples of spikeless neurones in the vertebrates are those in the retina, but I will bet they are present in cortex and olfactory bulb and probably quite widely.

I have exceptionally little confidence in such predictions about cells that will be found incapable of spiking. In contrast, I feel much more willing to bet that lots of cells will be found which, although capable of spiking, need not spike to exert physiological effects on other cells. Surely these are present in the cerebral cortex, indeed in each of the cortices, including cerebellar, tectal, and cochlear nuclear; also in thalamus, striatum, olfactory bulb, substantia gelatinosa, and elsewhere in the vertebrates, in octopus optic lobe, vertical lobe, and quite possibly in the basal lobe and subesophageal ganglion, in the arthropod brain and optic ganglion – in short, they are probably widespread.

I would like to give special attention to that large and diverse class of cells sometimes called paraneurones, which are traditionally not regarded as neurones but are at least closely related to neurosecretory neurones. They include pinealocytes, pancreatic and gastroenteric endocrine cells releasing gastrin, secretin, cholecystokinin, or somatostatin, basal granulated cells of bronchial epithelia, adrenal chromaffin cells, small intensely fluorescing (SIF) cells in sympathetic ganglia, melanocytes and secondary sense cells such as taste bud, Merkel's disc and hair cell receptors of the lateral line system, vestibule and cochlea. Among these, spikes have been found so far in several – in glucose-stimulated pancreatic β cells, adrenomedullary adenohypophyseal, and parafollicular cells (Fujita & Kobayashi, 1979) and in some hair cells, especially among electroreceptors (Bennett, 1967; Harder, 1968). But it seems likely that the general mode of response among paraneurones is graded. To enlarge briefly on electroreceptors, those of mormyrid fish called Knollenorganen exhibit a regular spike train (Bennett, 1967), which can be from a few tens per second to several hundred, based on an intrinsic, oscillatory state of the hair cell. In some species of mormyrids and probably in all gymnotoid electric fish the homologous receptor cells oscillate with a graded potential (Viancour, 1979); sometimes the graded potential looks spike-like in form. I am sure we do not as yet have the full story on these oscillatory receptor potentials but they have a considerable general neurobiological interest. For example, they may help to explain how electroreceptor afferent fibres can have frequency response or filter curves as steep as those of cochlear nerve fibres although there is no basilar membrane or mechanical resonance to attribute it to (see Russell, this volume).

Finally, we may ask what the upper size limits of spikeless neurones will be. The crab-leg stretch receptor has an axon 5–9 mm long in the species studied. One wonders whether the homologous axons in large crabs (> 5 times the carapace width) work the same way; if so, this figure might be multiplied several times. Note that Bush gives a length constant in 90-μm-diameter, 9-mm-long *Scylla* axons as 60 mm. It seems reasonable to suppose the system could still work even if the attenuation by decremental spread were 90% or more. I believe this is happening when a subthreshold slow potential injected into a follower cell of the lobster cardiac ganglion modulates the pacemaker cells at the other end of the ganglion 10 mm away (Watanabe & Bullock, 1960). This leads me to raise the old question of why spikes? What may be the functional advantages – emphasising the plural – of developing all-or-none impulses?

The functional significance of spiking

So far we have been looking for correlates, significance and predictability of spikeless neurones. To gain perspective it might be well to change our point of view and ask what are the correlates, significance and predictability of spiking neurones. In the Introduction Shepherd has addressed the issue already. Today, this question is much more difficult that it once was (Bishop, 1956). We can no longer say that spiking is mainly an adaptation to long distance signalling. Amacrine cells in the retina and in the lateral geniculate of the rat, as well as many short-axon intrinsic neurones in many ganglia and nervous centres have spikes. We should remember that most ganglia and major brain structures of most animals are well within the dimensions of the space constant of common axons, so that spikes gain little over decrementally spread potentials, in respect to conduction of signals. Accepting that over long distances – many millimetres or centimetres – spikes have a significance for faithful propagation, it seems clear that we have to look for another significance in the common intrinsic neurone where such distances are not involved. Remember, spikes were already well-known to unicellular organisms and the ancestors of Hydra, flat worms and other small invertebrates.

The proposition I want to put forward here is that there is a value to spikes over and above whatever advantages they offer to unicellular organisms and for long-distance conduction. I am thinking of the value in respect to encoding information. By introducing devices that encode and decode pulse trains, a wide dynamic range of signals becomes available represented in terms of numbers of pulses, intervals, distribution of intervals and derivatives of these with respect to time. These several forms of candidate codes may be advantageous in comparison with, say, the amplitude of a graded potential, perhaps in several ways at once. I can guess some advantages but I'm not prepared to evaluate them quantitatively. For example, there may be more independence of signal transmission from unwanted effects of temperature, of osmotic fluctuations, of d.c. or slowly fluctuating electric field potentials and of other unknown sources of 'noise' that would directly alter the amplitude of a graded potential but only indirectly alter intervals between spikes. Trains of spikes might give an advantage merely because the effects of these unwanted agencies are relatively confined to the loci of the encoding and the decoding. In addition it is possible that the spiking loci or decoding devices are less sensitive to these perturbations or have a narrower frequency pass band for them than would the simple amplitude-graded signal. Signal-to-noise ratio and dynamic range might be improved by the voltage-to-spike rate and spike rate to transmitter-release conversions, or if not simple spike rate, one or a combination of the other parameters inherent in a more or less regular train of impulses.

The functional significance of stratification of nervous tissue

I have referred to that special type of nervous architecture called cortical. It seems to me germane to the topic of this book, if we were to extrapolate from the single-cell level to that of organised arrays of cells, and look at the role of graded, local activity in interneuronal transactions on a larger scale. What can we expect, say in a cortex, if spikeless interactions are as common as I am betting? In order to pose a more specific question – though still a pretty slippery one – I propose to ask 'what might be the functional significance of stratification in nervous tissue?'

I will not dwell on the ontogenetic factors or significance, nor on the phylogenetic aspects. To emphasise that point let me grant for purposes of discussion that such structures are laminated because (in one sense of that term) they develop that way; there are good ontogenetic mechanisms that account for it. But I am asking, given stratification, does it have any functional consequences? What I would like is to imagine the functional difference between two cortices with the same variety of inputs, of cell types, of connections and of outputs but differing in that one is well stratified and the other is scrambled.

We tend to regard the more stratified, that is the more differentiated, tissue as more advanced, more derived, and capable of higher levels of function. The optic tectum of elasmobranchs is not as well developed as that of some teleosts, reptiles and birds. The neuropile masses of the optic lobes of arthropods are regarded as more advanced where there are more distinct strata and the same is true in the inner plexiform layer of vertebrate retinas and in comparing neocortex to the cerebral pallium of lower vertebrates. To be sure, some of this is due to a greater variety of cell types and connections in the so-called higher cases. But let us ask whether there might be any further factor or functional meaning to stratification per se.

In a recent symposium devoted to laminated structures (Creutzfeldt, 1976), the question of the functional meaning of lamination does not appear to have been discussed. Surely it is not so self-evident a question that an explicit proposal is trivial. Can we even be sure, as we are so inclined to say, that better lamination goes with more highly developed or complex function? Is the optic tectum of reptiles really more complex in function than that in mammals or than the cerebral cortex in mammals, which is not nearly as well stratified? The torus semicircularis of gymnotoid electric fish is much more laminated than its homologue, the inferior colliculus, in mammals, or the torus of most fishes, including mormyrid electric fish! In the following discussion I assume, as a *first* approximation, that function *is* more complex when comparing stratified structures with their unstratified homologues. A second approximation may compel us to qualify this. But the first approximation is the most needed and that calls for thinking not only of the mammalian cerebral cortex but of the many other stratified cortices and layered structures, including those in the cerebellum, tectum, olfactory bulb, torus semicircularis, dorsal cochlear nucleus, lateral line lobe, retina, and invertebrate plexiform layers such as those of the cephalopod and arthropod optic lobes. It is particularly important to compare structures in animals where stratification is less developed with equivalent structures in animals where it is more developed.

I am not aware of a functional correlate that has been proposed on the basis of comparing more and less stratified structures with some equivalence or homology. I put forward the following, in spite of difficulties based on ignorance, to give a target to shoot at, hoping thereby to elicit other targets.

We begin with a category of structures that tends to form sheets for whatever reasons. This tendency goes way back to early stages. The nerve net in coelenterate medusae and polyps is perforce in a well-defined layer between the epithelial cell bodies and the basement membrane or mesoglea (Bullock & Horridge, 1965). Time and again in more advanced nervous systems, though there are many exceptions, equivalent neurones have taken up a two dimensional extent and, in various degrees, a common orientation, giving then the basis for subdivision into strata.

Consider such an unstratified extended sheet of cells, several deep, of one or of a number of types. Consider that a basic principle of connectivity for at least some of the inputs or intrinsic interconnections is overlap of the set of receiving cells in this sheet to which each entering fibre projects, with those of neighouring fibres. There can be some private or 1:1 connections, but the following proposition depends more particularly upon the component of overlapping connections.

Now, could it be that stratification develops in such a pre-existing sheet of cells simultaneously with more and more dependence upon local circuits? As there develops confinement of the targets of the different types of arriving fibres to limited parts of the postjunctional neurones, there will be more differentiation of axonal terminals, dendrites and output messages. With the evolution of these forms of increased specification, a tendency to bring together into one stratum the corresponding parts of each of the overlapping cell types would provide several advantages. One is to shorten the aggregate length of the fibres in the pre- and post-junctional arbors in comparison to the theoretical, disordered situation where the target portions of the set of cells to which each input fibre would send projections, would be scattered through the depth of the cortex. In terms of volume saved alone, this factor might be significant in permitting a certain degree of differentiation. Another consequence would be to permit short axon cells to become more important and to reach more of the overlapping targets with their limited arbors. Perhaps most significantly, it would permit relatively more dependence on local interactions, especially those that involve more than two neurones. Local circuitry means not only connectivity mediated by graded, local events via defined junctions as in spikeless transmission, but also other, non-circuit forms of signalling. One such is the influence of local fields of current such as occur in the axon cap of Mauthner's neurone in goldfish. Another is the action of transmitters that diffuse through some volume of tissue, without requiring specialised synaptic contacts. Note that both pre- and post-junctional elements can be axonal, dendritic or somatic.

The differentiation in the axis normal to the cortical surface and its functional consequences must be added and permuted with those of differentiation in the plane of the cortex which varies within and among cortices from slight or gradual to marked, abrupt and local. Besides neighbour relations and declining connectivity with distance, discontinuities are now well-known such as the columns and modules of the cerebral neocortex (Szentagothai, 1975, 1978; Szentagothai & Arbib, 1974) and the two distinct systems of intersecting slabs in the visual cortex (Hubel & Wiesel, 1977). The result is a three-dimensional matrix of repeating elements that distributes certain input and output parameters at least in the plane of the cortex and probably others in the different depths, at right angles to that plane. Our knowledge is still quite primitive as to the functional parameters segregated with depth but I am sure it will come.

As a second approximation, at an advanced evolutionary stage, it seems reasonable to expect that a more advanced system in terms of the foregoing processes might *not* appear as well stratified as a less evolved system, by any particular method of preparation. With sufficient complexity a cortex where many cells span most or all of the layers could develop, not more or sharper strata as revealed by a given method, but more kinds of stratification, each requiring a different kind of criterion. Based on recent advances, the cerebral cortex, with many kinds of afferents and efferents and 7–10 times as many neurone types as the cerebellar cortex (types that are chiefly defined by connectivity), exemplifies this evolution. The cortex is not adequately described as a 6-layered sheet but somewhat better as a 4- to 6-layered sheet on each of half-a-dozen criteria that are probably not mutually predictive and by which the levels of the discontinuities are not quite the same. These include criteria based on the use of the electron microscope, on Golgi, Nissl and myelin staining methods, on histochemistry for each of many substances, such as succinic dehydrogenase, and on physiological measurements such as current source density of evoked potentials.

This last criterion deserves special comment. Experience is still limited but it is clear from profiles based on 20 or more recording depths through the cortex (Mitzdorf & Singer, 1978) that after computing the second spatial derivative, evoked potentials exhibit sharp boundaries at the edges of sources and sinks. These results are only from stimulating the primary afferent projection to a sensory cortex. There are several other specific and nonspecific inputs ending in different layers. No doubt, when the method is applied to responses evoked from other inputs to the same cortex, additional boundaries will be revealed. Evoked potentials in laminated cortices, which are just special cases of the broader class of compound field potentials in higher central levels – the electroencephalogram – may be manifestations of the enhanced local interactions, including interactions without spikes, that such stratified structures are postulated to emphasise. It is well to be reminded in this connection that, with only special exceptions, the slow, field potentials exemplified by evoked potentials as well as by ongoing brain waves, are 'options' or facultative emergents in the sense that they probably could not be explained or predicted from even a fairly extensive knowledge of unit spike activity. For example mesencephalic or diencephalic auditory structures may show no evoked potential at a time when units are relaying signals to higher centres (Biederman-Thorson, 1970; Bullock & Ridgway, 1972). I take this to be due to both anatomical and physiological factors that determine synchronisation and field geometry.

So much for this attempt to extrapolate the role of spikeless neurones into higher levels of ordered arrays. Now, if you will permit me one more and even bolder leap into interpretative imagination I would like to ask what would nervous tissue in organised higher centres look like in operation, if we could visualise all the forms of signalling, the transactions, the 'magic loom' as Sherrington called it?

Reasons for questioning the image of 'circuitry'

Let me list some reasons that make me dissatisfied with the current analog of circuitry and the associated terminology and mental imagery. Concepts are changing, to be sure, and we have seen a giant step in the recent emphasis on so-called 'local circuits' – those involving subthreshold activity and distances of the order of a few micrometers and less.

Still, 'circuits' comes from a domain of discourse, a technology, that carries a load of implied characteristics. It is a class of models that asserts that the operations in the living tissue can be represented by an equivalent diagram. The diagrams are dominated by three features: *a*, discrete one-way contacts called synapses; *b*, discrete fibres pre- and post-synaptically, effectively insulated between synapses; and *c*, discrete events that make up the signals – impulses and non-impulsive brief, graded transients. Even the newer emphasis on 'local circuits', as usually treated, is to a large degree congruent with such discrete models; the image of local circuits leans heavily on dendro-dendritic and reciprocal synapses which are not ipso facto a great departure and can readily be incorporated into circuit diagrams that work even distorted by changes in the length of lines between terminals. Elsewhere in science and engineering the terminology of circuits would not be used if there were many kinds of series and parallel transducers, converting and reconverting the forms of energy or if arrays of large numbers of transducers (e.g. synaptic contacts) have to be specified in respect to numbers, areas, spacing and three-dimensional distribution to represent the functional system. Likewise, we would not speak of a circuit if there were too many distinct forms of signals, especially non-electrical signals, and their generators, shapers, gates, valves and detectors. A thoroughly hybrid system with sufficient complexity is more likely to be called, for example as electro-optical-mechano-chemical device or system.

Consider some examples of our best-known current circuits – whether your favourite is in insect ventral ganglia, lobster stomatogastric ganglion, *Aplysia* visceral ganglia, the vertebrate retina, cerebellum, vestibulo-ocular reflex or ofactory bulb. Nobody pretends these diagrammatic circuits are complete representations of the real thing, but I would like to convert this intellectual honesty or caution into a stimulus for a new saltation in our mental habits of both imagery and terminology about the working nervous system, especially its grey matter and neuropile.

Here is a partial list of reasons that a new image is needed.

(i) We have to incorporate into our working scheme not only half-a-dozen 'classical' transmitters, with their non-interchangeable properties and dynamics but a dozen or more newer transmitters and 'modulators' – peptides, substance P, encephalins and the rest. We should also allow for a significant role in interneuronal influence of potassium, calcium, pH, oxygen and other simple substances, of substances long dismissed as metabolites, of purines, of hormones, even of proteins. A 'circuit' with symbols for each of these would be well on the way to a new form of representation.

(ii) The probability seems to increase yearly that transmitters sometimes operate effectively at distances of many micrometers, and not only at the loci of membrane apposition and density recognised with the electron microscope. The so-called varicosities along axons of locus coeruleus are just one example.

(iii) Transmitter release is now well established to be graded and to be effective even when the presynaptic event is less than a spike. The relation of presynaptic potential to transmitter released can be quite nonlinear.

(iv) Nearly continuous release of transmitter, gated by input, is accepted for rods and cones. It might well turn up elsewhere.

(v) An increasing variety of potentials must be incorporated into our scheme. Plateau potentials, slow and infraslow potentials that spread decrementally for long distances relative to the lengths of dendritic trees and axon arbors, local potentials that can overshoot, and regenerative hyperpolarising potentials are examples.

(vi) In many processes the direction of spread of signals is not fixed and neither direction can categorically be called antidromic. It depends on what is happening elsewhere and the relative timing.

(vii) The points along axons where regenerative action gives rise to spikes as well as the points were the safety factor falls below 1.0 and decrement sets in are also variable with conditions and timing. This may be particularly important near axon terminals, but probably also at some branch points and discontinuities in diameter.

(viii) In addition to the foregoing there is a long list of neuronal integrative variables that decisively determine the outputs of neurones as a function of their inputs in time and space and which are not symbolised in the 'circuit' image. These dynamic properties of each element belong in any scheme that undertakes to represent the working system. Just to mention a few, I am thinking of variable degrees of accommodation, after-potentials, iterativeness as function of depolarisation, spike invasion of dendrites, burst formation, facilitation, post-tetanic potentiation, autoinhibition, alternative modes of summing heterosynaptic potentials, interactions of electrical and chemical transmission and the like (Calvin, 1975; Calvin & Loeser, 1975; Calvin & Sypert, 1976; Bennett & Goodenough, 1978; Bullock, 1979). These examples are chosen from the neuronal level. Another list could come from the level of neural masses: e.g. synchronisation, recruitment, kindling and the like.

(ix) This blanket item will serve to embrace dendro-dendritic synapses, reciprocal synapses and the other arguments for so-called 'local circuits' put forward by Rakić (1975) and Schmitt, Dev & Smith (1976). There has been uneasiness over the term circuits since the beginning. The evidence is not so much in favour of circuits in a useful meaning of that term but of an integration, in three dimensional volume at least tens of μm in diameter, of graded signals and nonlinear responses of diverse sorts.

(x) We have moved gradually from more physiological to more anatomical factors. I would remind you now of further evidence for a complex volume integration of activity at many loci, not represented by circuitry. The lobster stomatograstric ganglion, where we have quite a complete list, has about 40 chemical synapses, functionally, and some 10^6 seen electron-microscopically (Selverston, Russell & Miller, 1976; King, 1976*a*, *b*). From EM observations, some 25000 contacts are shown to make one functional synapse, consistent in its properties, dynamics and effectiveness, as between lobsters and therefore reflecting a well specified distribution, although I do not suppose the individual contacts are specified.

(xi) We still have to find functional meaning for all the bottle-brushes, nests, calyces, climbing vines and other forms of terminations seen with the light microscope and emphasised by Cajal and for the architecture of layered dendrites and precisely angled axons like the parallel fibres of the cerebellar cortex at right angles to the Purkinje dendrite espaliers. The same need to find functional meaning is true for glomeruli, those three dimensional compound junctions of several neurones in a structured knot.

(xii) A still larger scale complex of terminals is the dense neuropile in the Mauthner's cell axon cap. Although not yet really understood, either functionally or structurally, this is clearly a precedent for field effects that might well be widespread, usually on a small scale in intimate thickets of neuropile.

(xiii) The last item in my present list is the abundance in transmission electron micrographs of what some microscopists call 'casual contiguity', that is, places where processes come close together without glia intervening and without specialisation. Although usually regarded as not representing functional junctions, I cannot help suggesting that in the aggregate these might mediate some weak, perhaps nonspecific synchronising influence in neuronal masses.

Could it be that we have been limited in our images of the working architecture by our tendency to accept a plausible picture such as those based on circuit diagrams and discrete elements, channeling our thoughts too narrowly? I believe this is what has happened. As I survey the current literature, with its exciting revelations in componentry on many fronts, it is clear that most workers, dealing with cellular mechanisms, and also those dealing with global aspects, such as evoked potentials in cognitive tasks, are likely to regard the questions raised here and the aim of an adequate framework for conceptualising the functioning grey matter as futile or at least premature. Certainly we are groping, overwhelmed by a plethora of phenomena. Hoping for simplifying principles, we constantly encounter principles that require adding levels of complexity. What once could be called a model is now a vague and highly personal image, probably more disparate among workers than ever before.

Let us try to think of a new analogue – not to claim it is a good representation of nervous tissue, but to disinhibit our mental habits. Picture a crowd at the football stadium. There are discrete units called people but more than a few forms of communication, contact and signalling. Voices yell at distant targets, known and unknown, whisper to neighbours in coded language, groan to no one in particular. Hats, hot dogs, confetti and bottles are thrown, some to specified targets (friend, customer, enemy or referee), some to no particular target. Hands, lips and other body contacts are used. Silver and paper act as specialised transmitters. Many other forms of emissions exert effects: effluvia, odours, radiant heat, and items that would be merely disturbing waste to some but are the assigned livelihoods of janitors and others. Note that targets of communications may be contiguous, nearby, or distant; more or less specified; specified individually or by class – for example any team fan or any member of a certain ethnic group. Note common driving stimuli can synchronise many, though usually not everybody. Note 'circuits' of various kinds: (i) local reflexes in and out of the same unit; (ii) local circuits of a few units such as Daddy getting cokes for the family, using both connections to specific individuals and to generic groups and, en route, to anybody he bumps into or who picks his pocket; (iii) wider group circuits with mixed specific and nonspecific types of communications as in spreading excitment when a scuffle or arrest is noticed, for a radius of, say 10–50 people, plus a longer range of both specific and nonspecific individuals as the TV camera swings in or the policeman uses his walkie talkie.

I propose that a 'circuit' in our context of nervous tissue is an oversimplified abstraction involving a limited subset of communicated signals and a seriously diagrammatic picture of the three dimensional topology; that in fact there are many parallel types of signals and forms of response, often skipping over neighbours in direct contact and acting upon more-or-less-specified classes of nearby or even remote elements. Thus the true picture of what is going on could not be drawn as a familiar circuit; and specified influence would not be properly called connectivity, except for a very limited subset of neighbours. Instead of the usual terms 'neutral net' or 'local circuit' I would suggest we think of a 'neural throng', that is, a form of densely packed social gathering with more structure and goals than a mob.

I don't want to undervalue in the slightest the importance, the primary necessity, of working out the circuitry in terms of connectivity, as a first-order priority. But just as a good map of a town cannot tell you the pattern of traffic, by foot, hand-cart, car and truck, and the very different sorts of responses over time and space, according to the nature of the materials and transactions, so a connectivity diagram in the brain is only a beginning.

My message is not intended to discourage or to dismay by shouting complexity but quite to the contrary to convey my feelings of excitement based on a widening appreciation, like Balboa discovering the Pacific, of the discoveries there are to be made, at each of many levels of integration. Neuronal influence without impulses, singly and in little or much specified arrays of units, on intimate targets and on neighbourhoods must be a significant part of the whole – or so it seems to me.

References

Bennett, M. V. L. (1967). Mechanisms of electroreception. In *Lateral Line Detectors*, ed. P. H. Cahn. Bloomington, Ind.: Indiana University Press.

Bennett, M. V. L. & Goodenough, D. A. (1978). Gap junctions, electronic coupling, and intercellular communication. *Neurosciences Res. Prog. Bull.* **16**, 373–486.

Biederman-Thorson, M. (1970). Auditory evoked responses in the cerebrum (field L) and ovoid nucleus of the ring dove. *Brain Research*, **24**, 235–45.

Bishop, G. H. (1956). Natural history of the nerve impulse. *Physiol. Rev.* **36**, 376–99.

Bremer, F. (1944). L'activité 'spontanée' des centres nerveuses. *Bull. Acad. roy. Med. Belgique*, **9**, 148–73.

Bullock, T. H. (1945). Problems in the comparative study of brain waves. *Yale J. Biol. and Med.* **17**, 657–79.

Bullock, T. H. (1947). Problems in invertebrate electrophysiology. *Physiol. Rev.* **27**, 643-64.

Bullock, T. H. (1948). Properties of a single synapse in the stellate ganglion of squid. *J. Neurophysiol.* **11**, 343–64.

Bullock, T. H. (1957). Neuronal integrative mechanisms. In *Recent Advances in Invertebrate Physiology*, ed. B. T. Scheer. Eugene, Oregon: University of Oregon Press.

Bullock, T. H. (1980). Reassessment of neural connectivity and its specification. In: *Information Processing in the Nervous System*, eds. H. M. Pinsker and W. D. Willis, Jr. New York: Raven Press, pp. 199–220.

Bullock, T. H. & Horridge, G. A. (1965). *Structure and Function in the Nervous Systems of Invertebrates*. San Francisco: W. H. Freeman & Co. (2 vols.).

Bullock, T. H. & Ridgway, S. H. (1972). Evoked potentials in the auditory system of alert porpoises to their own and artificial sounds. *J. Neurobiol.* **3**, 79–99.

Calvin, W. H. & Loeser, J. D. (1975). Doublet and burst firing patterns within the dorsal column nuclei of cat and man. *Exp. Neurol.* **48**, 406–26.

Calvin, W. H. & Sypert, G. W. (1976). Fast and slow pyramidal tract neurones: An intracellular analysis of their contrasting repetitive firing properties in the cat. *J. Neurophysiol.* **39**, 420–54.

Creutzfeldt, O. (1976). Afferent and Intrinsic Organization of Laminated Structures in the Brain. *Experimental Brain Research, Suppl. 1.* New York: Springer-Verlag.

Erber, J. & Menzel, R. (1977). Visual interneurons in the median protocerebrum of the bee. *J. Comp. Physiol.* **121**, 65–77.

Fain, G., Quandt, F. N. & Gerschenfeld, H. M. (1977). Calcium-dependent regenerative responses in rods. *Nature, Lond.* **269**, 707–10.

Fujita, T. & Kobayashi, S. (1979). Current views on the paraneurone concept. *Trends in Neuroscience*, **2**, 27–30.

Gerard, R. W. (1941). The interaction of neurones. *Ohio J. Sci.* **41**, 160–72.

Graubard, K., Raper, J. A. & Hartline, D. K. (1980). Graded synaptic transmission between spiking neurons. *Proc. Nat. Acad. Sci.*, **77**, 3733–5.

Gwilliam, G. F. (1963). The mechanism of the shadow reflex in Cirripedia I. *Biol. Bull.* **125**, 470–85.

Harder, W. (1968). Die Beziehungen zwischen Elektrorezeptoren, elektrischem Organ, Seitenlinienorganen und Nervensystem bei den Mormyridae (Teleostei, Pisces). *Z. vergl. Physiol.* **59**, 272–318.

Hubel, D. H. & Wiesel, T. N. (1977). Functional architecture of macaque monkey visual cortex. *Proc. R. Soc. Lond. Series B*, **198**, 1–59.

King, D. G. (1976*a*). Organization of crustacean neuropil. I. Patterns of synaptic connections in lobster somatogastric ganglion. *J. Neurocytol.* **5**, 207–37.

King, D. G. (1976*b*). Organization of crustacean neuropil. II. Distribution of synaptic contacts on identified motor neurons in lobster stomatogastric ganglion. *J. Neurocytol.* **5**, 239–66.

Maynard, D. M. (1966). Organization of central ganglia. In *Invertebrate Nervous System*, ed. C. A. G. Wiersma. Chicago: University of Chicago Press.

Mitzdorf, U. & Singer, W. (1978). Prominent excitatory pathways in the cat visual cortex (A 17 and A 18): A current source density analysis of electrically evoked potentials. *Exp. Brain Res.* **33**, 371–94.

Mori, K. & Takagi, S. F. (1975). Spike generation in the mitral cell dendrite of the rabbit olfactory bulb. *Brain Research*, **100**, 685–9.

Parry, D. A. (1947). The function of the insect ocellus. *J. Exp. Biol.* **24**, 211–19.

Rakić, P. (1975). Local circuit neurones. *Neurosciences Res. Prog. Bull.* **13**, 291–446.

Ripley, S. H., Bush, B. M. H. & Roberts, A. (1968). Crab muscle receptor which responds without impulses. *Nature, Lond.* **218**, 1170–1.

Schmitt, F. O., Dev, P. & Smith, B. H. (1976). Electrotonic processing of information by brain cells. *Science*, **193**, 114–20.

Selverston, A. I., Russell, D. F. & Miller, J. P. (1976). The stomatogastric nervous system: Structure and function of a small neural network. *Progress in Neurobiol.* **7**, 215–90.

Siegler, M. V. S. & Burrows, M. (1979). The morphology of local non-spiking interneurones in the metathoracic ganglion of the locust. *J. Comp. Neurol.* **183**, 121–47.

Sloper, J. J. & Powell, T. P. S. (1978*a*). Dendro-dendritic and reciprocal synapses in the primate motor cortex. *Proc. R. Soc. Lond. Series B*, **203**, 23–38.

Sloper, J. J. & Powell, T. P. S. (1978*b*). Gap junctions between dendrites and somata of neurones in the primate sensori-motor cortex. *Proc. R. Soc. Lond. Series B*. **203**, 39–47.

Strumwasser, F., Kaczmarek, K. & Viele, D. (1978). The peptidergic bag cell neurones of *Aplysia*: Morphological and electrophysiological studies of dissociated cells in tissue culture. *Neuroscience Abstracts*, **4**, 207.

Szentagothai, J. (1975). The 'module-concept' in cerebral cortex architecture. *Brain Research*, **95**, 475–96.

Szentagothai, J. (1978). The neurone network of the cerebral cortex: A functional interpretation. *Proc. R. Soc. Lond. Series B*, **201**, 219–48.

Szentagothai, J. & Arbib, M. A. (1974). Conceptual models of neural organisation. *Neurosci. Res. Progm. Bull.* **12**, 307–510.

Viancour, T. A. (1979). Electroreceptors of a weakly electric fish. II. Individually tuned receptor oscillators. *J. Comp. Physiol.*

Vowles, D. M. (1964). Models of the insect brain. In *Neural Theory and Modeling*, ed. R. F. Reiss, pp. 377–99. Stanford, Calif.: Stanford University Press.

Watanabe, A. & Bullock, T. H. (1960). Modulation of activity of one neuron by subthreshold slow potentials in another in lobster cardiac ganglion. *J. Gen. Physiol.* **43**, 1031–45.

Waterman, T. H. & Wiersma, C. A. G. (1954). The functional relation between retinal cells and optic nerve in *Limulus*. *J. exp. Zool.* **126**, 59–86.

Werblin, F. S. & Dowling, J. E. (1969). Organisation of the retina of the mud puppy. *Necturus maculosus* II. Intracellular recording. *J. Neurophysiol.* **32**, 339–55.

Young, J. Z. (1971). *Anatomy of the Nervous System of Octopus vulgaris*. Oxford: Clarendon Press.

INDEX

Figures in italics refer to diagrams, and those in bold type to general discussions of the topic.